Springer
Berlin
Heidelberg
New York
Hong Kong
London
Milan
Paris
Tokyo

Physics and Astronomy

ONLINE LIBRARY

springeronline.com

ADVANCES IN MATERIALS RESEARCH

Series Editor-in-Chief: Y. Kawazoe

Series Editors: M. Hasegawa   A. Inoue   N. Kobayashi   T. Sakurai   L. Wille

The series Advances in Materials Research reports in a systematic and comprehensive way on the latest progress in basic materials sciences. It contains both theoretically and experimentally oriented texts written by leading experts in the field. Advances in Materials Research is a continuation of the series Research Institute of Tohoku University (RITU).

T. Fukuda   P. Rudolph   S. Uda (Eds.)

Fiber Crystal Growth from the Melt

With 202 Figures

 Springer

Professor Dr. Tsuguo Fukuda
Institute of Multidisciplinary Research
for Advanced Materials
Tohoku University, 2-1-1 Katahira
Aoba-ku, Sendai 980-8577, Japan
e-mail: t-fukuda@tagen.tohoku.ac.jp
and
President
Fukuda X'tal Laboratory
c/o ICR 6-6-3 Minami-Yoshinari
Aoba-ku, Sendai 989-3204, Japan
e-mail: fukuda@fxtal.co.jp

Professor Dr. habil. Peter Rudolph
Institut für Kristallzüchtung
Max-Born-Strasse 2
12489 Berlin, Germany
e-mail: rudolph@ikz-berlin.de

Professor Dr. Satoshi Uda
Institute for Materials Research
Tohoku University
2-1-1, Katahira, Aoba-ku
Sendai 980-8577, Japan
e-mail: uda@imr.tohoku.ac.jp

Series Editor-in-Chief:

Professor Yoshiyuki Kawazoe
Institute for Materials Research, Tohoku University
2-1-1 Katahira, Aoba-ku, Sendai 980-8577, Japan

Series Editors:

Professor Masayuki Hasegawa
Professor Akihisa Inoue
Professor Norio Kobayashi
Professor Toshio Sakurai
Institute for Materials Research, Tohoku University
2-1-1 Katahira, Aoba-ku, Sendai 980-8577, Japan

Professor Luc Wille
Department of Physics, Florida Atlantic University
777 Glades Road, Boca Raton, FL 33431, USA

Library of Congress Cataloging-in-Publication Data: Fiber crystal growth from the melt/T. Fukuda, P. Rudolph, S. Uda (eds.). p. cm. – (Advances in materials research, ISSN 1435-1889; 6) Includes bibliographical references and index. ISBN 3-540-40596-8 (acid-free paper) 1. Crystal growth. 2. Crystal whiskers. I. Fukuda, Tsuguo. II. Rudolph, Peter, Dozent Dr. sc. nat. III. Uda, S. (Satoshi), 1955– . IV. Series. QD 921. F48 2004 548'.5–dc22 2003060695

ISSN 1435-1889
ISBN 3-540-40596-8 Springer-Verlag Berlin Heidelberg New York

© Springer-Verlag Berlin Heidelberg 2004
Printed in Germany

The use of general descriptive names, registered names, trademarks, etc. in this publication does not imply, even in the absence of a specific statement, that such names are exempt from the relevant protective laws and regulations and therefore free for general use.

Typesetting: Data conversion by Jürgen Grunert, Berlin
Cover concept: eStudio Calamar Steinen, Cover design: *design & production*, Heidelberg

Printed on acid-free paper   SPIN 10927007   57/3141/ba - 5 4 3 2 1 0

Preface

There are numerous well-established bulk crystal growth techniques for industrial use, like the Czochralski and Bridgman growth or hydro-thermal-synthesis. However, crystals grown by these methods need slicing, lapping, polishing and dicing before they are of practical use for manufacturing devices. Also, the fabrication loss is up to 50% of the initial weight of grown crystals. Hence, there is a serious challange for crystal growers to develop so-called shaped growth techniques offering ready-to-use crystals with form similar to the devices dimensions. The edge-defined film-fed growth (EFG) is one such technique already applied to the commercial growth of several shaped crystals such as sapphire tubes or silicon ribbons. For diverse applications, however, the most suitable shape possesses the fiber shape.

Single crystalline fibers became the subject of increasing interest because of their remarkable characteristics. Due to their very large length-to-diameter ratio composed with perfect crystallographic structure and chemical homogeneity, the mechanical and physical properties approach theoretical values, scarcely achievable by bulk growth methods. Fiber crystals are particularly well-suited for wave guiding in the IR region, tunable narrow-band filters and nonlinear optics due to their long interaction length and tight beam confinement. They are of increasing interest for room-temperature micro lasers and laser modulators, especially for generating of second and higher harmonics in the green, blue and violet region. Owing mainly to their high applicability in modern optics the growth of oxide fiber crystals has been achieved (sapphire, YAG, $LiNbO_3$, further niobates, BBO, TAG and numerous miscellaneous). At present, the growth of crystalline eutectic fibers with extremly high tensile strength is under forced development. Their ultra-high strength enables their application as reinforcing agents in structural components. The growth of semiconductor fibers is of fundamental capacity.

Today, two fiber melt growth techniques are well established - the micro floating zone by laser-heated pedestal growth (LHPG) and the micro pulling down (μ-PD) drawing from a die in the bottom of the melt crucible. Whereas LHPG produces fibers free from contact with a die, i.e., of lowest contamination level, μ-PD is favored by a diffusion-controlled growth mode from a large melt reservoir maintaining a very uniform fiber composition. In both methods dislocation-free and single-domain fibers with diameters in the region of 50 - 1000 μm have been grown. The μ-PD technique especially which could be described as an up-side-down EFG technique, shows some remarkable technological benefits such as simpler furnace structure, better shaping and fiber length possibilities, radial concentration profiling and deeming unnecessary the preparation of polycrystalline source rods, making this method predestined not only to fundamental material research but also to industrial applications.

In general, the growth of fiber crystals from the melt allows excellent fundamental studies on phase relations and diagrams, capillary stabilities with diminished gravity in value, growth kinetics (i.e., facetting phenomena), nearly unidirectional heat and mass transports, and dynamics of individual imperfections. Particularly, the brief and uncomplicated test of the crystallization behavior of new materials by fiber growth, even before the costly conventional bulk growth techniques is used, is of immense importance for research. Actually, one can say, that a growth apparatus for single crystal fibers belongs to each modern material science or crystal growth laboratory.

Since the early 1990s a micro-pulling-down method has been developed in Prof. Fukuda's research group at the Tohoku University in Sendai together with further developing the "Growth and Characterization of Single Crystal for Functional Devices" project which belonged to the Research for the Future Program of the Japan Society for the Promotion of Science (JSPS). This program is a joint effort of various distinguished scientists and engineers from all over the world and turned out to be an international interdisciplinary study specializing in the growth of fiber-shaped crystals or shape-controlled crystals. Any subject concerning the growth via the μ-PD method such as fundamental growth theory, furnace structure, practical growth, quality evaluation and functional characteristics of grown crystals has been compiled in this book based on the collective accomplishments of the project. This constitutes a magnificent contribution to the maturity of the pulling down principles from a meniscus shaper. Meanwhile, this method is applied by numerous international laboratories, e.g., University Claude Bernard Lyon, Sungkyunkwan University Suwon of South Korea, Institute of Crystal Growth in Berlin, among others. LHPG growth was developed in Feigelson's Laboratory at the Stanford University of California (ca. 1980) and is today used at the Pennsilvania State University, the National Central University of Taiwan, the University of Sao Paulo and the University of Aveiro in Portugal.

The present book is the first monograph on fiber crystal growth from the melt. Following a broad introduction to shaped crystal growth principles and application fields, the fundamentals of mass and heat transport, especially of the μ-PD mode, are developed. Then follow special chapters describing material selection, growth specifics, technologies and results of grown dielectric fiber crystals mainly used for laser and non-linear optics. The growth and use of eutectic fibers with high-tensile strength are added. Each chapter is written self-consistently in order to provide a well-rounded overview on the theory and practise of the material class. Well-known international specialists in the fields of fundamentals, simulation, methodics, analysis and application of shaped and fiber crystals have been arranged as authors. The editors thank them for their prompt readiness to contribute recent results to this treatise on rapidly developing branch of crystal growth and material science.

The book will interest readers in the fields of materials science, crystal growth, physics, chemistry, crystallography, optics, non-linear optics, laser optics, mechanics, and engineering.

The authors are greatful to the JSPS and its 161st committee of "Science and Technology of Crystal Growth" for their support and cooperation and they also thank all who contributed to the project of "Growth and Characterization of Single Crystal for Functional Devices."

The editors express their thanks to Dr. C. Ascheron and Mr. C.-D. Bachem both from the Springer Verlag for their excellent cooperation during the technical book prepartion phase and to Mr. J. Grunert for formatting the manuscript.

August 2003

Tsuguo Fukuda
Peter Rudolph
Satoshi Uda

Contents

List of Contributors

Georges Boulon
Physical Chemistry
of Luminescent Materials
Claude Bernard /Lyon1 University
CNRS UMR 5620, Bat.A.Kastler
10 rue Ampere
69622 Villeurbanne Cedex, France
boulon@pcml.univ-lyon1.fr

Valery I. Chani
Department of Materials Science &
Engineering
McMaster University
1280 Main Street West,
Hamilton, Ontario, L8S 4L7, Canada
vchani@home.com

Boris M. Epelbaum
University of Erlangen-Nürnberg
Dept of Materials Science 6,
Martensstr. 7
D-91058 Erlangen, Germany
boris.epelbaum@ww.uni-erlangen.de

Tsuguo Fukuda
Institute for Materials Research
Tohoku University
2-1-1 Katahira, Aoba-ku
Sendai, 980-8577, Japan
fukuda@fxtal.co.jp

C.W. Lan
Department of Chemical Engineering
National Taiwan University
Taipei, Taiwan
cwlan@ccms.ntu.edu.tw

Kheirreddine Lebbou
Physical Chemistry
of Luminescent Materials
Claude Bernard /Lyon1 University
CNRS UMR 5620, Bat.A.Kastler
10 rue Ampere
69622 Villeurbanne Cedex, France
lebbou@univ-lyon1.fr

Peter Rudolph
Institut für Kristallzüchtung
Max-Born-Str. 2
12489 Berlin, Germany
rudolph@ikz-berlin.de

Satoshi Uda
Electronics Device Research
and Development Center
Mitsubishi Materials Co., Ltd.
2270 Yokoze, Chichibu,
Saitama 368-8503, Japan
uda@imr.tohoku.ac.jp

Akira Yoshikawa
Institute for Materials Research
Tohoku University,
2-1-1 Katahira, Aoba-ku
Sendai, 980-8577, Japan
yoshikawa@imr.edu

List of Symbols

(taken from chapter 2)

English Symbols

C_L	–	solute concentration in the liquid
$C_{L(i)}^j$	–	concentration of the jth species in the liquid at the interface
C_L^I	–	solute concentration in zone I of the μ-PD system
C_L^{II}	–	solute concentration in zone II of the μ-PD system
$C_{L(x)}$	–	solute concentration in the liquid at x in the coordinate system with the origin at the solid-liquid interface
C_S	–	solute concentration in the solid
C_0	–	initial solute concentration in the liquid
C_{Lmz}	–	steady-state concentration at $x = L_{mz}$
$\overline{C}$	–	normalized solute concentration ($= C / C_\infty$)
C_∞	–	solute concentration in the bulk liquid
c_p	–	specific heat at constant pressure
D	–	diffusion constant
D_{eff}	–	effective diffusion constant
d_{cry}	–	diameter of crystal
d_{cap}	–	diameter of capillary channel
E	–	interface electric field
E_r	–	radial electric field
E_c	–	charge-separation effect-related electric field
E_t	–	Seebeck-effect-related electric field
$\left(E_{c_L}\right)_r$	–	charge-separation-driven electric field along the radial direction
$\left(E_{t_L}\right)_r$	–	Seebeck-effect-induced electric field along the radial direction
e	–	electronic charge
g	–	solidified melt fraction
G_L	–	temperature gradient in the liquid in the axial direction
G_{L_r}	–	radial temperature gradient in the liquid
G_S	–	temperature gradient in the solid in the axial direction
H	–	molten zone height
ΔH	–	latent heat
J_J	–	solute flux in zone "J", $J =$ I, II
J^j	–	flux of the jth species in the liquid
k_0	–	equilibrium solute partition coefficient
k, k_{eff}	–	effective solute partition coefficient
k_{E_0}	–	field-modified equilibrium solute partition coefficient
k_E	–	field-modified effective solute partition coefficient

K_L	–	thermal conductivity of liquid
K_S	–	thermal conductivity of solid
k_B	–	Boltzmann's constant
L_{cap}	–	position of the entrance of the capillary channel in the coordinate system with the origin at the solid-liquid interface
L_{mz}	–	position of the upper end of the molten zone in the coordinate system with the origin at the solid-liquid interface
ΔM	–	magnitude of the imbalance of the solute mass flow at the zone boundary
Ma	–	Marangoni number
q_β^j	–	field-driven flux term for the jth species ($\beta = L, S$)
r	–	radial distance from the axis of symmetry
R^*	–	crystal radius
R	–	normalized radius ($= r / R^*$)
R_1	–	crucible radius
R_2	–	crystal radius
S_J	–	cross-section of zone J, $J =$ I, II
T	–	temperature
$u(z)$	–	convection term working along the z-direction
u	–	vector flow velocity of the melt
u_z	–	axial component of the vector flow velocity
u_r	–	radial component of the vector flow velocity
V, V_{cry}	–	growth rate
V'_{cry}	–	transient growth rate from the steady-state growth rate, V_{cry}
$\overline{V}_{cap}$	–	mean velocity of fluid in the capillary channel
$\overline{V}_{cry}$	–	mean growth rate
V^J	–	fluid velocity in zone J, $J =$ I, II
V_E	–	vector field-modified effective velocity
V_{E_z}	–	axial component of the vector field-modified effective velocity
V_{E_r}	–	radial component of the vector field-modified effective velocity
V_{E_L}	–	field-modified effective growth velocity for the liquid
V_{E_S}	–	field-modified effective growth velocity for the solid
V^*	–	critical velocity
V_r	–	radial growth velocity
$V_{E_r}^{R^*}$	–	V_{E_r} at $r = R^*$
Z	–	normalized one-dimensional direction parameter ($= z / \delta_c$)
$Z(r)$	–	meniscus profile of molten zone with radius r
z^j	–	valence of the jth species

Greek Symbols

α	–	constant
α_i	–	crystallization EMF
α_L	–	thermoelectric coefficient of liquid
α_s	–	thermoelectric coefficient of solid
β	–	constant
δ_c	–	thickness of the solute diffusion boundary layer
ϕ	–	angle between the meniscus and the growth axis
ϕ_0	–	critical angle between the meniscus and the growth axis
ϕ_d	–	diameter of a fiber
$\Delta\phi$	–	thermoelectric potential difference
γ	–	constant
γ_{ij}	–	surface energy between phase i and j
$\overline{\eta}$	–	dynamical viscosity
λ	–	constant
ρ_s	–	density of crystal
ρ_L	–	density of liquid

1 What Do We Want With Fiber Crystals? An Introductory Overview

Peter Rudolph

The characteristics and wide field of advanced applications of micro single crystals in elongated form with small diameters in the μm–mm region are reviewed. Melt growth methods for production of fiber crystals are classified and discussed. After discussion of some growth fundamentals, selected results and utilization of oxide, eutectic, semiconductor and metal fibers are presented. The special suitability of fiber crystal growth for fundamental research is emphasized.

1.1 Introduction

Single crystalline fiber crystals have become the subject of intense study in recent years because of their remarkable characteristics. Usually, one means by fibers any materials in elongated form having a small diameter in the μm to mm region. The term fiber describes all types of materials that fit the definition, i.e. filaments, wires and whiskers. It is possible to grow such crystal shapes from melt, solution and vapor. Nowadays, there is an enormous fund of literature on fiber crystal growth. In particular, the growth of whiskers from the vapor or by the vapor–liquid–solid mechanism, showing nearly perfect single crystalline properties and being excellent objects of fundamental studies, is discussed in numerous original papers and monographs like those of Levitt [112], Maslov [121] and Givargizov [76]. This review, however, confines itself to melt growth methods only. At present, there are more than 200 original publications dealing with this field.

The first publications on fiber growth date back to the beginning of this century. Surprisingly, it was Czochralski [33] working alone who in 1917 pulled the first metallic single crystalline fibers ("wires") from the melt with diameters from 200 μm upwards. In order to study the stretch behavior of monocrystals, somewhat later, in 1922, von Gomperz [77] grew metallic fibers through the orifice in a mica disk floating on the melt surface. This was the "hour of birth" of shaped crystal growth from a die by using the principle of stablilized meniscus capillary which is of greatest importance for fiber growth methods today.

After intense studies of shaping on Ge by Stepanov [158] and Gaule and Pastoré [74], the first major push at the beginning of the 1970s was concerned with one of the extra-ordinary characteristics of fiber crystals – their *ultrahigh strength,* yielding > 1 GPa in EFG-grown sapphire fibers, for example (La Belle and Mlavsky [102–104]). This is due to their crystalline perfection and small dimensions, which minimize the occurrence of the defects that are responsible for the low strength of materials in bulk form [112]. This property favors fibers as *reinforcing agents* in structural components.

The development of optical waveguiding in the 1970s activated the further growth of single crystalline fibers for diversified applications (Haggerty [81], Burrus and Stone [18]). The *broad transmission window*, *high melting point*, and *resistance to chemical attack* of many crystalline materials make them attractive for *energy delivery*, particularly in hostile thermochemical environments (Fejer et al. [85]). Single crystalline fibers are also well suited to *nonlinear optical interactions*, whose efficiencies can be greatly enhanced by the *long interaction lengths* and *tight beam confinement* available in *guided wave structures* (Feigelson [52]). On the other hand, they are applicable for second-order interactions like *harmonic generation*, frequency mixing, parametric oscillation and electro-optic modulation (Magel et al. [117, 118]). In particular, $LiNbO_3$ single crystalline fibers are of favorable shape for the production of *photorefractive hologram storage* (Yu et al. [200–201], Yin et al. [181–187]). Because of the longitudinal dimension, Bragg gratings of a large number of periods can be synthesized within the fiber, which is suitable to develop *narrow-band tunable filters* with bandwidth as narrow as 0.01 nm [183, 186, 187]. Moreover, the fiber configuration has another advantage. By using it as a *lasing element* it removes heat very efficiently because of the short distance between the pumping region and surrounding temperature. Thus, it keeps the *laser rod at a low temperature*, which is preferable for increasing output (Ishibashi [89]). These superior characteristics of fiber crystals are complemented by their *well-adapted as-grown shape* similar to many kinds of optical device profiles. As a result, crystal machining and preparation costs are minimized drastically.

It was a logical consequence that in numerous international laboratories special research programs of single crystalline fiber growth were started, among them at Stanford University of California (c. 1980), Pennsylvania State University (c. 1992), Tohoku University in Sendai (1993), Nat. Central University of Taiwan (c. 1994), Université Claude Bernard, Lyon (c. 1995), University of Sao Paulo (c. 1996), Sungkyunkwan University Suwon of South Korea (ca. 1997), and the Institute of Crystal Growth in Berlin (1998).

The scientific approach to the fiber growth process immediately brought to light its specific fundamental qualities too. Because of the very small crystal diameter only diminishing values or even complete *absence of dislocations* was observed. Inoue and Komatsu [88] first showed this tendency in thin KCl rods and attributed this behavior to the drastically reduced thermomechanical stress. Moreover, fibers offer the best conditions for the *out-growth of defects* even if the glide directions are only slightly tilted away from the growth axis. Further, fibers usually grow by the *diffusion controlled growth mode*. The small melt zones and high growth rates produce an *effective distribution coefficient near to unity* leading, as a result, to uniform axial component distribution. This fact also makes it possible to study phase relations with incongruent melting behavior by using a feeding melt reservoir with the same composition as the solid to be expected. Nowadays, it is the increasing practice of several laboratories, like IKZ Berlin, to follow the pioneering work of Fukudas laboratory and apply the fiber growth method even for fundamental research of the crystallization specifics of new multicomponent materials (Fig.1.1). First the growth kinetics, composition stability and crystallinity become clear by testing in a low-cost, material- and time-saving fiberpulling apparatus before the material is recommended (or rejected) for Czochralski growth. This approach is very effective for materials research, especially in university and institute laboratories

Finally, a new exciting development proves to be the fiber growth of single crystalline *eutectic composites*, like $Al_2O_3/Y_3Al_5O_{12}$ (Yoshikawa et al. [194–197]) or Al_2O_3/ZrO_2 (Lee et al. [110]), with *superior high-temperature strength properties* exceeding those of eutectic bulk crystals markedly (Fig. 1.2). This is due to the exponential increase of the tensile strength with decreasing diameter, especially in the region below 100 μm (Ström-Olsen et al. [159]), and the structure refinement at high growth rates. The rapidly growing number of related publications since 1999 shows that this research direction is developing very promisingly.

In the following, fiber melt growth methods, tested materials and their applications will be introduced. Only characteristic features can be summarized. For more complete details see reviews of Feigelson [50–52] and Rudolph and Fukuda [148].

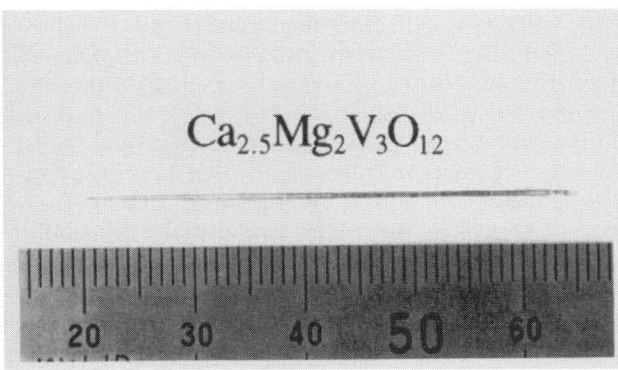

Fig. 1.1. $Ca_{2.5}Mg_2V_3O_{12}$ – a possible new material for optical insulator applications first tested by fiber growth before the Czochralski growth was settled (courtesy of Fukuda Laboratory).

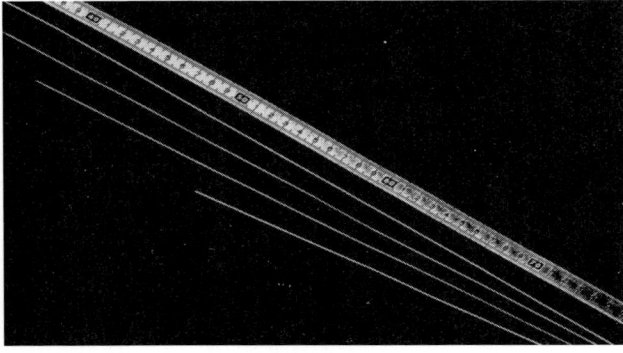

Fig. 1.2. As-grown $Al_2O_3/Y_3Al_5O_{12}$ eutectic fibers (Rudolph, Yoshikawa and Fukuda [149]).

1.2 Methods

Basically, there are two ways to form single crystalline fibers from the melt (i) unidirectional solidification, including micro zone melting, in a capillary tube, or (ii) the containerless pulling of a filament from a melt meniscus. However, despite good diametric constancy, usually, case (i) is scarcely used for fiber growth due to the limited heater length and the complicated fiber exposition after the growth. Feigelson [51] found few papers dealing with solidification experiments of low-melting oxides and semiconductors in thin tubes. Moreover, the unidirectional crystallization of organic optically active materials in very thin glass tubes with inner diameters of 20–50 µm has been described [8]. These authors applied wet-ting capillaries to raise the molten material into those when were inserted verti-cally into a melt volume placed within the heater below. The inverted Bridgman–Stockbarger technique was used for the solidification process. Recently, this crys-tallization method was applied by Kurlov et al. [101a] to crystallize a bundle of sapphire fibers within capillaries in vertical 65 mm long Mo blocks dived in the melt. The fiber extraction was achieved by dissolving the Mo matrix in an acid mixture. After that the molybdenum sediment was fully regenerated.

Compared with that, it is the common practice to pull crystalline fibers from a free melt meniscus (case ii). One has to note, however, that it is not possible to grow crystals with very small diameters from a conventional voluminous Czochralski crucible, not only because of the acting convective and temperature oscillations in the melt affecting meniscus demolition, but mainly due to the in-herent capillary instability, typical for small aspect ratios between crystal and cru-cible radius (R/R_{cr}). Tatarchenko [166] showed theoretically that there is no capil-lary stability in the conventional Czochralski arrangement. Certainly, this instability can be restrained by increasing the crystal radius above the capillary constant ($R > a$) but this case is far from the fiber growth condition due to the mil-limeter-dimension of a. In a physical sense one can say that the meniscus is me-chanically fixed only at its upper periphery, i.e. at the growing interface, by the ac-tion of the three-phase (vapor–solid–liquid) equilibrium. In contrast, the below two-phase (vapor–liquid) contour is labile. Its fixation and, hence, the achieve-ment of total capillary stability is only possible by anchoring at a solid body with a given radius R_{sh}. Solid boundaries offer feed rods of the same material to be grown (like in the case of zone floating) or dies made of a different material (like in the case of a swimming coracle or Stepanov, EFG and pendant drope shapers). Using such measures there are no further obstacles to the growth of thin fibers from the melt (Fig. 1.3). However, one has to be aware that the total stability is lost if $R/R_{sh} < 0.5$ [166]. Note that the first unsupported fiber pulling experiment of Czochralski [33] succeeded even though a small crucible (melt "puddle") was used. Also Ohnishi and Yao [127] grew $LiNbO_3$ fibers by the so-called "micro-Czochralski" (µ-CZ) technique because in this case the meniscus was supported at a microprotuberance of a heated platinum wire covered by a skin of melt.

Excluding the individual (but quite fascinating) µ-CZ tests, we can classify melt growth fiber techniques into the following two main categories: (1) *micro floating zone methods*, and (2) *pulling techniques from a shaper* (i.e. die). Whereas the methods of (1) vary by different principles of heating and heat focusing (laser beam, lamp mirror furnace) as well as floating directions (up- or downward) the variants of (2) differ by up- or downward pulling (drawing) from a nonwetting shaper (Fig. 1.3). In the following these versions will be discussed in detail.

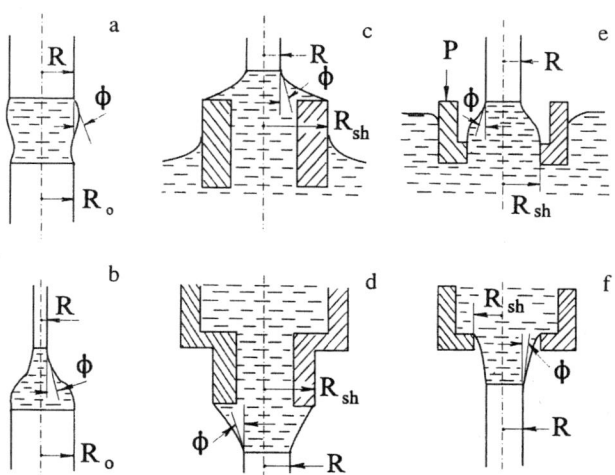

Fig. 1.3. Principles of capillary stable crystal growth conditions: (a) zone floating ($R = R_0$); (b) pedestal growth ($R < R_0$); (c) pulling upward from wetting shaper; (d) drawing downward from wetting shaper. (e) Pulling upward from nonwetting shaper; (f) drawing downward from nonwetting shaper (R-crystal radius, R_0-feed rod radius, R_{sh}-shaper radius, Φ-growth angle, P-pressure)

1.2.1 Micro Floating Zone Methods

The functioning principle of diameter reduction of the growing crystal during the float-zone process is known from the first pedestal growth experiments of Poplawski [138] who provided focused energy by an arc image furnace. Further, the well located heating behavior of a laser beam was successfully demonstrated for floating zone growth by Eickhoff and Gürs in 1969 [40] and Gasson and Cockayne in 1970 [73]. On the basis of these experiences the "laser heated pedestal growth" (LHPG) method was first developed by Haggerty in 1972 [81] as mentioned by Feigelson in his review paper [50]. He pointed out that because of its great versatility and the broad range of materials to be studied, this method was chosen and considerably cultivated at the Stanford Center for Materials Research in the USA.

Figure 1.4 demonstrates the miniature pedestal growth (MPG; so called by Fejer et al. [54]) of a single crystal fiber. A tightly focused CO_2 laser, emitting 10.6 µm radiation, is the heat source used to melt the refractory material. The source rod may be fabricated from a single crystal, polycrystalline, sintered or pressured powder material. A seed rod is used to determine the crystallographic orientation of the fiber to be grown. Growth proceeds by simultaneous upward (or downward) translation of the seed and source rods with the molten zone positioned between them. The laser focal spot, and consequently the zone height, remains fixed during fiber growth. To achieve an axially symmetric irradiance a reflaxicon as an optical element was incorporated in the laser beam wich consists of an inner cone surrounded by a larger coaxial one [53]. The combination of the reflaxicon with a parabolic mirror produces a minimum spot size of 30 μm. For the growth of oxide fibers, having a high absorption at 10.6 μm by photons (the absorption coefficient of sapphire close to its melting point is 4000 cm^{-1}, for example), it takes about 30–60 W to melt a rod of about 1 mm in diameter. Compared to that, for silicon, the absorption of which is only 1000 cm^{-1} near the melting point, a power of 100 W has been estimated for the same diameter [95]. Very high axial temperature gradients in the range greater then 10^3 K cm^{-1} have been measured along growing oxide fibers [30, 170]. The fiber-to-source rod diameter ratio is set by the mass balance to be the square root of the source rod-to-fiber translation rate (1.1). Typical fiber growth rates range from 1 to 10 mm min^{-1} (0.1–100 mm min^{-1}) with diameter reductions of approximately three. Such system can produce fibers 3–1700 μm in diameter and up to 20 cm in length. When the source rod is less than 100 μm, however, the power required to form and maintain the molten zone is so small that it is difficult to maintain laser stability. Using an automated fiber-diameter control system, diameter variations of less than 0.1 % have been achieved [50, 52].

Until now similar LHPG arrangements, at times somewhat modified in the laser beam reflection system [31, 130] or pulling-down direction [35, 163], have been used by numerous authors (see Sect. 1.4.1 and Table 1.1). Halogen lamp heating or lamp-laser combinations were applied by Kawakami et al. [94] and Adachi et al. [1], respectively. In order to dam the convection in the melt a completely different heater principle was tested by Koh et al. [97] for the growth of oxide rods using an alive sieve-like metallic plate immersed in the floating liquid zone.

Ardila et al. [5, 7] developed a modified LHPG system for growth in an isostatic gaseous environment at large pressures in the range up to 10 atm.

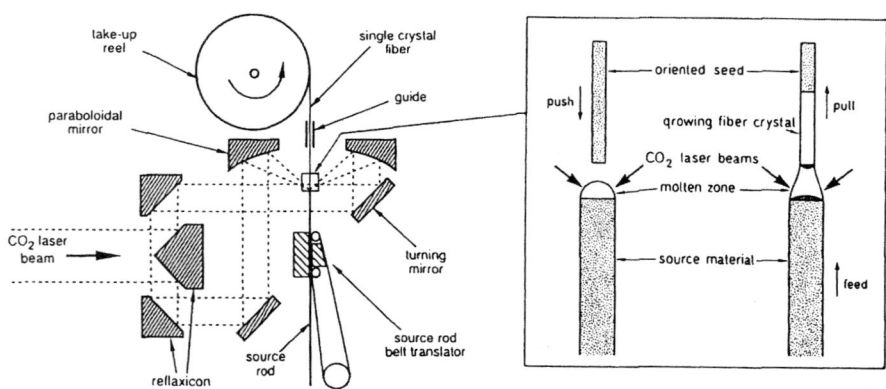

Fig. 1.4. Scheme of the laser-heated miniature pedestal growth (Fejer et al. [53,54])

A special fascinating feature of the LHPG is the relatively easy possibility of achieving periodically poled ferroelectric fibers in order to apply them for quasi-phase-matched second-harmonic generation (Fejer et al. 1989 [55], Magel et al. 1990 [118], Brenier et al. 1997 [13]). One can create during the growth periodically reversed ferroelectric domains with boundaries following the growing interface shape by modulating the freezing interface position, either by rotating an asymmetric heat input produced by a rotating mask or by periodically modulating the laser heating power. Obviously, the large axial temperature gradient and high growth rate, together with the high Curie temperature, are responsible for the formation of well-defined domain walls near the crystallizing interface [118] (see Sect. 1.4.1, Fig. 1.13). Chen et al. [30] discussed the influence of alternating composition gradients on domain formation.

1.2.2 Pulling Techniques from a Die

Upward Pulling from a Meniscus Shaper

In the 1920s von Gomperz [77], Mark et al. [120], Linder [113], and Hoyem [83] grew the first metallic crystals through a hole in a floating mica disk, primarily, to prevent oxidation of the melt surface. Much later, 1959–1960, well-aimed studies on the role of surface tension in the pulling of shaped single crystals were started by guiding the melt through a capillary (die) by Stepanov [158] and Gaulé and Pastore [74]. Using a nonwetting die, dived into a melt, depending on the pressure P, a stable concave, cylindrical or convex meniscus contour can be formed (Fig. 1.5). Although Tsivinskii et al. [169] already in 1966 demonstrated the wetting regime too (they grew thin Ge strips by using a tungsten shaper forming a wetting angle of $\Theta \approx 36°$), later a special branch of shaped crystal growth and metallurgy from nonwetting dies ($\Theta > 90°$) was cultivated in Russia. Mainly germanium ribbons and aluminum profiles were pulled from graphite and coated cast-iron shapers, respectively (see the monographs of Antonov et al. [3] and Maslov [121]). Egorov et al. [37] grew 21 germanium rods (4 mm in diameter each) at the same time from a special multicapillary die (they also later need these "growth in sections" from a wetting tungsten shaper for pulling YAG, sapphire and alexandrite crystal rods well-suited for the production of bearings [39]). However, there is almost no information on upward fiber pulling from nonwetting dies with diameters in μm dimensions. Only Koh et al. [98] tested this principle for the growth of 800 μm-thick $Bi_{1-x}Sb_x$ mixed crystal fibers. They used a long nonwetting capillary tube of graphite which was pressed in a hull with the melt. As a result, the melt was injected to the capillary top for meniscus shaping quasi-hydraulically.

In 1967, La Belle and Mlavsky [102] reported on the growth of sapphire filaments from a wetting die made of molybdenum ($\Theta \approx 30°$). In the first instance they used the Gomperz principle [77] whereupon the crystal was pulled through a disk-like die located on the surface of the melt. But at once they improved the technique by a capillary tube attached to the bottom of the crucible (Figs. 1.6a and b) in order to prevent melt surface deplation and, hence, the continuous lowering of the orifice as the filament is grown [104, 132]. They observed that this growth system displays a considerable degree of self-stabilization and very long fibers of some dozen meters in length and of nearly constant diameters between 100 and 200 μm could be grown over a range of pulling rates 25–50 mm min^{-1} (Fig. 1.6c). As can be seen from Fig. 1.6b the shape and stability of the below meniscus contour is fixed by the edge (or area) of the die top. Thus, the authors patented this technique as "Edge-defined Film-fed Growth (EFG)" (Swedish Patent 325 552 from 1970). In the period that followed the principle was widely used for the growth of silicon and sapphire ribbons and tubes from graphite ($\Theta \approx 30°$) and molybdenum dies, respectively (see the review of Tatarchenko [166]).

The EFG method with upward pulling direction is now scarcely used for growing fiber crystals. Thin LiNbO$_3$ rods with diameters of about 5 mm and an in situ doped core region have been grown by Shimamura et al. [153]. They were the first to use a double-die principle to ascend different doped melts to the top edge.

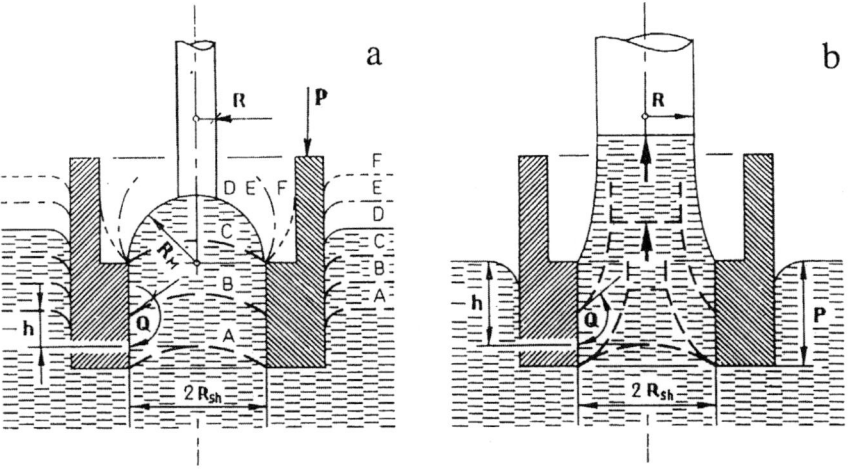

Fig. 1.5. The principle of shaped crystal growth from a nonwetting die (after Rudolph [145]): variation of the meniscus height by (**a**) the pressure *P* on the die, and (**b**) the pulling strength, i.e. rate (-wetting angle, *R* , R_M, R_{sh}-crystal, meniscus and shaper radius, respectively; *h*-height of capillary depression).

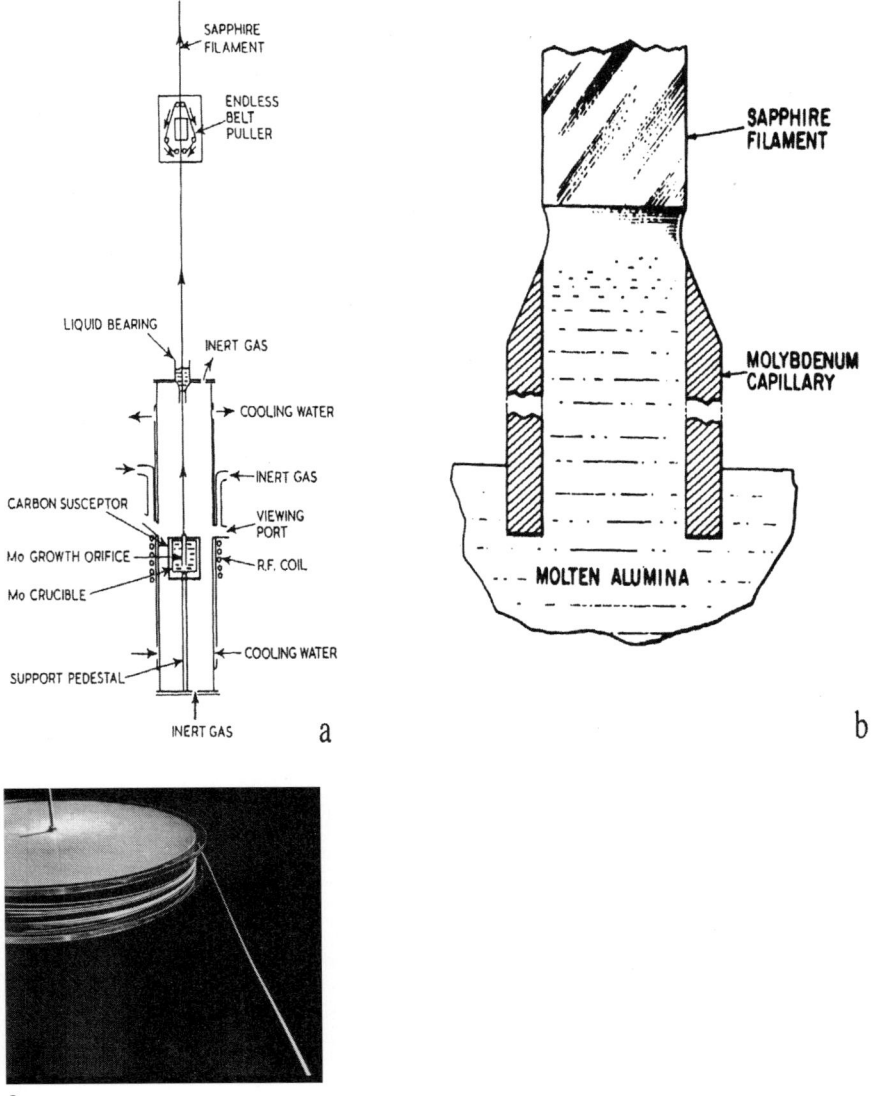

Fig. 1.6. The production of sapphire filaments after Pollock [132]: (**a**) schematic diagram of the growth apparatus; (**b**) the EFG principle used in the arrangement; (**c**) rolling up of the filaments (La Belle and Mlavsky [104]).

Pulling Down from a Meniscus Shaper

First Kim et al. in 1976 [95] and somewhat later Ricard [142] proposed the inverted Stepanov (IS) or IEFG technique to draw silicon sheets from a nonwetting silica ($\Theta = 87°$) or wetting graphite slot in the crucible bottom (Fig. 1.7). Continuous material charging from the top was demonstrated by Bell [10] and Ricard [142] using a polycristalline feed rod or pellets, respectively (Fig. 1.7a). At first, it was assumed that in the nonwetting case, a supplementary force, e.g. a gravitational head or differential gas pressure, should be required to push the liquid through the slot [10]. But later it was shown by numerous pulling down experiments that a seed, touched in the hanging melt drop, is able to draw the meniscus outward (see Fig. 1.3f and Epelbaum et al. [42] for example).

Mimura et al. [122] described for the first time in 1980 an inverted drawing technique for the growth of fiber crystals (KSR-5) consisting of (i) a melt container, (ii) an extra heated long capillary at the container bottom, and (iii) a wetting or nonwetting shaper. This construction was helpful in deepening the axial temperature gradient. A modified drawing down apparatus was also used for the growth of MgO-doped LiNbO$_3$ fibers by Oguri et al. [126]. They applied an alive conical shaped Pt heater for accommodation and melting of the feed charge with an orifice at the cone tip for the seeding and drawing procedure.

Since 1992, in the framework of the "Micro Crystal Project", at the Fukuda Laboratory of Tohoku University, the development and broad application of single crystalline fiber growth by the "micro pulling down" (µm-PD) method was established (see the introduction of Rudolph [147]). The method, described for the first time by Yoon and Fukuda [188–193], consists a directly resistivity heated Pt crucible with the starting melt having a micro nozzle with a capillary in the bottom. After a thin seed crystal is touched to the hanging melt droplet at the nozzle the fiber is pulled down with velocities in the range 0.5–20 mm min^{-1} (Fig. 1.8a).

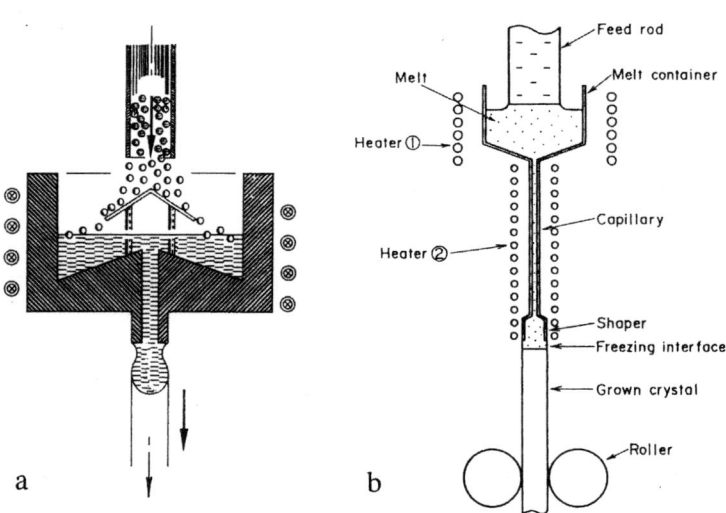

Fig. 1.7. Schematic diagrams of pulling downward growth systems by inverted Stepanov, i.e. inverted EFG, principles: (**a**) Si growth from a wetting graphite slot (after Ricard [142]); (**b**) KRS-5 fiber drawing from a shaper (after Mimura et al. [122, 123]).

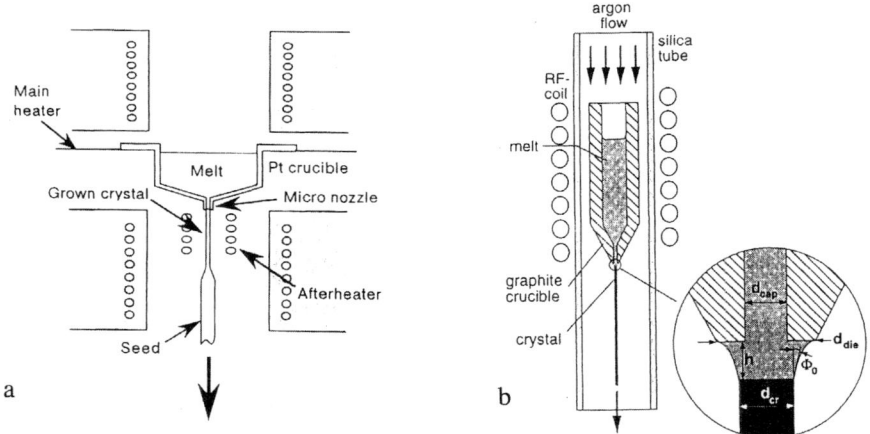

Fig. 1.8. Schematic diagram of the μ-PD growth apparatus with (**a**) alive heated crucible (after Yoon et al. [188]) and (**b**) indirect r.f. heated melt container (after Schäfer et al. [152]).

Very deep axial temperature gradients were measured yielding in LiNbO$_3$ fibers about 1500 K cm^{-1} [188]. Depending on the container materials indirect r.f. heating is also used (Fig. 1.8b). Constant fiber diameters between 50 μm and several mm can be grown by the μ-PD method. A double-die for in situ cladding of the drawing fibers was tested successfully by Epelbaum et al. [45].

The stable growth of thinner fibers is described by Yoshikawa et al. [195]. To overcome the capillary instability if $R/R_{sh} < 0.5$ (see Sect. 1.3.3) an additional shaping element (thin iridium wire) was inserted into the crucible orifice. If its end is located slightly below the bottom opening the central region of the meniscus is fixed and very thin fibers can be grown from the melt skin covering the inserted wire tip.

In general, a hanging meniscus combines capillary stability by die anchoring with the positive gravity effect and damped disturbances of the growing interface due to suppressed convection in this region. Additionally it was found by Uda et al. [174] that under such conditions in thermoelectric materials the radial influence of the interface electric field is reduced. As a result a more uniform distribution of dopants or solvents in the crystal can be obtained than with pulling upward methods (see also Chen and Hu [31]).

1.3 Selected Fundamentals

Basic considerations, showing the characteristics of fiber growth by the LHPG technique, were first discussed by Feigelson in 1986 [51]. Here only some important relationships will be given. For more details see the special chapter on the theory of fiber growth of Uda within the given book.

1.3.1 Conservation of Mass

The dimensional stability of a growing fiber from a shaped meniscus or pedestal requires that the zone length and volume be kept constant. Neglecting the density difference between melt and crystal the condition for steady state growth of a fiber with constant diameter is

$$R^2 v = R_{sh}^2 v_{sh} \quad \text{or} \quad R = R_{sh} \sqrt{\frac{v_{sh}}{v}} \tag{1.1}$$

with R, R_{sh} the radius of crystal and shaper (i.e. capillary or feed rod), and v, v_{sh} the pulling rate and mean flow velocity in the capillary channel (i.e. push rate of the feed rod), respectively.

1.3.2 Balance of Heat Transfer

At the crystallization front the heat-balance equation should be satisfied:

$$\rho_S \Delta H_f v + k_L G_L = k_S G_S \tag{1.2}$$

where ρ_S is the density of the solid, ΔH_f the latent heat of fusion, k_L and k_S the thermal conductivities of the liquid and solid phases, respectively, and G_L and G_S the temperature gradients in the solid and liquid phases at the crystallization front (meniscus height). Numerous papers deal with the determination of the values of G_L and G_S by measurements ([31, 32, 170] for LHPG; [174, 188–193] for μ-PD) and numeric calculations ([101] for FZ Si bulk growth; [100] for fiber growth). Even very large axial temperature gradients G_S, which increase with decreasing fiber diameter (Fig. 1.9a), are responsible for the very high fiber pulling rates used (1.2).

A characteristic length z_d (decay length) along the fiber axis over which the temperature drops to the surrounding value was formulated by Korpela et al. [100] considering coupled radiation-conduction heat transfer:

$$z_d = R \left[-\frac{1}{2} Pe + \left(\frac{1}{4} Pe^2 + 2Bi + 16\sigma_{St} n^2 T_s^3 k_P R^2 \Big/ k \right)^{1/2} \right]^{-1} \tag{1.3}$$

with R-crystal radius, Pe-Peclet number (= $r\,c\,v\,R\,/\,k$), r-reflectivity, c-specific heat, v-pulling velocity, k-thermal conductivity, Bi-Biot number (= $h\,R\,/\,k$), h-convective heat transfer coefficient, σ_{St}-Stefan-Boltzmann constant, n-index of refraction, T_s-surrounding temperature, and k_P, k_R-Planck and Rosseland (Fig. 1.9b) mean absorption coefficient, respectively.

As can be seen from (1.3), and Fig. 1.9b, z_d increases as the Biot and Peclet numbers (i.e. fiber diameters) become smaller and the absorptivity of the fiber lowers. Thus, low absorptivity (oxides) prevents the cooling of the fiber by radiation. But it doesn't mean that in materials with high absorbing power (semiconductors, metals) a smaller decay has to be expected. In this case the higher thermal conductivity k affects the value of z_d markedly. The authors showed that for very thin fibers neither the emissivity of the phase front, nor the reflectivity of the surface are of evident importance.

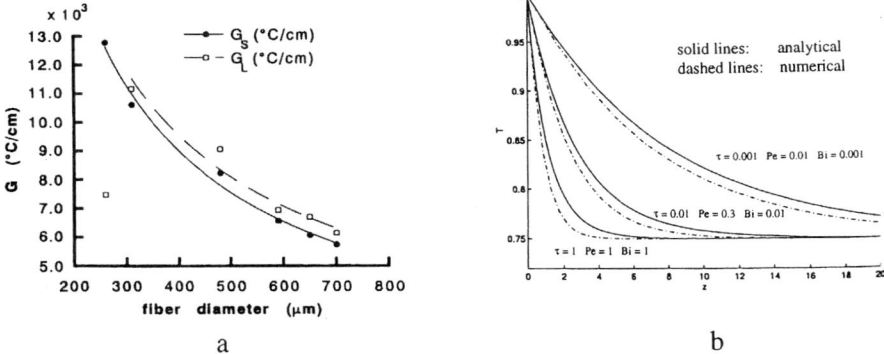

a   b

Fig. 1.9. Axial temperature distribution. (**a**) Calibration curve of measured temperature gradients G_S and G_L in LHPG LiNbO$_3$ fibers versus fiber diameter after Uda and Tiller [170]. (**b**) Non-dimensional axial temperature distributions in a cylindrical rod after Korpela et al. [100] (τ-optical thickness $= R\sqrt{k_p k_r}$, Pe , Bi - Peclet and Biot numbers, see text).

1.3.3 Capillary Stability

Fig. 1.3 compiles the principles of capillary shaping in different fiber growth methods. As can be seen, a uniform cross-section obtains if the growth angle ϕ is constant and equals ϕ_0 which forms by the thermodynamic equilibrium condition at the three-phase interface line and, hence, is independent of the growth parameters like pulling velocity, diameter or zone height. The value of ϕ_0 differs from zero and yields for Si 11° {111}[145,165], YAG 8° {100} [51] and sapphire 17° {0001} [35], for example.

Deviations of the actual growth angle ϕ from ϕ_0 by perturbations of the meniscus height or radius lead to nonstationary variations of the fiber diameter as

$$\frac{dR}{dt} = v \tan(\phi - \phi_0) \neq 0 . \tag{1.4}$$

Note that Shimamura et al. [154] found during pulling down growth of silicon fibers that in this case a somewhat different growth angle ϕ of 9° is required (instead of $\phi = \phi_0 = 11°$) to guarantee a steady state with constant fiber diameter. They concluded that this effect is due to gravity, i.e. hydrostatic pressure.

Kim et al. [96] studied the maximum stable zone length float-zone growth of small-diameter sapphire and silicon crystals (250–1500 μm) under the condition that the crystal radius equals the feed rod radius ($R = R_0$). They showed that for such dimensions, where the Bond numbers Bo are much smaller than unity, the maximum stable zone length l_{max} and the zone stability are controlled primarily by the surface tension ($Bo = \rho g d^2/4 \gamma \ll 1$, with ρ the density of the melt, g the acceleration due to gravity, d the diameter and γ the surface tension). This result is in good agreement with the theory according to which l_{max} yields

$$l_{max} = \pi d \quad \text{(micro zone floating with } R = R_0\text{)}. \tag{1.5}$$

Tatarchenko [166] showed that for the pedestal growth technique, where $R < R_0$, the theoretical capillary stability is kept until $R > R_0/2$. In the practical case of fiber growth by LHPG even a slightly higher ratio of $\frac{1}{2} - \frac{1}{3}$ can be used (Feigelson [51]). Tang et al. [164] found empirically that stable LHPG growth of oxides occurs for a zone length of

$$l_{max} = \tfrac{3}{2}(R + R_0) \quad \text{(micro pedestal zone growth)}. \tag{1.6}$$

Young and Heminger [198] presented a model of the time-depending behavior of a LHPG arrangement for a feed-to-crystal reduction ratio of 2 after which the change and disturbation decay of the fiber diameter depend very sensitively on the perturbation levels of the pulling rate, feed velocity and heating temperature as well as on coupled mechanisms. They found that the material parameters, such as Peclet number, Stefan number and density ratio between the solid and melt, also play an important role in setting the response of the fiber radius. Moreover, they supposed that the oscillatory Marangoni convection in the LHPG zone could set the growth angle vibrating, leading to diameter fluctuations of the growing fiber.

For the case of fiber drawing from a nonwetting die the maximal meniscus height h_{max} can be determined from the estimation of Egorov et al. [38] as

$$h_{max} \approx R_{sh} = 2\gamma/P \quad \text{(drawing from non-wetting die)} \tag{1.7}$$

with P the mechanical pressure on the die (or hydrostatic pressure of the surrounded melt).

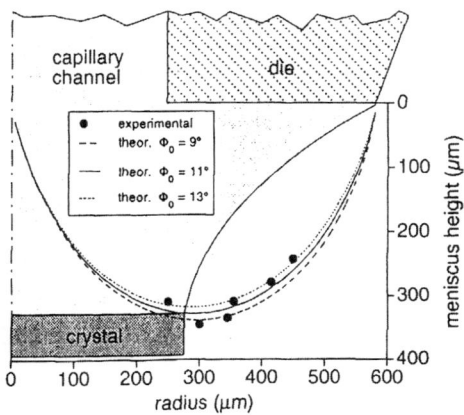

Fig. 1.10. Meniscus height and corresponding diameter of pulling-down Si fibers, calculated from (1.8), and experimentally observed values [152].

For the edge-defined growth from a wetting die, also applicable for the case of μ-PD of Si from a graphite nozzle, the relationship between meniscus height and crystal radius is given by Tatarchenko [165] as

$$h_{max} = R\cos\phi_0 \left[\cosh^{-1}\left(\frac{R_{sh}}{R\cos\phi_0} \right) - \cosh^{-1}\left(\frac{1}{\cos\phi_0} \right) \right]$$ (1.8)

(μ-PD and EFG growth).

Schäfer et al. [152] found good agreement between (1.8) and experimental meniscus heights for μ-PD growth of $Si_{1-x}Ge_x$ fibers (Fig. 1.10).

Capillary stable crystals with constant diameter d in the range $0.5\,d_{sh} < d < 0.7\,d_{sh}$ were grown without a special diameter-control system. Rudolph et al. [149] studied the maximal meniscus height for the stable μ-PD growth regime of $Al_2O_3/Y_3Al_5O_{12}$ eutectic fibers. Yoshikawa et al. [195] calculated the meniscus shape for a modified μ-PD arrangement with inserted wire in the crucible orifice to stabilize the growth conditions for very thin fibers.

1.3.4 Segregation and Axial Component Distribution

In general, fiber growth techniques from the melt imply quite favorable conditions for the axially homogeneous component distribution, like high pulling rate (i.e. effective distribution coefficients of nearly unity), small solidifying volume and small melt zones. Additionally, the high axial temperature gradient and nearly unidirected heat flow ensure the morphological stability of the growing interface.

To a good approximation for single-pass micro zone melting with identical crystal and feed diameters ($R = R_0$) the distribution of a given concentration C_S along the fiber axis z can be described by the well-known Pfann relation [49]. Foulon et al. [61] used for the pedestal growth technique, i.e. LHPG, the precise formula of Pfann considering the differences between crystal and feed radius

$(R < R_0)$, crystal and feed speed $(v > v_0)$ and theoretical (ρ) and real compactness (ρ_0) of the (mostly sintered) feed rod $(\rho > \rho_0)$

$$C_s = C_0 \left[\frac{R_0^2 v_0 \rho_0}{R^2 v \rho} - \left(\frac{R_0^2 v_0 \rho_0}{R^2 v \rho} - k_0 \right) \exp\left(-\frac{\pi R^2}{V} v_0 k_0 t \right) \right] \tag{1.9}$$

with C_s and C_0 the concentration in the crystal and feed rod, respectively, k_0 the equilibrium distribution coefficient, l the zone length, V the volume of the molten zone, and t the time.

The authors showed that, in accordance with (1.9), after the starting transient good uniformity of the axial component distribution is available. However, as Erdei et al. [49] emphasized, in the LHPG method very accurate preparation of the feed rod with high composition uniformity and compactness is required to prevent the translation of feed rod inhomogeneities to the fiber crystal. Hence, numerous authors reported on double-pass LHPG in order to improve the homogeneity and compactness of the feed rod before the final growth is provided (especially if complex material systems are used; see [75], for example). This sensitive coupling mechanism between fiber and feed rod quality seems to be, obviously, one of the principial problems of micro floating zone growth.

For the μ-PD method the starting material is completely molten and mixed. Thus, no special measures of homogenization are required if the composition is scaled exactly before the melting process. Further, the volume ratio of the bulk melt to growing single crystal fiber is very large so that, compared to LHPG, the solute concentration charge in the bulk melt is very small for a considerable length of the fiber. A quasi-diffusion controlled regime can be obtained if the capillary of the feeding nozzle is thin and long enough to satisfy the ratio $v_{cap} >> D / L_{cap}$ (v_{cap}-melt speed in the capillary ≈ crystal pulling speed v; D-diffusion coefficient of the given solvent in the melt; L_{cap}-capillary length), wellknown from the double-crucible Czochralski method.

The characteristics of the axial distribution in fibers grown by the μ-PD method are analyzed in more detail by Uda et al. [173] and Lan et al. [106] and are discussed by Uda in Chap. 2 of this book.

1.4 Fiber Materials and their Applications

1.4.1 Oxides

Sapphire

Crystalline sapphire fibers are of special interest due to their extremely superior mechanical, chemical and optical properties. Because of the high melting point (2053 °C), excellent chemical resistivity and wide transparency range (0.24–4 μm) single-crystalline fibers are well suited for sensor applications, i.e. thermometry, in high-temperature or chemically hostile processes (a thermometer, based on emission from a doped tip guided through a sapphire fiber achieves mK resolution [54]). Sapphire fibers are also used in medical power delivery systems operating at the 2.936 µm Er:YAG wavelength where silica glass fibers are highly absorbing [92]. As already mentioned above interest began with sapphire wires in the 1960s [102] and was concerned with the high tensile strength and elasticity which mode them usuable as structural reinforcement in composite materials, for example [125]. There is still an increasing demand for continuous pulling of very thin pure sapphire, mixed ceramic fibers, e.g. $CaO \cdot Al_2O_3$ and Al_2O_3/ZrO_2 [159, 110, 111], and crystalline eutectic fibers, e.g. $Al_2O_3/Y_3Al_5O_{12}$ [46, 194–196] for application in structural components with improved mechanical and thermal performance (see also Sect. 1.4.2). Finally, very thin Cr-doped Al_2O_3 (ruby) rods are applicable as micro lasers operating continuously at room temperature without active cooling [20].

La Belle and Mlavsky [102–104] grew continuous sapphire filaments in the c- and a-axis directions with high pulling speeds (up to 20 cm min^{-1} for a-direction) using the EFG technique with a molybdenum shaper. Flexible single crystalline wires of diameters in the range of 100–500 µm, spooled up onto a rotating drum, have been obtained. The single crystallinity with small number of low-angle grain boundaries was ascertained by X-ray analysis. The first fibers, having a diameter of about 100 µm, showed a high tensile strength of 5–10 times higher than Verneuil grown single crystals. The detailed fracture strength analysis of these filamentaries is described by Pollock and Bailey [133, 136]. At a diameter of 250 µm average tensile strengths of 2.75 GPa and 2.4–2.9 GPa of c- and a-axis fibers were measured at room temperature, respectively. Whereas the c-axis samples retained over 30 % of the room temperature value when tested at 1325 °C the a-axis cystals showed only a slight fall in strength with increasing temperature (80 % of the room temperature strength at 1300 °C). However, a sensitive correlation between the orientation accuracy and strength has been found in a-axis fibers. If the seed was slightly off a-axis or a misorientation developed during growth, c-plane facets (ridges) occurred at the filament surface leading to a fall in tensile fracture strength below 2 GPa by their stress concentration effect (an enhanced thermomechanical stress at growing facets can be assumed considering the unsteadiness of isotherms in this region; see the monograph of Antonov et al. [3]). Measuring the transmission quality Pollock [132] observed numerous 1 µm micro voids to lie in patterns inside the fibers which are generated at the liquid-solid interface and varied with the growth rate and temperature. In particular, at higher pulling speeds (> 6 cm min^{-1}) the voids are arranged in substructural cells elongated along the growth axis reflecting cellular interface morphology, i.e. the phenomena of thermal and/or constitutional supercooling. If the growth rate exceeded 10 cm min^{-1} dendritic structures with voids in a zigzag pattern have been observed by Pollock and Bailey [135]. In the first instance the authors attributed the generation of voids to the shrinking effect during solidification due to the large difference in density between liquid and solid Al_2O_3. Later Jegorov et al. [90] added further possible origins of micro voids, which they obtained in sapphire EFG ribbons. They attributed them to bubble generation mechanisms by chemical reactions with the molybdenum die and by melt dissociation both generating several gas products like MoO, MoO_2, AlO, AlO_2 and O_2. In fact, the situation was markedly improved by using a tungsten die with substantially higher thermal conductivity (1.0 W cm^{-2} K^{-1}) than molybdenum (0.4 W cm^{-2} K^{-1}) which stabilized the uniformity of the growing interface and filament morphology by enhanced dissipation of heat of fusion. Void-free filaments were grown at rates up to 3.0 cm min^{-1}. As a result, the light pipe efficiency and average tensile strength (3.3 GPa) were improved markedly [134].

Morscher and Sayir [125] investigated the effect of temperature on the bend survival radius of industrially produced c-axis-oriented EFG sapphire fibers. They concluded, however, that although single-crystalline fibers possess excellent creep resistance, their use as reinforcements in composites will be severely limited if the application requires even mild curvatures (bend radii less than 10 mm for 40 μm diameter). Therefore, to improve their applicability in structural components, during the last decade the target has been somewhat changed from single-crystalline to polycrystalline or amorphous fiber structures. Namely, it was shown that in such fibers enhanced elasticity and much higher pulling rates, i.e. markedly more productivity, can be achieved without loss of tensile strengths. Recently, using a novel technique of melt extraction, introduced by Marginder and Mobley [119] for metallic filaments, in which the material is cast directly from a laser heated droplet by introducing a sharpened high-speed rotating (1.5–65 m s^{-1}) molybdenum wheel into the liquid, continuous CaO·Al$_2$O$_3$ fibers with constant diameters down to 10 μm and tensile strengths of 3 GPa were obtained [2]. Elongated transmission behavior up to a wavelength of 6 μm was observed in fibers of the same alloy, which grew crystalline at Al$_2$O$_3$ contents of 81.5–100 % using the melt spun technique by extrusion from an orifice in the bottom of the crucible (Wallenberger et al. [175, 176]). The crystalline phase was found to be the metastable δ-Al$_2$O$_3$ phase caused by the rapid cooling of the molten jet. In order to stabilize the fiber diameter by an in situ pyrolytic surface reaction, propane was introduced below the crucible to form a carbon-rich skin of about 800 nm to act as a sheath against the appeareance of diameter oscillations by Rayleigh waves [177].

In the middle of the 1970s the LHPG production of ruby rods (Cr:Al$_2$O$_3$) for micro lasers was started by Burrus and Coldren [19], Takagi and Ishii [163], and Burrus and Stone [20]. Takagi and Ishii [163] analysed micro void generation for inverted LHPG and found that absolutely void-free sapphire filaments could be grown at drastically reduced growth rates (≤ 0.8 mm min^{-1}) or with a well-focused laser beam, i.e. the highest temperature gradient at the growing interface.

Burrus and Stone [20] pointed out that for a ruby rod of conventional dimensions and relatively high Cr^{3+} concentration the absorption of the pump light produces a very large temperature rise in the bulk increasing the laser threshold substantially. Compared to that, the application of ruby fibers increases the surface-to-volume ratio of the laser crystal and, hence, the effectiveness of natural cooling very obviously. For the first time they grew thin ruby laser crystals with a diameter of 40 μm and low Cr^{3+} concentration (0.02 wt %) by multistep LHPD (in order to reduce the chromium content successively) which continuously operated in a room-temperature air environment with an incident pump power of 32 mW. Compared to pure sapphire fibers in chromium doped materials the growth rate had to be reduced to below 10 mm min^{-1}. A regrowth process, of the surface region only, of a 0.5 wt %-doped sample by the same CO$_2$ laser with reduced power produced a quasi-cladded fiber with a steep radial Cr^{3+} gradient, generated by thermally induced outdiffusion of chromium [19]. This structure showed radial refractive index modulation with improved waveguide, i.e. laser translation efficiency.

In order to produce waveguides for optical power delivery systems, long sapphire fibers of lengths over 2 m and high constant diameter of 110 µm were grown by using the LHPG with accuracely controlled laser power within 0.5 % and a two-step reduction process (Jundt et al. [92]). The fibers showed scatter losses less than 0.2 dB m^{-1} in the wavelength region of 458–1064 nm. An absorption loss of 0.88 dB m^{-1} for 2936 nm light was measured for a fiber grown in an atmosphere of pure oxygen. The authors attributed this effect to reduced OH incorporation. Phomsakha et al. [130] improved the transmittance of LHPG sapphire fibers by using He as inert gas ambience.

YAG

The first Nd-doped YAG ($Y_3Al_5O_{15}$) fiber crystals for room temperature cw laser operation were grown by the LHPG method in the middle of the 1970s by Burrus and Stone [18, 160]. Starting with a platinum wire as quasi-seed the obtained rods were successively regrown until a diameter as small at 50 µm and a length of about 20 cm was achieved. At a growth rate of 5 mm min^{-1} micro crystals free of cracked cores, typically obtained in conventionally grown Nd:YAG crystals, were pulled for the first time. After dissection of the fibers in to small laser segments of 5–10 mm lengths and polishing the ends, they were placed in an external flat-mirror resonator and end-pumped by a krypton laser [18] or an (Al,Ga)As LED [160]. It was also for the first time that a Nd:YAG laser operated cw without a heat sink (due to the large surface-to-volume ratio of the fiber) pumped with a single LED. The lowest threshold was observed at a diode drive current of 45 mA where the emitted optical power was about 7 mW.

Fejer et al. [54] reported on monolithic miniature Nd:YAG oscillators with mW thresholds and single-transverse mode output having low scatter losses on the order of only 10^{-2} dB cm^{-1}. In order to enhance the light-guiding behavior a low melting compatible glass of higher refractive index was coated onto Nd:YAG single crystalline fibers by Digonnet et al. [34]. In addition, to the very high propagation loss of 8·10^{-2} dB cm^{-1} such cladded fibers of 40 µm diameter and 4 mm length also exhibited a superior lasing slope efficiency of 35 % and a cw output power of 65 mW with 17 mW operation at 1.32 µm.

These pioneering experiments with Nd:YAG micro laser crystals were very helpful to delve into the sensitive correlation between laser efficiency and fiber morphology. An optimum fiber length of ≤ 5 mm and diameter of 80–100 µm were ascertained if a LED luminous area of 85 µm was used. Whereas for larger lengths and diameters the mode filling factor and diode radiance are reduced, respectively, for smaller areas an increasing loss results from extension of the high-field region to the fiber surface.

Meanwhile well-aimed LHPG experiments for the production of YAG micro lasers were established [89]. In order to incorporate dopants, usually dopants containing sintered ceramics source rods were used [168]. However, Ishibashi et al. [89] did not adopt this method due to the danger of air bubbles in ceramics. They used undoped YAG rods on which two thin coats made of co-dopants (first CaO then Cr_2O_3) were deposited (Ca^{2+} ions act as a charge compensator to change the octahedrally coordinated Cr^{3+} ions to tetrahedrally coordinated Cr^{4+}). Both Cr and Ca ions were incorporated into the crystal during the micro zone floating process in the range of 0.01–0.04 wt %. To avoid facet growth and to give maximum laser gain, the orientation was 15° from the [100] to [110] direction. For a fiber length of 12 mm and a diameter of 320 µm the maximum cw output power was 68 mW four an input of 1.8 W and wavelength of 1.06 µm. A crystal loss less than $2 \cdot 10^{-3}$ cm^{-1} was estimated.

Recently, the growth of Nd- and Yb-doped YAG miniature lasers by the µ-PD method has also been started [25, 26]. Nd:YAG fibers as large as 550 mm in length were grown from a micro nozzle in the bottom of an Ir crucible by heating with a RF generator. To avoid of the crucible oxidation the growth arrangement was located into a vertical fused silica tube with a flowing Ar atmosphere. The best reproducibility was found for pulling down rates of 2–5 mm min^{-1} and fiber diameters of 500–800 µm. As already predicted above theoretically (Sect. 1.3.4) the effective segregation coefficient of Nd^{3+} was found to be 0.8, i.e.near to one, at a pulling rate of 5 mm min^{-1}. For Yb^{3+} a value even closer to unity (= 0.95) was obtained.

KRS-5

Thallium bromoiodite [Tl(Br,I), named KRS-5] and thallium bromide (TlBr) crystals are very promising materials for light transmission from 0.6 µm in the visible to approximatly 35 µm in the infrared. It is well known that conventional silica glass has infrared cutoffs in the range 4–5 µm where for KRS-5 a transmission loss minimum of 10^{-2}–10^{-5} dB km^{-1} has been estimated. These facts stimulated interest in the production of IR waveguides, especially for delivery of the CO_2 laser (10.6 µm) to the target. In 1978 Pinnow et al. [131] first prepared polymer clad polycrystalline KRS-5 and thallium bromide fibers with core diameters in the range 75–500 µm by a high-pressure ram extrusion process. The high extrusion rates of several centimeters per minute made possible the creation of fibers several meters long with extremly high flexibility at room temperature. A total loss of 10 % per meter was observed at 10.6 µm. In order to overcome the scattering loss from grain boundaries of polycrystalline fibers, two years later Mimura et al. [122] grew single-crystalline KRS-5 fibers 0.6–1 mm in diameter and 1–2 m in length by pulling down from an inverted EFG capillary in the bottom of the crucible, which was continuously charged by a feed rod from above. But, unfortunately, the transmission behavior of these quite interesting fibers was not reported.

LiNbO₃

Without doubt, lithium niobate (and related compounds) is one of the most investigated oxide crystal materials be cause of its large nonlinear optical coefficients and high piezoelectrical performance which makes it attractive for a wide range of applications such as in nonlinear optical devices, waveguides, surface acoustic wave devices and photorefractive data stores. Hence, is no wonder that most fiber growth experiments continue to deal with this material. Numerous original papers have been published in current journals since the beginning of the 1980s. Two growth methods are mainly applied to produce dislocation-free single-crystalline LiNbO$_3$ fibers with diameters down to 10 μm – (i) LHPG (Fejer et al. [53–56]; Feigelson [50–53]; Luh et al. [115, 116]; Foulon et al. [61], Yu and Yin [199]; Yin et al. [181–187]), and (ii) μ-PD (Fukuda et al. [69], Yoon et al. [188–193], Shur et al. [156]) including the drawing down techniques of Oguri et al. [126]. The μ-CZ method was used by Ohnishi and Yao [127].

Intensive studies on the correlation between LHPG growth conditions and LiNbO$_3$ fiber quality, especially their ferroelectric domain structure were first demonstrated by Luh et al. [115, 116]. With a diameter reduction ratio from feed rod to fiber of 2.5 to 1 and typical growth rates from 1 to 3 mm min^{-1}, perfect single-crystalline fibers with diameters ranging from 100 to 800 μm were grown routinely. The crystalline quality of μ-PD fibers with a diameter between 60 and 800 μm was analysed by Yoon et al. [188, 191, 192]. They found a sensitive dependence of the dislocation density on the fiber diameter. Whereas in thin fibers with diameters up to 500 μm no dislocations and grain boundaries were ascertained at 800 μm, a steep increase of the dislocation density, revealed by X-ray transmission topography, was sometimes noted.

As was already observed by the LHPG experiments of Fejer et al. [53] the high crystal perfection is reflected by polyhedral growth elements (facets) on the fiber surface. As sketched in Fig. 1.11a characteristic 3-fold symmetry of the cross-section, showing three growth ridges separated by an angle of 120°, appears if the fibers are grown along the c-axis (see also [127]). Two ridges are charcteristic for a-axis fibers (Fig. 1.11b) reflecting the crystallographic glide mirror plane in this orientation. If the seed is oriented exactly the ridges run smoothly and continuously down the length of the fiber.

Although the observation of Fejer et al. [53] that the noncircular nature of the cross-section does not introduce additional optical loss, Yoon et al. [190] later pointed out that this is true only so long as the diameter of the exciting laser mode does not coincide with or exceed the fiber diameter (this is in accordance with the observations of Ishibashi [89] for YAG fibers where growth facets were found as an obstacle for efficient laser oscillation). In general, for this reason and with the aim of increasing the nonlinear optical interaction length it is quite favorable to concentrate the active laser beam in the central region of the fiber. Sudo et al. [161] demonstrated this structure by refractive index cladding for LiNbO$_3$ single-crystal fibers of diameters 65–90 μm. A MgO layer was deposited onto the fiber surface producing Mg ion indiffusion during the annealing process at 1050 °C for 40 h. A 9 μm thick cladding periphery, enriched by about 20 mol % MgO with a near abruptly decreased refractive index, was obtained (Fig. 1.12a). Shimamura et al. [153] and Rudolph et al. [146] reported the in-situ Nd, Cr core doping of 1.5–4 mm thick LiNbO$_3$ rods applying the double die EFG technique. A sharp core region of 2 mm diameter was measured by radial ESMA scanning (Fig. 1.14b). Epelbaum et al. [45] succeeded in in situ clad doping in much thinner LiNbO$_3$ crystals with a diameter of 700 μm by the double-die μ-PD method. The radial distribution of the doping element (Mn), demonstrated in Fig. 1.12c, shows a 40–60 μm doped outer area. The authors attributed the small enrichment of Mn in the crystal core to the radial electric field, directed from the rim to the core and generated at the interface during growth in high temperature gradients, which drives the Mn ions to the center of the fiber. This inherent transfer effect was carefully studied by Uda and Tiller [170] for Cr- and by Uda et al. [174] for Mn-doped $LiNbO_3$ fibers grown by the LHPG and μ-PD methods, respectively. They also showed that the significant thermoelectric power, generated at the growing interface, affects the distribution and clustering of the intrinsic melt species (among others LiO^- and O^{2-}) supersaturated enough to nucleate micro bubbles which impair the optical transmission behavior of the fibers [171, 172]. Hence, for both undoped and doped $LiNbO_3$ fiber growth medium temperature gradients in the molten zone are recommended. But, in general, in the μ-PD process of middle and larger diameters (≥ 800 µm) the characteristic inversion of the radial temperature gradient in the main volume of the zone leads to field compensation and more homogeneous distribution of the dopants (Uda et al. [174]).

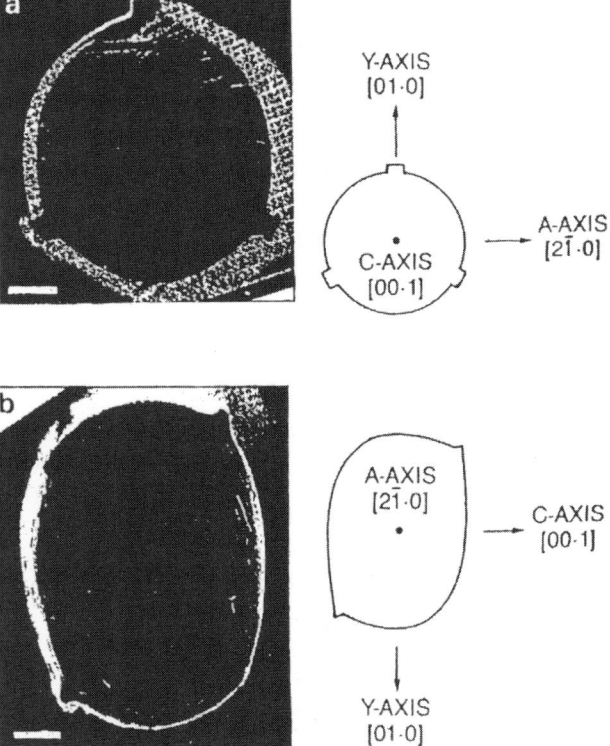

Fig. 1.11. Typical morphologies of single crystalline $LiNbO_3$ fiber cross sections perpendicular to (**a**) the c- and (**b**) the a-axis (Luh et al. [115]).

The high axial electric field, generated in growing $LiNbO_3$ fibers, is responsible for the appeareance of the characteristic ferroelectric domain arrangement (Luh et al. [115, 116, 126, 188]. In comparison with bulk Czochralski crystal fibers, single-domain crystals were grown along the c-axis. Luh et al. [115] explained this effect by the temperature gradient during fiber growth being significantly larger than those for Czochralski growth. In fact, Uda and Tiller [170] measured by a high spatial resolution pyrometer axial temperature gradients of 1500–12000 K cm^{-1} for $LiNbO_3$ fiber growth by LHPG. This result is in good agreement with ~ 5000 K cm^{-1} along the fiber surface analyzed by thermal imaging radiometry by Chen and Hu [32]. For μ-PD growth, Yoon and Fukuda [188] ascertained an axial temperature gradient of 2000–3000 K cm^{-1} by using a growing-in micro thermocouple. Such very high temperature differences generate an electric field on the order of 1 V cm^{-1}. This is much greater than the poling field used to pole Czochralski crystals during or after growth. Unlike the random domain patterns found in a-axis bulk Czochralski crystals, a-axis oriented fibers have a bidomain structure with the domain boundary along the fiber axis and parallel to the c-face. This alignment, observed with the convex shape of the growing interface, reflects the Curie isotherm morphology due to the local proximity of both in the very steep temperature gradient. In fact, the position of the domain boundary is controllable during growth by modifying the growing interface curvature. As Feigelson [52] proposed, this unexpected phenomenon may used for fabrication of unique fast double-throw switches, Q-switches for fiber lasers and mode converters.

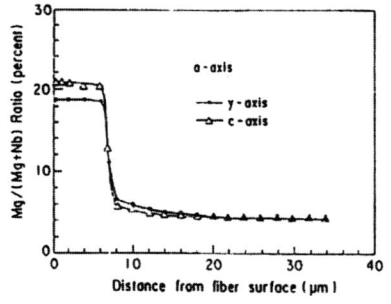

a

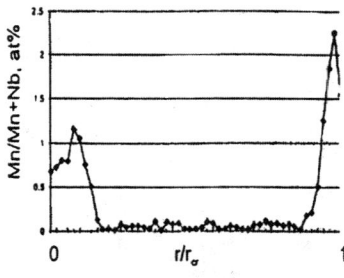

c

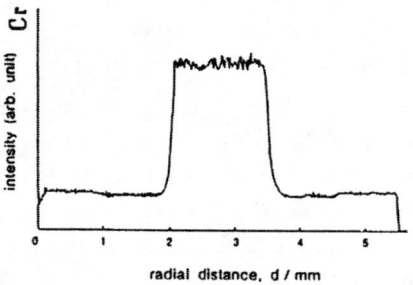

b

Fig. 1.12. Radial concentration profiles of in-situ cladded $LiNbO_3$ fibers doped with (**a**) MgO [161], (**b**) Cr [146] and (**c**) Mn [45]

Foulon et al. [61] and Ferriol et al. [58] characterized the optical properties of simultaneously Yb^{3+}- and MgO-doped as well as Sc_2O_3-doped LHPG fibers. Whereas the Yb^{3+} ions are very suitable for laser diode pumping due to the absence of up-conversion, excited-state absorption or concentration quenching (because Yb^{3+} has only a two-level state), the presence of MgO or Sc_2O_3 decreases the photorefractive damage. A resonably high value of the stimulated emission cross-section was obtained in Yb^{3+}-doped $Sc^{3+}:LiNbO_3$ and $Mg^{2+}:LiNbO_3$ fibers. It was shown that the coincidence between the wavelength of the laser emission (near 1060 nm) and the wavelength of the non-critical birefringent phase matching at room temperature can be achieved by choosing the host composition (i.e. Li_2O content) which is relatively well controllable by the starting composition of the feed rod due to the high growth speed and, hence, the effective distribution coefficient near to unity. Good homogeneity of the MgO distribution for a doping level between 0 and 6 mol % was obtained. However, it was found by spectroscopic investigations of the co-doped fibers that the addition of MgO was not so favorable for emission as a similar content of Sc_2O_3.

Quasi-phase-matched (QPM) $LiNbO_3$ fibers with periodically alternating ferro-electric domains (Fig. 1.13) were produced by the LHPG method using a periodically perturbing growth rate by asymmetric heat input or modulated heating power (Magel et al. [117, 118], Fejer et al. [55, 56], Jundt et al. [93]). The samples, having a diameter of 250 µm, were Mg-doped (5 % in the source rod). Very accurate domain periodicities between 2 and 7 µm were achieved. Brenier et al. [13] succeeded in periodic poling of MgO-doped fibers with an alternating electric field induced by two electrodes parallel to the growth axis. Such crystals are usuable for waveguiding with increased efficiency of nonlinear optical interactions and QPM supported frequency doubling of the 808–870 nm (Al,Ga)As diode laser and 1064 nm Nd:YAG laser line by second harmonic generation (SHG), for example. Fig. 1.14 shows the first blue emission spectrum of a SHG QPM micro laser based on a $Mg:LiNbO_3$ fiber and reported by Magel et al. [118]. The d_{33} nonlinear coefficient was addressed ($\lambda = 814.5$ nm) by polarizing the x-propagating fundamental wave lineary along the crystal z-axis. Then the d_{22} coefficient was addresed ($\lambda = 850.5$ nm) by rotating the polarization $90°$ so that it was parallel to the y-axis.

Jundt et al. [93] demonstrated QPM room-temperature frequency doubling of the Nd:YAG laser to generate blue (467 nm), green (532 nm) and red (660 nm) light using a-axis $Mg:LiNbO_3$ fiber crystals periodically poled with domain widths of 2.29, 3.47 and 6.31 µm. A 1.24 mm long sample generated 1.7 W of green power from an input of 4.2W. Fejer et al. [55] found no significant reduction in the throughput of periodically poled fibers by photorefractive damage although this effect for antiparallel domains had been expected.

Very hopeful results on µ-PD growth of Er-doped $LiNbO_3$ fibers were reported by Shur et al. [156]. Er^{3+} ions show an optical intra-4f transition around 1.5 µm being in coincidence with the low-loss window of a standard optical telecommunication fiber. Over the growth length of 30–35 mm a nearly uniform axial distribution of the dopant was obtained, well suited for combination with integrated optics.

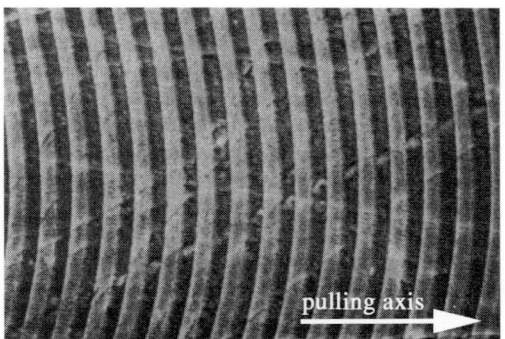

Fig. 1.13. Magnified section of a quasi-phase-matched (QPM) LiNbO$_3$ fiber with periodically alternating ferroelectric domains (Fejer et al. [55]).

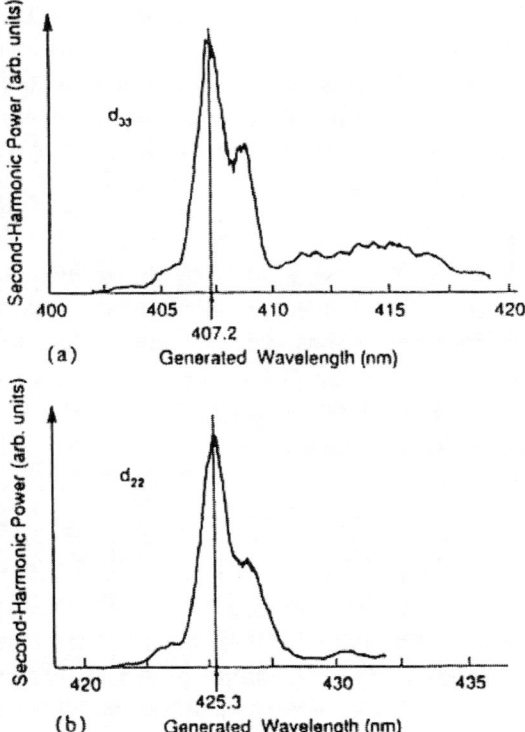

Fig. 1.14. SHG power versus wavelength in a Mg:LiNbO$_3$ micro laser for d_{33} (**a**) and d_{22} (**b**) (Magel et al. [118])

Recently, LHPG-grown photorefractive LiNbO$_3$ fibers were successfully applied as a tunable narrow-band filter by Yin et al. [181–187] (Fig. 1.15). Such filters play a significant role for spectral imaging, wavelength division multiplexing, phase array antenna signal processing, multichannal communication networking, among others. For that an incident light beam is launched into the fiber at one end, one beam of specific wavelength that matches the Bragg grating is reflected, and the rest of the light of other wavelengths will be transmitted through the fiber filter. Due to the electro-optic effect the reflected wavelength can be tuned by an electric field contacted to the fiber jacket. Because of the longitudinal dimension of the fiber this filter has a very narrow bandwidth (Fig. 1.16). Yin [186] reported tuning speeds in the ns-range and a bandwidth as narrow as 0.01 nm, being several orders faster than the conventional acousto-optic tunable filter.

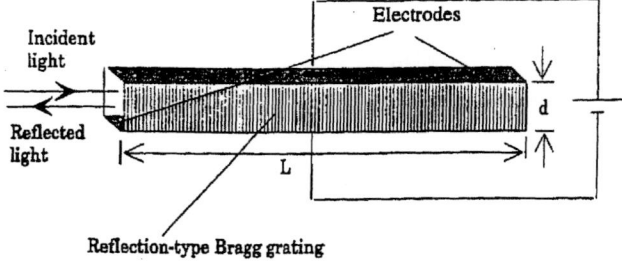

Fig. 1.15. Operation of a tunable fiber filter (L – fiber length, d – fiber diameter). An electric field is applied to influence the refractive index change (after Yin et al. [183]).

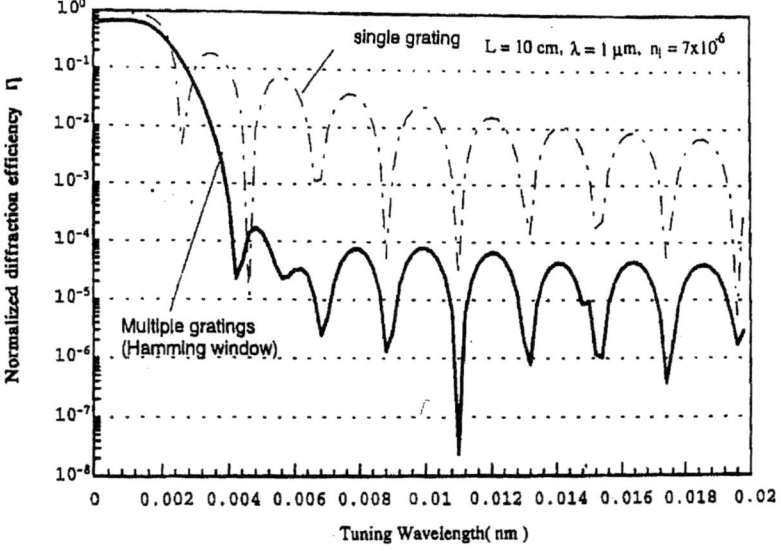

Fig. 1.16. Wavelength response for the tunable $LiNbO_3$ fiber filter (L – fiber length, λ - recording wavelength, n_1 – amplitude modulation index of the refractive index n_0 [183].

Further Niobates (KN, KLN, KLTN, KTN, KNN, BNN and SB)

Similar to $LiNbO_3$, also $KNbO_3$ (KN) is a well-known ferroelectric material for nonlinear optic and photorefractive applications. However, it melts incongruently. Therefore, usually, the crystals have to be grown from a K_2O-rich nonstoichiometric flux using TSSG, for example, with very low growth rates (< 0.01 mm min^{-1}). For the first time, Chani et al. [21, 23, 24] reported on single crystalline KN rods with a diameter of 2 mm, grown from a melt-solution with high K_2O excess at relatively high growth rates of 0.5–1 mm min^{-1} by using the pulling-down method from a larger capillary channel with diameter of 2 mm in order to ensure the supply of constant flux composition. Similar to TSSG bulk crystals the fibers were single phase and showed very flat cubic planes expressing a four-fold symmetry corresponding to the [100] orientation. Chani et al. [22, 23] refer to a serious problem with KN fiber growth – the formation of the second phase $K_4Nb_6O_{17}$ which prevents the growth of relatively large crystals with high pulling rate.

Potassium lithium niobate $K_3Li_{2-x}Nb_{5+x}O_{15+2x}$ (KLN, $0.15 < x < 0.5$) is a very suitable material for frequency doubling of (Al,Ga)As diode lasers at 300 K and presents several material advantages. It is uniaxially negative (crystallizing in the tetragonal tungsten bronze structure) and shows high nonlinear coefficients and damage threshold. No depoling occurs by mechanical shock or temperature increase. Moreover, the SHG wavelength can be adjusted to the laser wavelength of the activator ion by varying the composition. So far, neither Czochralski nor solution growth (TSSG) has succeeded in producing crack-free bulk material due to the segregation induced composition inhomogeneities along the growth axis. The large difference between the liquidus and solidus lines of the tetragonal phase requires, at low-rate bulk growth, methods differing melt composition with markedly lower Nb_2O_5 content than in the desired solid phase. Thus, for this material high-speed diffusion controlled fiber growth with an effective distribution coefficient equal to unity is of advantage.

The first single-crystalline KLN fibers with diameters up to 500 µm were grown by the µ-PD technique (Yoon and Fukuda [189–192]). Until now this method has been cultivated for KLN fiber growth and recently tested under industrial conditions [69, 85, 87]. The X-ray Rocking curve, taken from µ-PD fibers, grown with a pulling-down rate of 0.2 mm min^{-1}, showed a FWHM value of 17 arcsec [86, 87].

In 1997 the LHPG growth of Nd^{3+}-doped KLN fibers with similar diameters was reported by Ferriol et al. [57, 58] and Adachi et al. [1].

Generally, in both methods high crystallinity was achieved and the appearence of specific crystal habits can be observed. Whereas c-oriented growth is mostly characterized by a nearly round or slightly polygonal cross-section crystals along one of the a-axes always show rectangular cross-sections. In this case the fiber surface consists of perfect (001) and (010) facets elongated parallel to the growth axis (Fig. 1.17). However, the flatness roughens if the melt composition decreases below $x < 0.3$. Yoon et al. [191] did not find any impairment of the laser efficiency by the crystal habit so long as the dimension of the cross-section exceeded the diameter of the exciting laser mode [≈ 100 µm at (Al,Ga)As laser].

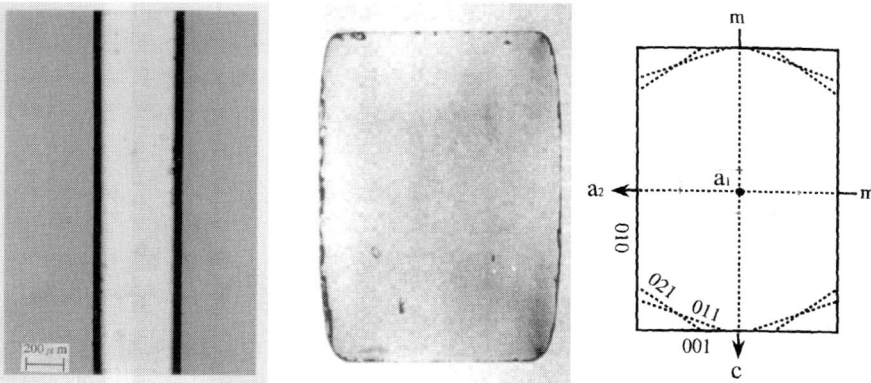

Fig. 1.17. Morphology of a KLN fiber crystal (x = 0.3) grown along the a_1-axis: the (001) facet along the fiber length (left); cross-section of an as-grown single-crystal fiber (middle); and the sketch of the cross-section with characteristic crystallographic planes (right) [191].

Both the μ-PD and LHPG methods produce KLN fibers with axial composition homogeneity. Second-harmonic blue lasers with a wavelength from 400 to 460 nm and 466 nm were generated in segments, cut from μ-PD and LHPG fibers, respectively. The phase-matching quality in μ-PD crystals was controlled by the starting melt composition and pulling rate (Imai et al.). In principle, the efficiency was better than for Czochralski pieces. However, the average conversion efficiency of fibers grown with a lower rate (0.2 mm min^{-1}) was higher (0.03 % W^{-1}, i.e. 15 % of calculated) than of that grown at a higher rate of 1.3 mm min^{-1} (0.006 % W^{-1}, i.e. 3 % of calculated). The authors attributed this effect to an observed radial composition inhomogeneity leading to a parabolic changed phase-matching wavelength by radial refractive index variation whereby the difference between the center and edge increases with the growth rate. They concluded that further optimization of the axial temperature gradient and growing interface shape is required.

Foulon et al. [63] and Ferriol [59] found that at x = 0.24 in LHPG fibers the wavelength for noncritical second-harmonic generation phase matching was found to be near 930 nm. This corresponds to a second harmonic generation in the blue range at 465 nm.

The replacement of Nb^{5+} with Ta^{5+} in KLN crystals improves the transparency spectra in the blue light region, decreases the absorption coefficient, decolorizes the material, and shifts the absorption edge towards a shorter wavelength. Moreover, such mixing crystals of the type $K_3Li_2(Ta_xNb_{1-x})_5O_{15}$ (KLTN) show a higher chemical resistivity. The first μ - PD fibers of length 10 mm with x ≤ 0.2 and excellent axial compositional homogeneity were grown by Chani et al. in 1998 [22].

A modified LHPG method with cooling argon gas flow was applied for the growth of <100>-oriented $KTa_{1-x}Nb_xO_3$ (KTN) crystal fibers with diameters of about 600 μm [86]. This material can be used in optical modulators and voltage controlled photorefractive devices. In comparison to bulk melt growth techniques, where only very slow growth rates and considerable potassium dissociation take place, a 50 times faster fiber pulling rate (0.1–0.4 mm min^{-1}) could be achieved by LHPG. The potassium evaporation was compensated by adding excess potassium to the source rod in order to form a K_2O-rich molten zone. To avoid subsequent effect of constitutional supercooling by an increased temperature gradient the argon gas stream was supplied to the as-grown crystal rod in situ. However, as a result an enrichment of crystal imperfections and Nb content in the center region were measured during these first tests.

$K_2NdNb_5O_{15}$ (KNN) and Ba_2NdNbO_6 fiber crystals of 500 µm diameter were grown by the LPHG method with pulling rates of > 1.5 mm min^{-1} and 0.45–1.0 mm min^{-1}, respectively (Qi et al. [140]). The higher speed for KNN crystals was preferred due to significant potassium evaporation. The occupancy mechanisms of the Nd^{3+} ions in both materials were measured by optical absorption and photoluminescence.

Also $Ba_2NaNb_5O_{15}$ (BNN) exhibits large optical nonlinearities. Doping with Fe or Mo leads to phase holographic storage. However, industrial applications of this material have not yet been made due to the difficulties of growth of high-quality bulk crystals as with the Czochralski method. Beyond microtwinning the main problem arises from the large thermal expansions of the c-axis during cooling of the as-grown bulk crystals leading to cracks [67]. For the first time crack- and microtwin-free fiber crystals of this material were grown using the LHPG method with pulling rates between 0.5 and 0.66 mm min^{-1} by Foulon et al. [62, 67]. Identical results were also recently obtain by the µ-PD growth technique (Lebbou et. al [107]). It was found that there is a positive effect of rare earth addition (Yb^{3+}) on the stability of the structure solving the problems of twinning and cracking very effectively [107, 108]. In LHPG Nd^{3+}-doped BNN fibers the stimulated emission cross-section at 1060 nm has been found to be resonably high [64, 65, 68]. Hence, such BNN fibers are very practicable for self-doubling micro laser arrangements in the green.

In 1988 Feigelson [52], and in 1989 Yamamoto and Bhalla [179] reported the first LHPG results of incongruent melting $Sr_xBa_{1-x}Nb_2O_6$ (SBN) single-crystalline fibers with diameter of 500 µm. No compositional striations were found as in Czochralski crystals. Such material is a promising medium for holographic data storage too. The use of fiber bundles in this application is particularly attractive because each individual fiber can store many (10–30) holograms with little crosstalk between adjacend fibers [178]. Comparing with the Czochralski technique, where diameter instabilities are typical, first stable growth perpendicular to the c-orientation (the highest photorefractive performance) was achieved by the LHPG method, whereby a quality improvement with decreasing diameter and growth rate (0.5 mm min^{-1}) was ascertained [178].

Extensive studies on the correlation between LHPG conditions and dislocation density in undoped and Ce-doped SBN fiber crystals were done by Yamamoto et al. [180]. They found that the growth from single-crystalline feeds produced fibers with far fewer dislocations than crystals from ceramic starting rods. Generally, under stable growth conditions the core was dislocation-free and after a transient of about 10 mm a rapid decrease of the dislocation density of the outer region with increasing fiber length took place. No dependence was observed between dislocation density and growth rate. The authors attributed the tendency of the central area to be dislocation-free to the analyzed convex shape of the growing interface. This shape is supported by Marangoni convection where the convection cells are concentrated in the outer zone region. Hence, air bubbles in ceramic feeds, stirred by convection, can disturb the growing interface, i.e. induce strains and, therefore, dislocations, in the periphery only. Also the possible influence of the gravitational bouyancy, leading to the enrichment of the bubbles at the upper growing interface, was discussed. For this reason a preferred downward directed growth was proposed. A core effect with a radial optical inhomogeneity, correlating with a Ba enrichment (Sr deficiency) in the centre, was also found and analyzed by Galambos et al. [70] and Erdei et al. [49]. They attributed such radial segregation effects to a variable diffusion boundary layer supported by an intensive convection flow in the laser-heated molten zone. In order to minimize this harmful phenomenon the complexe segregation interplay between Ba, Sr and Nb_2O_3 needs to be considered. A well-specified congruent composition of the feed rod combined with the pulling-down regime can probably minimize the core effect in SBN fibers.

$SrAl_{0.5}Nb_{0.5}O_3$ (SAN) and $SrAl_{0.5}Ta_{0.5}O_3$ (SAT) fibers were grown by Guo et al. [79,80] applying the LHPG method (Table 1.1).

BBO

β-BaB_2O_4 (BBO) is a material of highest importance for nonlinear optics because of its excellent parameters (6 times larger effective SHG coefficients than in KDP, wide transparency from 190 to 3500 nm, large birefringence, low dispersion, high damage threshold and good mechanical properties). These qualities allow phasematching for harmonic generation from 200 to 1500 nm [164]. However, one of the challenges for crystal growth is the existence of two crystalline phases, α (centrosymmetric) and β (noncentrosymmetric), with a first-order phase transition 170 °C below the melting point of the high-temperature α phase ($T_m^{\alpha} = 1098$ °C). Therefore, crystals of the interesting β phase can be grown only from meltsolution in order to lower the liquidus temperature below the phase transformation temperature. Tang et al. [164] first grew 20 mm-long single crystalline BBO fibers of high optical quality by the LHPG method from B_2O_3 (≥ 40 wt %) and Na_2O (≥ 3 mol %) solvents. Small amounts of these components were placed on the top of the pure BBO source rod. The molten zone was then formed by direct fusion using the CO_2 laser. The oriented seed was obtained from a previously grown TSSG crystal. Typical growth rates were in the range of 0.1–0.5 mm min^{-1}.

The predominant type of defects found in some fibers were inclusions caused by constitutional supercooling. Therefore, the authors concluded that more stable growth conditions with a smooth growing interface by constant melt-solution composition, laser power and slower growth velocity must be used.

TAG

Terbium aluminum garnet $Tb_3Al_5O_{12}$ (TAG) is a very suitable material for applications in optical communication systems. It can be used as a Faraday isolator at visible and IR wavelengths. However, the growth of bulk crystals with high Verdet constant is difficult due to the incongruent melting behavior. Ganschow et al. [71] succeeded in TAG fiber growth by using the μ-PD method. They investigated the most favorable melt composition among the starting charges of $(1-x)Al_2O_2 + x$ Tb_2O_3. The highest quality, i.e. transparency, was obtained for $x = 0.317$. The microcrystals were 700 µm in diameter and up to 100 mm in length. However, numerous eutectic inclusions were still presented indicating the necessity of further optimization of the pulling process.

In order to stabilize the garnet phase of TAG Chani et al. [27, 28] grew µ-PD fibers co-doped with the smaller lanthanide ion Lu^{3+} substituting Tb^{3+}. After Pawlak et al. [129], however, the smaller amount of terbium ions decreases the Faraday effect. They stabilized the garnet phase by replacing the aluminum ion by scandium in the octahedral site. The µ-PD crystals were 1–3 mm in diameter and 80 mm in length. The highest pulling-down rate was 5 mm min^{-1}. The best crystals were grown from a melt composition $\{Tb_{2.8}Sc_{0.2}\}[Sc](Al_3)O_{12}$. A Faraday rotation angle of 30° for a 2 mm long sample was measured.

Miscellaneous

Many more fiber growth experiments were carried out with different oxides (including HT_c materials) using the LHPG and µ-PD methods. Already in 1988 Feigelson [52] compiled 25 different oxide materials which were grown in fiber shape. Today this value is nearly doubled. Due to their large abundance an extensive discussion of each material would exceed the volume of the present review. Please see Chap. 3 and 4, where numerous oxide fiber crystals are discussed in detail. Table 1.1 summarizes the most important materials and results not reported in the above sections.

Table 1.1. Selected dielectric fiber crystals, growth parameters and applications (materials already discussed and referred to above are not more considered)

Material	Method	Growth rate (mm/min)	Diameter (µm)	Qualities	Applications	Author (Year)	Ref.
$MgAl_2O_3$	LHPG	1.67	100–150	single crystalline, faceting	Reinforcement tensile stregth: 2.14 GPa	Sigalovsky et al. (1993)	[157]
Mg_2SiO_4	LHPG	1	700	single crystalline	micro laser	Jia et al. (1991)	[91]
$Li_{1-x}Nb_{1-x}WO_3$	LHPG	0.58–0.17	500–600	single crystalline	Self-doubling laser	Foulon et al. (1996)	[66]
Gd_2O_3 $GdAlO_3$	LHPG	0.5	~1000	single crystalline	micro laser upconversion laser	Brenier et al. (1996–98)	[12–14]
$GdCa_{(1-x}M_x)_4$ $(BO)_3$	µ-PD	1–2	800	single crystalline high hardness	nonlinear optics	Ganschow et al. (2001)	[72]
$Ca_3Nb_xGa_yO_{12}$	µ-PD			single crystalline	IR micro laser	Brenier et al. (1999)	[16]
$Ca_8La_2(PO_4)_6O_2$	µ-PD			single crystalline	micro laser	Boulon et al. (2001)	[11]
$ZnLiNbO_4$	LHPG	8–54	700–1000	bi- and twinned crystals	blue and green SHG, laser	Ferriol et al. (1999)	[60]

$BaTiO_3$	LHPG	5–15	100	single crystalline	magneto-optics	Saifi et al. (1986)	[150]
			~1000		photore-fractivity	Lan et al. (1997)	[105]
$BaTi_{1-x}Zr_xO_3$	LHPG	0.16–0.25	1000	single crystalline	piezoelec-tricity	Yu and Yin (2001)	[199]
$Ba_{0.77}Ca_{0.23}TiO_3$	LHPG	0.35–0.65		single crystalline	photorefracti-vity	Barbosa et al. (2001)	[9]
$BaMg_{1/3}Ta_{2/3}O_3$	LHPG	0.5–1	300–1000	single crystalline	microwave devices $(T_m \sim 3000°C!)$	Guo et al. (1994)	[79]
$SrAl_{0.5}Ta_{0.5}O_3$	LHPG	0.5–1	500	single crystalline	microwave devices YBCO fitting	Guo et al. (1995)	[80]
$Bi_{12}TiO_{20}$ $Bi_{12}SiO_{20}$	LHPG	0.2–0.3	650–1000	single crystalline, facetting	photorefrac-tivity, data storage	Prokofiev et al. (1994)	[139]
$ScTaO_4$	LHPG	0.5–1	300–1000	single crystalline, me-tastable phase	ferroelectric-ity	Elwell et al. (1985)	[41]
$Y_{2-x}Sc_xO_3$	LHPG	0.5–2	300–700	single crystalline, in-clusions	IR upconver-sion, IR laser detection	Tissue et al. (1991)	[167, 168]
Sr_2RuO_4	LHPG	0.3–1	800	single crystalline, cracks (≥1mm/min)	supercon-ductor (at 1K)	Ardila et al. (1997)	[4]
$SrVO_3$	LHPG	0.5	600	single crystalline	supercon-ductor	Ardila et al. (2000)	[6]
$LaLuO_3$ $Lu_4Al_2O_9$	LHPG			single crystalline	scintillators	Zhang et al. (1997)	[203]
YVO_4	LHPG	5–30	750–1500	crystalline, second phases	laser host, phosphors	Erdei et al. (1993 - 94)	[47, 48]
	HP-LHPG		500·400	single crystalline		Ardila et al. (2001)	[7]
$Ca_3(VO_4)_2$	LHPG			single crystalline		Ribeiro et al. (2000)	[141]
$Bi_2Sr_2CaCu_2O_8$	LHPG	1.5–4.8	200	texture with small crys-tallites	high T_c super-conductivity	Gazit et al. (1988)	[75]
	μ-PD	0.4	1500	microcrystal plates		Lebbou et al. (2000)	[109]
$Bi_2CaSr_2Cu_2O_x$	LHPG	0.19		texture with small crys-tallites	high T_c super-conductivity	Brody et al. (1989)	[17]
$PbWO_4$	μ-PD	0.15–0.45	700	single crystalline, stoichiome-try drift	acusto-optics ionic induc-tors	Epelbaum et al. (1997)	[45]
$Ca_3(Li,Nb,Ga)_5O_{12}$	μ-PD	3–18	800	single crystalline	magneto-optics	Yu et al. (1997)	[200]
$Tb_3Ga_5O_{12}$	μ-PD	1.5–4.0	400–1200	crystalline	Faraday isola-tor	Chani et al. (2000)	[29]

NaCa$_2$Mg$_2$V$_3$O$_{12}$	µ-PD	6	800–1200	single crystalline	fundamentals	Yu et al. (1997)	[201]
CsI	pulling down	5–6	700–1000	crystalline, micro striations	IR waveguides at 10 µm	Okamura et al. (1980)	[128]
CsBr	pulling down	5–10	700–2000	crystalline, teflon cladding	IR waveguides at 10 µm	Mimura et al (1981)	[123]

1.4.2 Eutectics

Generally, the interest in eutectic compositions arises from two basic considerations, namely, that the total material behavior E results from (i) the intrinsic addition of the partial component properties E_A and E_B (like density ρ_A, ρ_B or strength σ_A, σ_B etc.) as

$$E_\Sigma = x_A \cdot E_A + (1 - x_A) \cdot E_B \tag{1.10}$$

with x the amount (molar fraction) of phase A in the matrix B, and (ii) the dot product of the partial physical derivatives $(dY/dX)_A$ and $(dZ/dY)_B$, e.g. tensor properties, as

$$E_\Pi = (dY/dX)_A \cdot (dZ/dY)_B = k_1 \cdot k_2 \cdot A \cdot B \tag{1.11}$$

where the value Y, produced by an extrinsic input value X in phase A, on its part intrinsically causes an effect Z in phase B (the generation of an electrical potential Z by a magnetic field X if phase A is piezomagnetic and phase B is piezoelectric, for example). Whereas the first characteristic in (1.10) is of high practical importance and already used in numerous well-known applications like strengthened, thoughened or resilianced structural components, the effect of (1.11) is of a more fundamental character due to the lack of well-suited eutectic systems and mutual component solubility, reducing markedly the structural and energetical efficiencies k_1 and k_2.

In case (i) the shape (aspect ratio) and structural perfection are important factors. Usually, the sum properties can be considerably improved in very thin, long and crystallographically perfect eutectic specimens. From this point of view eutectic fibers with perfect component arrangements (lamellar or fibrous) should be of special interest. However, from a study of the literature it can be concluded that even the combination of structural perfection and fiber shape has not yet been explored completely and, maybe, it is only at the beginning stage. In 1988 Feigelson [52] speculated on the production of fiber diameters comparable with the size of eutectic phases leading to self-cladding structures if the surface energy of one phase shows a significantly larger value than that of the other one and, hence, will form up at the crystal periphery. Recently, Yoshikawa et al. [195] tested this idea for the first time experimentally by using the Al$_2$O$_3$/Y$_3$Al$_5$O$_{12}$ eutectic composition.

They grew μ - PD fibers of 120–150 μm diameter with very low pulling rates in the range 0.04–0.15 mm min^{-1}. In fact, they observed the tendency of the YAG phase to concentrate in the core area and of the sapphirephase at the periphery.

The production of bulk eutectic structures with embedded whiskers in a matrix has been well known for a long time (Salkind et al. [151]). For instance, the Al-Al$_3$Ni system is made of Al$_3$Ni whiskers embedded well-oriented in the Al matrix by a uniaxial crystallization process. Further, an unidirectional solidified eutectic alloy of the copper-chromium system contains about 1.6 % by volume of aligned chromium whiskers in the copper matrix. Dissolving the matrix with acid the exposed individual chromium fibers of some millimeters in length show very high structural perfection and exhibit excellent tensile strengths. A similar approach was later demonstrated by Kawakami et al. [94]. They crystallized sintered rods of MgAl$_2$O$_4$/Mg$_2$SiO$_4$ eutectics with a diameter of 5 mm by use of a floating zone apparatus with halogen lamp mirror heating. A micro structure colony was obtained at a growth rate of 2 mm min^{-1} consisting of very thin (2 μm) spinel fibers. After the Mg$_2$SiO$_4$ matrix was chemically separated in 25 wt % NaOH aqueous solution, bundles of <100>-oriented MgAl$_2$O$_4$ whiskers were exposed. Although this method produces extremly thin fibers, which are interesting for fundamental research, the above discussed composite sum or product qualities of eutectica are not fulfilled due to phase separation after growth. Therefore, the question arises of which composite properties will appear in directional solidified eutectic fibers? What will occur if the diameter of the growing rod approaches the eutectic needle dimension? This question is currently under special investigation.

Pollock and Stormont [137] f irst reported on small eutectic Li-CaF$_2$ and LiF-NaCl ribbons (6 mm wide, 1.5 mm thick) grown by the EFG method from a nickel die with pulling rates up to 6.67 mm min^{-1}. Whereas a well-alternated lamellar structure was observed for LiF/CaF$_2$, embedded needles of LiF were found in the NaCl matrix. In both composites a considerable degree of unidirectional uniformity along the whole ribbon length has been obtained at rates of up to 4 mm min^{-1}. The measurements of the interlaminar spacing λ showed a dependence on the growth velocity v as $\lambda \sim v^{-0.55}$ consistently for both systems and predicted by the theoretical relationship $e^{-0.5}$. Somewhat later the observations were confirmed by Rogalskii et al. [143] using the same growth technique for pulling of thin NaCl/LiF rods with a diameter of 1 mm. Both results demonstrated emphatically the productive power of shaped micro crystal growth because only under such conditions can strong controlled unidirectional heat transfer, i.e. perfect alignment of the eutectic phases, take place (see the monograph of Rudolph [145]). The authors proposed the application of thin NaCl/LiF rods for waveguiding in the region 0.2–15 μm whereby the light is transmitted within the NaCl matrix and collimated at the embedded NaF or LiF needles.

Since 1998 at Fukuda's Laboratory the μ-PD growth of eutectic fibers of oxide composites has been under very active development. Such compositions, like sapphire/YAG, for example, show ultrahigh tensile yield strength, flexibility, and chemical resistivity up to very high temperatures. Filaments can be used as excellent constructional elements in high-temperature systems (e.g. engines) even within oxidizing atmospheres. It is well known that the tensile strength increases with decreasing fiber diameter, decreasing phase component dimensions (dispersivity) and increasing crystallinity. Therefore, high-velocity pulling-down fiber growth proves to be an especially suitable technique. Yoshikawa et al. [194–196], Epelbaum et al. [46] and Rudolph et al. [149] grew sapphire/YAG fibers of diameters down to 150 μm and lengths up to 500 mm having crystalline eutectic microstructures with interlamellar spacing of $\lambda \sim v^{-1/2}$ (v is the crystallization rate; Fig. 1.18). At a growth rate of 10 mm min^{-1} the value of λ yields about 1 μm. Such fibers show a tensile strength of 576.5 MPa at 1500 °C, about three times as high as that of Bridgman bulk material. In order to further improve the temperature performance, Durbin et al. [36] grew sapphirre/YAG fibers with pulling rates up to 20 mm min^{-1} giving a uniform microstructure with characteristic size as low as 0.15 μm and tensile stress of ~ 600 MPa. Lee et al. [110] clearly demonstrated the increase of tensile strength of sapphire/ZrO$_2$ fibers with a pulling rate (Fig. 1.19). Al$_2$O$_3$/Y$_3$Al$_5$O$_{12}$/ZrO$_2$ ternary eutectic fibers, grown with a pulling rate of 15 mm min^{-1}, show an extreme hardness of 17.4 GPa and tensile strength of 970 MPa at 1200 °C [111]. The maximum high-temperature value of 624 MPa at 1500 °C was obtained in sapphire/Tm$_3$Al$_5$O$_{12}$ fibers [197]. For more details see Chap. 6 of this book.

It should be noted that the industrial production of eutectic fibers with a high-strength amorphous structure by super-high-speed pulling methods, like inviscid melt spinning (e.g. Wallenberger et al. [176, 177], Mitchell et al. [124], and Sung et al. [162]) or melt extraction (Ström-Olsen et al. [159]), is already well known and comprehesively discussed in the literature. In particular, eutectic vitreous calcia-alumina (CaO/Al$_2$O$_3$) fibers, extruded with rapid solidification through a 300 μm diameter orifice in the bottom of a crucible by a nitrogen pressure of 241 kPa to the top of the melt, show enhanced tensile strengths above 1 GPa at room temperature. However, the disperse phase arrangement in such fibers is far from the above discussed crystalline strongly aligned, i.e. highly anisotropic, ones.

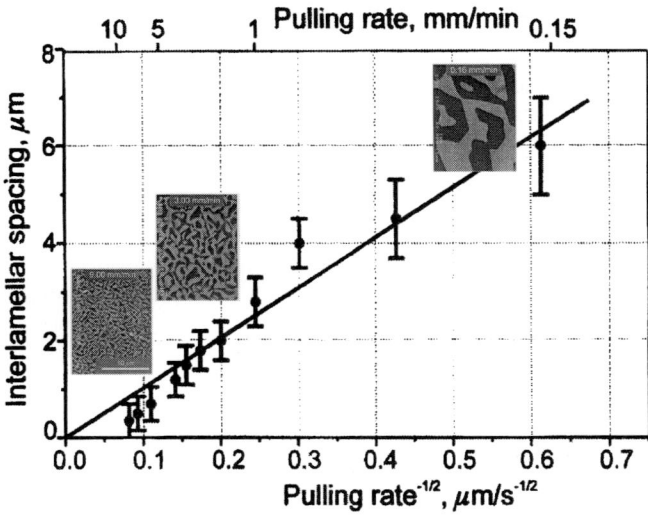

Fig. 1.18. Relation between microstructure and pulling rate of sapphire/YAG fibers after Yoshikawa et al. [194–196].

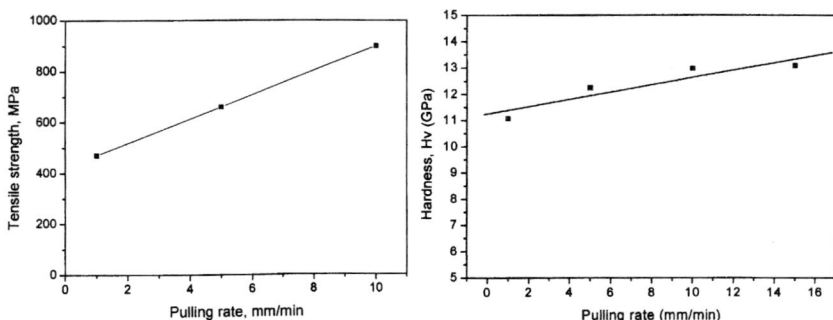

Fig. 1.19. Hardness and tensile strengths of sapphire/ZrO$_2$ eutectic fibers at 300 K vs. pulling rate after Lee et al.[110].

1.4.3 Semiconductors

Until now there have been very few reports on the melt growth of long semiconductor fibers with diameters in the μm region. This is, primarily, due to the lack in demand (we will not refer here to the very applicable field of low-dimensional semiconductor needle-shaped whiskers, grown from the vapor and solution or by the VLS mechanism, about which there exists an extensive special literature; see Levitt [112] or Givargizov [76], for example). Of course, it is possible that large-scale foils, assembled or woven from silicon or Ge$_{1-x}$Si$_x$ fibers with constant circular cross sections, could be adopted in photo-voltaics due to their excellent absorption ability. However, at present one can only speculate.

It has been known for a long time that various semiconductor profiles can be grown by shaped crystal growth techniques from the melt. In the 1960s the production of germanium rods (diameter 2–3 mm) and small ribbons (0.3 mm thick and some mm in width) was introduced by the Stepanov technique [3, 121]. Also silicon ribbons for photovoltaics have been grown since the 1070s by the EFG, inverted Stepanov technique or ribbon-to-ribbon growth with a laser heated floating zone [10]. But the first mention of LHPG Si and Ge fibers with diameters of 200 μm is found in Feigelson [52] and proposed as a model substance and for IR waveguiding, respectively. However, no growth and quality details were given.

In 1995 Riemann and coworkers [204] grew dislocation-free Si fibers with diamter of 400μm and length up to 120 cm by the floating zone method with a modified RF-coil. The fiber pulling was provided upward from a stabilised melt pedestal with velocities of 15–25 mm min^{-1}.

In 1996 Shimamura et al. [154] published the results of growth and characterization of single-crystalline silicon fibers (200 μm in diameter and 15 cm in length) grown by the μ-PD method from a nozzle in the bottom of a graphite crucible. Pulling rates up to 5 mm min^{-1} have been applied. To prevent oxidation the growth chamber, made of silica tube, was flushed with high-purity argon gas. The growth direction was always <111>. Also the Si fiber surfaces typically exhibited narrow ridges for perfect bulks. However, in these first fibers some dislocations and sub-grains occasionally appeared. Later, Epelbaum et al. [42, 43] tested the conditions for successful growth of defect-free thinner Si fibers with diameter less than 150 µm. It was shown that in the case of graphite crucibles the silicon melt is always saturated with carbon leading, by assistance of CO and SiO gases, to the chemical formation of harmful SiC particles which are able to plug the nozzle channel and enter the growing fiber crystal. In order to reduce this effect the argon stream was led over the melt surface passing the crucible from the top through vents in its wall close above the melt. As a result the melt surface could kept free of SiC particles and the effective lifespan of the crucible was increased. Further, different types of crucible-nozzle assembly where investigated to reduce the fiber diameter as much as possible. For instance, fibrous graphite was inserted in the 300 µm thick capillary nozzle to narrow the melt outlet. However, melt meniscus diameters less than 100 µm affected the growth stability considerably. Stable conditions were found for larger meniscus diameters and heights less than 20–25 µm if the pulling velocities were not more than 1.5 mm min^{-1}.

The distribution analysis of µ-PD Si$_{1-x}$Ge$_x$ fibers was described by Schäfer et al. [152], Koh et al. [99], Uda et al. [173] and Epelbaum et al. [43]. They showed that at high flow velocity of the melt through the die capillary, which was achieved by a larger crystal diameter (900 µm) reducing the RF heating power, segregation conditions with $k_{eff} = 1$ can be established and fibers of constant axial composition can be grown (Fig. 1.20). Due to the slower back-diffusion rate of the solvent component from the meniscus zone through the thin capillary to the bulk melt in the crucible (sect. 1.3.4.) and because of the very small convection influence, there is, de facto, no decay but always a nearly saturated Ge concentration at the interface (Uda et al. [173]). A uniform radial homogeneity was also ascertained if the diameter ratio between crystal and capillary was ≤ 2.8.

Koh et al. [99] demonstrated the modulation capability of the µ-PD method and produced Si$_{1-x}$Ge$_x$ fibers with linear and sinusoidal x variations along the growth axis by controlling the crystal pulling rate. Such tailored composition profiles are not only of interest from a fundamental point of view but, maybe, also suitable for functional gradient elements.

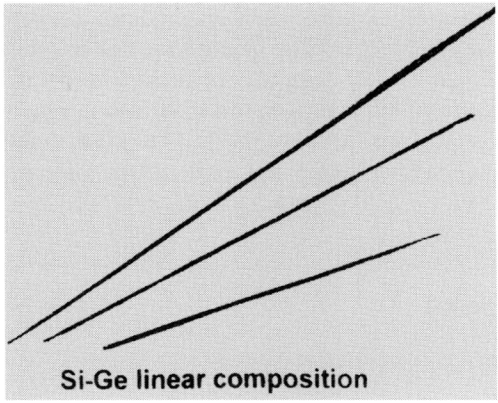

Si-Ge linear composition

Fig. 1.20. Si$_{1-x}$Ge$_x$ fibers grown by the µ-PD method at Fukuda's Laboratory (courtesy of Shimamura).

1.4.4 Metals

It was the demand for single-crystalline metallic fibers for basic reasearch (crystal structure, strength tests, electrical resistivity, thermo-electricity, etc.) which inspired the development of shaped growth methods from the melt in the 1920s. For the first time Czochralski [33] grew thin Sn, Pb and Zn fibers with diameters from 200 μm to 1 mm and lengths up to 19 cm from a small melt reservoir. Von Gomberz [77] improved the meniscus stability by insertation of a mica disk, floating on the melt surface and pierced with a circular hole through which the crystalline fiber was drawn. He reported on the growth of monocrystalline 35 cm-long Pb, Zn, Sn, Al, Cd and Bi wires having diameters between 100 μm and 1.5 mm. The same method was also used, mainly for growth of single crystalline Zn and Cu fibers, by numerous other authors (Mark et al. [120], Lindner [113, 114], Hoyem and Tyndall [83, 84]). At first the crystals were started on copper wires and later on seeding crystals whereby the first use of a crystal fragment as a seed can be attributed to Grüneisen and Goens [78].

In order to overcome the low surface tension-to-density ratio of the melt the first zone floating process with pedestal arrangement was tested in 1962 by Poplawsky [138] with ferrite crystals (Fe_2O_4, $MnFe_2O_4$, $NiFe_2O_4$), usable in microwave devices. RF heating was not applicable because of the high ferrite resistivity. Therefore, the zone melting energy was provided by an arc image heater. In 1986 Co, Fe and Co-Fe fibers were grown by the LHPG method (Hayashi et al. [82]). For effective absorption of the laser energy a sapphire rod and the metallic feed rod of Co or Fe were placed parallel to one another and the laser beam was focused directly onto the sapphire rod. The diameters of the Co and Fe single crystals, grown with a rate of 1.8 mm min^{-1}, were in the range from 50 to 500 μm. The Co-Fe alloys were also single, but with reduced quality in spite of uniform composition distribution.

Generally, at present the production of single-crystalline metal fibers is of low importance. Also the conventional pulling techniques from the melt, which had been practiced for some years in fiber metallurgy, are no longer used due to the low production rates and material compatibility problems. Today modified super-high-speed drawing methods, producing polycrystalline, amorphous and disperse alloy wires with velocities of 1.5–15 m s^{-1}, are adopted. In these techniques, mostly, a rotating disk brought into contact with a molten metal surface in a container or with a pendant drop (see Maringer and Mobley [119]). Diameters in the range of 25–125 μm and quasi-continuous fibers, spooled up onto drum, are typical.

1.5 Conclusions and Outlooks

It has been shown that single-crystalline fibers have become the subject of increasing interest because of their remarkable characteristics. Due to their very large length-to-diameter ratio composed with perfect crystallographic structure and chemical homogeneity the mechanical and physical properties approach the theoretical values, which was scarcely achievable by bulk growth methods of multi-component systems. Their ultrahigh strength enables them to be used as reinforcing agents in structural components and has led to the important technical branch of high-speed filament spinning. Fiber crystals are particularly well suited for waveguiding in the IR region, tunable narrow-band filters and nonlinear optics due to their long interaction length and tight beam confinement. They are increasingly used as room-temperature micro lasers and laser modulators, especially for generation of LD pumped second and higher harmonics in the green, blue and violet regions.

Single-crystalline fibers can be grown from the melt either by micro melt zone floating or by pulling from a micro shaper (die). Today two related techniques are well established – laser heated pedestal growth (LHPG) and micro pulling down (μ-PD) drawing from a die in the bottom of the melt crucible. Whereas LHPG produces fibers free from contact with a die, i.e. of lowest contamination level, the μ-PD method is favored by a diffusion controlled growth mode from a large melt reservoir maintaining a very uniform fiber composition. In both methods dislocation-free and single-domain fibers with diameters in the region 10–1000 and 60–1000 μm have been grown, respectively.

Due to the demands of modern optics, until now, the growth of oxide fiber crystals has been practised mainly (sapphire, YAG, KRS-5, $LiNbO_3$, further niobates, BBO, TAG and numerous miscellaneous crystals). At present, the growth of crystalline eutectic fibers with extremly high tensile strength is under forced development. The growth of semiconductor fibers is still of fundamental importance. The high-speed pulling of polycrystalline metallic filaments was well established in production a long time ago.

Of course, there are still new technological problems to be solved. Firstly, it seems to be very difficult to grow super-thin fibers with diameters below 50 μm from the melt having perfect single-crystalline structure and high diameter constancy. This is understandable from the discussed principle of capillary stability after which, for the case of LHPG, very small feed rods must be combined with an extremely uniform laser beam, and for growth methods from a shaper, infinitely small dies would be required. Secondly, in mixed crystal systems a certain radial inhomogeneity of component distribution has been frequently reported. This demands, as pointed out by the author, an improved synchronization between temperature gradient, growth direction and velocity, as well as careful scaling of the starting composition. Thirdly, if the melt evaporates incongruently there are some composition stability problems due to the large surface-to-volume ratio of molten micro zones or menisci. For this case, in addition to the requirement of high growth speeds, an in situ vapour pressure controlling system, already well established in the bulk growth of compound semiconductors, should to be quite considerable for future experiments.

In general, the growth of fiber crystals from the melt allows excellent fundamental studies on phase relations and diagrams, capillary stabilities with diminished gravity in value, growth kinetics (i.e. facetting phenomena), nearly unidirectional heat and mass transports, and dynamics of individual imperfections. In particular, the brief and uncomplicated test of the crystallization behavior of new materials by fiber growth, even before costly conventional bulk growth techniques used, is of immense encouragement for reseach. Actually, one can say that growth apparatus for single-crystal fibers belongs in each modern laboratory of material science or crystal growth.

References

1   M. Adachi, K. Miyashita, Z. Chen: Growth and Optical Properties of KLN Tungsten-Bronze Crystals, Films and Fibers. In: *Proc. Int. Symp. on Laser and Nonlinear Optic Mat. in Singapore, November 3–5, 1997*, ed. T. Sasaki (Jap. Soc. Prom. Sci., Data Storage Inst., Singapore 1997) p. 196.

2   M. Allahverdi, R.A.L. Drew, P. Rudkowska, G. Rudkowski, J.O. Ström-Olsen: Mat. Sci. Engin. **A207**, 12 (1996).

3   P.I. Antonov, L.M. Zatulovskii, A.S. Kostygov, D.I. Levinzon, S.P. Nikanorov, V.V. Peller, V.A. Tatarchenko, V.S. Juferev: *Production of Shaped Monocrystals and Forms by Stepanov Method (in Russian)*. (Nauka, Leningrad 1981).

4   D.R. Ardila, M.R.B. Andreeta, S.L. Cuffini, A.C. Hernandes, J.P. Andreeta, Y.P. Mascarenhas: J. Crystal Growth **177**, 52 (1997).

5   D.R. Ardila, J.B. Andreeta, C.T.M. Ribeiro, M.S. Li: Rev. Sci. Instruments **70**, 4606 (1999).

6   D.R. Ardila, J.P. Andreeta, H.C. Basso: J. Crystal Growth **211**, 313 (2000).

7   D.R. Ardila, A.S.S. de Camargo, J.P. Andreeta, L.A.O. Nunes: J. Crystal Growth **233**, 253 (2001).

8   R. Balasubramanian, K.B.R. Varma, A. Selvarajan: J. Crystal Growth **151,** 140 (1995).

9   L.B. Barbosa, D.R. Ardila, J.P. Andreeta: J. Crystal Growth **231**, 488 (2001).

10   A.E. Bell: RCA Review **38**, 109 (1977).

11   G. Boulon, A. Collombet, A. Brenier, M.T. Cohen Adad, A. Yoshikawa, K. Lebou, J.H. Lee, T. Fukuda: Adv. Functional Mat. **11**, 263 (2001).

12   A. Brenier, A.M. Jurdyc, H. Verweij, M.T. Cohen Adad, G. Boulon: Optical Mat. **5**, 233 (1996).

13   A. Brenier, G. Foulon, M. Ferriol, G. Boulon: J. Physics D-Applied Physics **30**, L37 (1997).

14   A. Brenier: Chem. Phys. Lett. **290**, 329 (1998).

15   A. Brenier, G. Boulon: J. Luminescence **82**, 285 (1999).

16   A. Brenier, G. Boulon, K. Shimamura, T. Fukuda: J. Crystal Growth **204**, 145 (1999).

17   H.D. Brody, J.S. Haggerty, M.J. Cima, M.C. Flemings, R.L. Barns, E.M. Gyorgy, D.W. Johnson, W.W. Rhodes, W.A. Sunder, R.A. Laudise: J. Crystal Growth **96**, 225 (1989).

18   C.A. Burrus, J. Stone: Appl. Phys. Lett. **26**, 318 (1975).

19   C.A. Burrus, L.A. Coldren: Appl. Phys. Lett. **31**, 383 (1977).

20   C.A. Burrus, J. Stone: J. Appl. Phys. **49**, 3118 (1978).

21   V.I. Chani, K. Shimamura, T. Fukuda: Flux Growth of $KNbO_3$ and Related Crystals by Pulling-Down Method. In: *Proc. Int. Symp. on Laser and Nonlinear Optic Mat. in Singapore, November 3–5, 1997*, ed. T. Sasaki (Jap. Soc. Prom. Sci., Data Storage Inst., Singapore 1997) p. 301.

22   V.I. Chani, K. Nagata, T. Kawaguchi, M. Imaeda, T. Fukuda: J. Crystal Growth **194**, 374 (1998).

23   V.I. Chani, K. Nagata, T. Fukuda: Ferroelectrics **218**, 9 (1998).

24   V.I. Chani, K. Shimamura, T. Fukuda: Cryst. Res. Technol. **34**, 519 (1999).

25   V.I. Chani, A. Yoshikawa, Y. Kuwano, K. Hasegawa, T. Fukuda: J. Crystal Growth **204** , 155 (1999).

26   V.I. Chani, Y. Yoshikawa, Y. Kuwano, K. Inaba, K. Omote, T. Fukuda: Mat. Res. Bull. **35**, 1615 (2000).

27   V.I. Chani, A. Yoshikawa, H. Machida, T. Fukuda: J. Crystal Growth **212**, 469 (2000).

28   V.I. Chani, A. Yoshikawa, H. Machida, T. Fukuda: Mat. Sci. Engin. **B75**, 53 (2000).

29   V.I. Chani, A. Yoshikawa, H. Machida, T. Satoh, T. Fukuda: J. Crystal Growth **210**, 663 (2000).

30   J.C. Chen, Q. Zhou, J. Hong, H. Wang, N. Ming, D. Feng, Ch. Fang: J. Appl. Phys. **66**, 336 (1989).

31   J.C. Chen, C. Hu: J. Crystal Growth **149**, 87 (1995).

32   J.C. Chen, C. Hu: J. Crystal Growth **158**, 289 (1996).

33   J. Czochralski: Zs. Phys. Chem. **92**, 219 (1917).

34   M.J.F. Digonnet, C. Gaeta, D. O'Meara, H.J. Shaw: IEEE J. Lightwave Tech. **5**, 642 (1987).

35   A.B. Dreeben, K.M. Kim, A. Schujko: J. Crystal Growth **50**, 126 (1980).

36   S.D. Durbin, A. Yoshikawa, K. Hasegawa, J.-H. Lee, B.M. Epelbaum, T. Fukuda, Y. Waku: Mat. Res. Soc. Symp. Proc. **581**, 577 (2000).

37   L.P. Egorov, E.S. Okun, L.M. Zatulovskii, P.M. Chaikin, V.V. Guljajev, V.Ju. Zivinskii, S.I. Levinzon, Ju. Smirnov, G.V. Satchkov: Izv. AN SSSR, ser. fizicheskaja **37**, 12 (1973).

38   L.P. Egorov, L.M. Zatulovskii, D.Ja. Kravezkii, P.M. Chaikin, E. Sidjakin: Izv. AN SSSR, ser. fizicheskaja **40**, 1361 (1976).

39   L.P. Egorov, L.M. Zatulovskii, D.Ja. Kravetzkii, B.B. Pelz, E.A. Freiman, P.M. Chaikin, I.E. Bereizna: Izv. AN SSSR, ser. fizicheskaja **43**, 1947 (1979).

40   K. Eickhoff, K. Gürs: J. Crystal Growth **6** (1969) 21.

41   D. Elwell, W.I. Kway, R.S. Feigelson: J. Crystal Growth **71**, 237 (1985).

42   B.M. Epelbaum, K. Shimamura, S. Uda, J. Kon, T. Fukuda: Cryst. Res. Technol. **31**, 1077 (1996).

43   B.M. Epelbaum, P.A. Gurjiyants, K. Inaba, K. Shimamura, S. Uda, J. Kon, T. Fukuda: Jpn. J. Appl. Phys. **36** , 2788 (1997).

44   B.M. Epelbaum, K. Inaba, S. Uda, T. Fukuda: J. Crystal Growth **178** , 426 (1997).

45   B.M. Epelbaum, K. Inaba, S. Uda, K. Shimamura, M. Imaeda, V.V. Kochurikhin, T. Fukuda: J. Crystal Growth **179**, 559 (1997).

46   B.M. Epelbaum, A. Yoshikawa, K. Shimamura, T. Fukuda, K. Suzuki, Y. Waku: J. Crystal Growth **198/199**, 471 (1999).

47   S. Erdei, F.W. Ainger: J. Crystal Growth **128**, 1025 (1993).

48   S. Erdei, G. Johnson Jr., F.W. Ainger: Cryst. Res. Technol. **29** , 815 (1994).

49   S. Erdei, L. Galambos, I. Tanaka, L. Hesselink, L.E. Cross, R.S. Feigelson, F.W. Ainger, H. Kojima: J. Crystal Growth **167**, 670 (1996).

50   R.S. Feigelson: 'Growth of Fiber Crystals'. In: *Crystal Growth of Electronic Materials*, ed. E. Kaldis (Elsevier, Amsterdam 1985) p. 127.

51   R.S. Feigelson: J. Crystal Growth **79**, 669 (1986).

52   R.S. Feigelson: Mat. Sci. Engin. **B1**, 67 (1988).

53   M.M. Fejer, J. Nightinale, G.A. Magel, R.L. Byer: Rev. Sci. Instrum. **55**, 1791 (1984).

54   M.M. Fejer, J. Nightingale, G. Magel, R. Byer: Laser Focus/Electro Optics, October (1985), p. 60.

55   M.M. Fejer, G.A. Magel, E.J. Lim: SPIE – Nonlinear Optical Properties of Materials **1148**, 213 (1989).

56   M.M. Fejer: *Guided Wave Nonlinear Optics*, ed. D.B. Ostrowski, R. Reinisch (Kluwer, Amsterdam 1992) p. 133.

57   M. Ferriol, G. Foulon, A. Brenier, M.T. Cohen-Adad, G. Boulon: J. Crystal Growth **173**, 226 (1997).

58   M. Ferriol, A. Dakki, M.T. Cohen-Adad, G. Foulon, A. Brenier, G. Boulon: J. Crystal Growth **178** , 529 (1997).

59   M. Ferriol, G. Boulon: Mat. Res. Bull. **34**, 533 (1999).

60   M. Ferriol, Y. Terada, T. Fukuda, G. Boulon: J. Crystal Growth **197**, 221 (1999).

61   G. Foulon, M. Ferriol, A. Brenier, M.T. Cohen-Adad, G. Boulon: Chem. Phys. Lett. **245**, 555 (1995).

62   G. Foulon, A. Brenier, M. Ferriol, M.T. Cohen-Adad, G. Boulon: Chem. Phys. Lett. **249**, 381 (1996).

63   G. Foulon, A. Brenier, M. Ferriol, G. Boulon: J. Physics D – Appl. Phys. **29**, 3003 (1996).

64   G. Foulon, M. Ferriol, A. Brenier, M.T. Cohen Adad, G. Boulon: Acta Physica Polonica **A 90**, 63 (1996).

65   G. Foulon, M. Ferriol, A. Brenier, G. Boulon, S. Lecocq: Europ. J. Solid State and Inorg. Chem. **33**, 673 (1996).

66   G. Foulon, A. Brenier, M. Ferriol, A. Rochal, M. T. Cohen Adad, G. Boulon: J. Luminescence **69**, 257 (1996).

67   G. Foulon, M. Ferriol, A. Brenier, M.T. Cohen-Adad, M. Boudeulle, G. Boulon: Optical Mat. **8** , 65 (1997).

68   G. Foulon, M. Ferriol, A. Brenier, M.T. Cohen-Adad, M. Boudeulle, G. Boulon: 'Nonlinear Single-Crystal Fibers of Undoped or Nd^{3+}-Doped Niobates'. In: *Proc. Int. Symp. on Laser and Nonlinear Optic Mat. in Singapore, November 3-5, 1997*, ed. T. Sasaki (Jap. Soc. Prom. Sci., Data Storage Inst., Singapore 1997) p. 167.

69   T. Fukuda, Y. Terada, K. Shimamura, V.V. Kochurikhin, B.M. Epelbaum, V.V. Chani, 'Crystal Growth of Oxide and Fluoride Materials for Optical Applications'. In: *Proc. Int. Symp. on Laser and Nonlinear Optic Mat. in Singapore, November 3–5, 1997*, ed. T. Sasaki (Jap. Soc. Prom. Sci., Data Storage Inst., Singapore 1997) p. 203.

70   L. Galambos, S. Erdei, I. Tanaka, L. Hesselink, L.E. Cross, R.S. Feigelson, F.W. Ainger, H. Kojima: J. Crystal Growth **167**, 660 (1996).

71   S. Ganschow, D. Klimm, B. M. Epelbaum, A. Yoshikawa, J. Doerschel, T. Fukuda: J. Crystal Growth **225**, 454 (2001).

72   S. Ganschow, A. Klos, D. Klimm, P. Reiche, A. Pajaczkowska. In: *Abstracts of the 13th Int. Conf. on Crystal Growth (ICCG-13), 30.07. – 04. 08. 2001 in Kyoto*, (Doshisha Univ., Kyoto 2001) p. 335.

73   D.B. Gasson, B. Cockayne: J. Mat. Sci. **5**, 100 (1970).

74   G.K. Gaulě, J.R. Pastore: *Metallurgy of Elemental and Compound Semiconductors – Boston 1960*. (Interscience, New York 1961) p. 201.

75   D. Gazit, R.S. Feigelson: J. Crystal Growth **91**, 318 (1988).

76   E.I. Givargizov: *Rost nitevidnykh i plastinchatykh kristallov iz para* (Nauka, Moscow 1977).
77   E. von Gomperz: Zs. f. Physik **8**, 184 (1922).
78   E. Grüneisen, E. Goens: Zs. f. Physik **12**, 58 (1923).
79   R. Guo, A.S. Bhalla, L.E. Cross, J. Appl. Phys. 75, 4704 (1994).
80   R. Guo, A.S. Bhalla, J. Sheen, F.W. Ainger, S. Erdei, E.C. Subbarao, L.E. Cross, J. Mat. Res. **10**, 18 (1995).
81   J.S. Haggerty: *Final Report NASA-CR-120948* (May 1972).
82   S. Hayashi, W.L. Kway, R.S. Feigelson: J. Crystal Growth **75**, 459 (1986).
83   A.G. Hoyem, E.P.T. Tyndall: Phys. Rev. **33**, 81 (1929).
84   A.G. Hoyem: Phys. Rev. **38**, 1357 (1931).
85   M. Imaeda, K. Imai, T. Fukuda: 'KLN single crystal fiber growth by micro-pulling-down method'. In: *Proc. Proc. Int. Symp. on Laser and Nonlinear Optic Mat. in Singapore, November 3–5, 1997*, ed. T. Sasaki (Jap. Soc. Prom. Sci., Data Storage Inst., Singapore 1997) p. 188.
86   T. Imai, S. Yagi, Y. Sugiyama, I. Hatakeyama: J. Crystal Growth **147**, 350 (1995).
87   K. Imai, M. Imaeda, S. Uda, T. Taniuchi, T. Fukuda: J. Crystal Growth **177**, 79 (1997).
88   T. Inoue, H. Komatsu: Kristall und Technik **14**, 1511 (1979).
89   S. Ishibashi, K. Naganuma, I. Yokohama: J. Crystal Growth **183**, 614 (1998).
90   L.P. Jegorov, L.M. Zatulovskii, D.I. Kravetskii, B.B. Pelts, E.A. Freiman, P.M. Chaikin, I.E. Berezina: Izv. AN SSSR, s. fizicheskaja **43**, 1947 (1979).
91   W. Jia, L. Lu, B.M. Tissue, W.M. Yen: J. Crystal Growth **109**, 329 (1991).
92   D.H. Jundt, M.M. Fejer, R.L. Byer: Appl. Phys. Lett. **55**, 2170 (1989).
93   D.H. Jundt, G.A. Magel, M.M. Fejer, R.L. Byer: Appl. Phys. Lett. **59**, 2657 (1991).
94   S. Kawakami, T. Yamada, S. Sakakibara, H. Tabata: J. Crystal Growth **154**, 193 (1995).
95   K.M. Kim, G.W. Cullen, S. Berkman, A.E. Bell: *Silicon Sheet Growth by the Inverted Stepanov Technique, Quarterly Progress Report No. 1.* (ERDA/JPL/ 954465 - 76/1, June 1976).
96   K.M. Kim, A.B. Dreeben, A. Schujko: J. Appl. Phys. **50**, 4472 (1979).
97   H.J. Koh, Y. Furukawa, P. Rudolph, T. Fukuda: J. Crystal Growth **149**, 236 (1995).
98   H.J. Koh, P. Rudolph, T. Fukuda: J. Crystal Growth **154**, 151 (1995).
99   H.J. Koh, N. Schafer, K. Shimamura, T. Fukuda: J. Crystal Growth **167**, 38 (1996).
100   S.A. Korpela, J. Ni, A. Chait, M. Kassemi: J. Crystal Growth **165**, 455 (1996).
101   H.K. Kuiken, P.J. Roksnoer: J. Crystal Growth **47**, 29 (1979).
101a   V.N. Kurlov, V.M. Kiiko, A.A. Kolchin, S.T. Mileiko: J. Crystal Growth **204**, 499 (1999).
102   H.E. La Belle Jr., A.I. Mlavsky: Nature **216**, 574 (1967).
103   H.E. La Belle Jr., A.I. Mlavsky: 'Growth of Continuous Whiskers from the Melt'. In: *Whisker Technology*, ed. A. P. Levitt (John Wiley, New York 1970) p. 121.
104   H.E. La Belle Jr., A.I. Mlavsky: Mat. Res. Bull. **6**, 571 (1971).
105   C.W. Lan, J.C. Chen, J.Y. Chang, C.C. Sun, H.F. Yau, M.W. Chang: 'An Integrated Research of the Growth of Bulk and Fiber Crystals for Nonlinear Optics Applications'. In: *Proc. Int. Symp. on Laser and Nonlinear Optic Mat. in Singapore, November 3–5, 1997*, ed. T. Sasaki (Jap. Soc. Prom. Sci., Data Storage Inst., Singapore 1997) p. 180.
106   C.W. Lan, S. Uda, T. Fukuda: J. Crystal Growth **193**, 552 (1998).

107   K. Lebbou, A. Yoshikawa, T. Fukuda, M.Th. Cohen-Adad, G. Boulon, A. Brenier, M. Ferriol: Mat. Res. Bull. **35**, 1277 (2000).

108   K. Lebbou, H. Itagaki, A. Yoshikawa, T. Fukuda, F. Carillo-Romo, G. Boulon, A. Brenier, M.Th. Cohen-Adad: J. Crystal Growth **210**, 655 (2000).

109   K. Lebbou, A. Yoshikawa, M. Kikuchi, T. Fukuda, M.Th. Cohen-Adad, G. Boulon: Physica C **336**, 254 (2000).

110   J.H. Lee, A. Yoshikawa, S.D. Durbin, D.H. Yoon, T. Fukuda, Y. Waku: J. Crystal Growth **222**, 791 (2001).

111   J. H. Lee, A. Yoshikawa, T. Fukuda, Y. Waku: J. Crystal Growth **231**, 115 (2001).

112   A.P. Levitt: 'Introductory Review'. In: *Whisker Technology*, ed. A.P. Levitt (John Wiley, New York 1970) p. 1.

113   E.G. Linder: Phys. Rev. **26**, 486 (1925).

114   E.G. Linder: Phys. Rev. **29**, 554 (1927).

115   Y.S. Luh, R.S. Feigelson, M.M. Fejer, R.L. Byer: J. Crystal Growth **78**, 135 (1986).

116   Y.S. Luh, M.M. Fejer, R.L. Byer, R.S. Feigelson: J. Crystal Growth **85**, 26 (1987).

117   G.A. Magel, E.J. Lim, M.M. Fejer, R.L. Byer: Optics News, December (1989), p. 20.

118   G.A. Magel, M.M. Fejer, R.L. Byer: Appl. Phys. Lett. **56**, 108 (1990).

119   R.E. Maringer, C.E. Mobley: J. Vac. Sci. Technol. **1**, 1067 (1974).

120   H. Mark, M. Polanyi, E. Schmid: Zs. f. Physik **12**, 58 (1923).

121   V.N. Maslov: *Vyraschtschivanije profilnikh poluprovodnikovykh monokristallov* (Metallurgija, Moscow 1977).

122   Y. Mimura, Y. Okamura, Y. Komazawa, Ch. Ota: Jpn. J. Appl. Phys. **19**, L269 (1980).

123   Y. Mimura, Y. Okamura, Y. Komazawa, Ch. Ota: Jpn. J. Appl. Phys. **20**, L17 (1981).

124   B.S. Mitchell, K.Y. Yon, S.A. Dunn, J.A. Koutsky: J. Non-Crystalline Solids **152**, 143 (1993).

125   G.N. Morscher, H. Sayir: Mat. Sci. Engin. **A 190**, 267 (1995).

126   H. Oguri, H. Yamamura, T. Orito: J. Crystal Growth **110**, 669 (1991).

127   N. Ohnishi, T. Yao: Jpn. J. Appl. Phys. **28**, L278 (1989).

128   Y. Okamura, Y. Mimura, Y. Komazawa, Ch. Ota: Jpn. J. Appl. Phys. **19**, L649 (1980).

129   D. A. Pawlak, Y. Kagamitani, A. Yoshikawa, K. Wozniak, H. Sato, H. Machida, T. Fukuda: J. Crystal Growth **226**, 341 (2001).

130   V. Phomsakha, R.S.F. Chang, N. Djeu: Rev. Sci. Instrum. **65**, 3860 (1994).

131   D.A. Pinnow, A.L. Gentile, A.G. Standlee, A.J. Timper, L.M. Hobrock: Appl. Phys. Lett. **33**, 28 (1978).

132   J.T.A. Pollock: J. Mat. Sci. **7**, 631 (1972).

133   J.T.A. Pollock: J. Mat. Sci. **7**, 649 (1972).

134   J.T.A. Pollock: J. Mat. Sci. **7**, 787 (1972).

135   J.T.A. Pollock, J.S. Bailey: J. Mat. Sci. **9**, 323 (1974).

136   J.T.A. Pollock, J.S. Bailey: J. Mat. Sci. **9**, 510 (1974).

137   J.T.A. Pollock, R. Stormont: J. Mat. Sci. **9**, 508 (1974).

138   R.P. Poplawsky: J. Appl. Phys. **33**, 1616 (1962).

139   V.V. Prokofiev, J.P. Andreeta, C.J. De Lima, M.R.B. Andreeta, A.C. Hernandes, J.F. Carvalho, A.A. Kamshilin, T. Jaaskelainen: J. Crystal Growth **137**, 528 (1994).

140   X. Qi, H.G. Gallagher, T.P.J. Han, B. Henderson, R. Illingworth, I.S. Ruddock: Chem. Phys. Lett. **264**, 623 (1997).

141 C.T.M. Ribeiro, D. R. Ardila, J. P. Andreeta, M. S. Li: Adv. Mat. Optics and Electronics **10**, 9 (2000).
142 J. Ricard in: *6. Int. Conference on Crystal Growth (ICCG-6), Extended Abstracts*, Vol. III (Moscow 1980) p. 150.
143 G.I. Rogalskii, S.P. Nikanorov, V.V. Peller, A.G. Ambrok: Izv. AN SSSR, ser. fizicheskaja **44**, 340 (1980).
144 P. Rudkowski, G. Rudkowska, J.O. Ström-Olsen: Mat. Sci. Engin. **A 133**, 158 (1991).
145 P. Rudolph: *Profilzüchtung von Einkristallen*. (Akademie Verlag, Berlin 1982).
146 P. Rudolph, K. Shimamura, T. Fukuda: Cryst. Res. Technol. **29**, 801 (1994).
147 P. Rudolph: III-Vs Review **9**, 27 (1996).
148 P. Rudolph, T. Fukuda: Cryst. Res. Technol. **34**, 3 (1999).
149 P. Rudolph, A. Yoshikawa, T. Fukuda: Jpn. J. Appl. Phys. **39**, 5966 (2000).
150 M. Saifi, B. Dubois, E.M. Vogel, F.A. Thiel: J. Mat. Res. **1**, 452 (1986).
151 M.J. Salkind, F.D. Lemkey, F.D. George: 'Properties of Whisker Composites'. In: *Whisker Technology*, ed. A. P. Levitt (John Wiley, New York 1970) p. 343.
152 N. Schäfer, T. Yamada, K. Shimamura, H.J. Koh, T. Fukuda: J. Crystal Growth **166**, 675 (1996).
153 K. Shimamura, N. Kodama, T. Fukuda: J. Crystal Growth **142**, 400 (1994).
154 K. Shimamura, S. Uda, T. Yamada, S. Sakaguchi, T. Fukuda: Jpn. J. Appl. Phys. **35**, L793 (1996).
155 K. Shimamura, T. Yamada, S. Sakaguchi, T. Fukuda: Jpn. J. Appl. Phys. **35**, L793 (1996).
156 J.W. Shur, W.S. Yang, S.J. Suh, J.H. Lee, T. Fukuda, D.H. Yoon: J. Crystal Growth **229**, 223 (2001).
157 J. Sigalovsky, J.S. Haggerty, J.E. Sheehan: J. Crystal Growth **134**, 313 (1993).
158 A.V. Stepanov: Zh. technich. fiziki **29**, 381 (1959).
159 J.O. Ström-Olsen, G. Rudkowska, P. Rudkowski, M. Allahverdi, R.A.L. Drew: Mat. Sci. Engin. **A179/180**, 158 (1994).
160 J. Stone, C.A. Burrus, A.G. Dentai, B.I. Miller: Appl. Phys. Lett. **29**, 37 (1976).
161 S. Sudo, I. Yokohama, A. Cordova-Plaza, M.M. Fejer, R.L. Byer: Appl. Phys. Lett. **56**, 1931 (1990).
162 Y.M. Sung, S.A. Dunn, J.A. Koutsky: Ceramics International **21**, 169 (1995).
163 K. Takagi, M. Ishii: J. Mat. Sci. **12**, 517 (1977).
164 D.Y. Tang, R.K. Route, R.S. Feigelson: J. Crystal Growth **91**, 81 (1988).
165 A.V. Tatarchenko: J. Crystal Growth **37**, 272 (1977).
166 A.V. Tatarchenko: 'Shaped Crystal Growth'. In: *Handbook of Crystal Growth, Vol. 2*, ed. D. T. J. Hurle (Elsevier Science, Amsterdam 1994) p. 1015.
167 B.M. Tissue, L. Lu, L. Ma, W. Jia, M.L. Norton, W. Yen: J. Crystal Growth **109**, 323 (1991).
168 B.M. Tissue, W. Jia, L. Lu, W.M. Yen: J. Appl. Phys. **70**, 3775 (1991).
169 S.V. Tsivinskii, Yu. Koptev, A.V. Stepanov: Soviet Physics – Solid State **8**, 449 (1966).
170 S. Uda, W.A. Tiller: J. Crystal Growth **121**, 93 (1992).
171 S. Uda, W.A. Tiller: J. Crystal Growth **121**, 155 (1992).
172 S. Uda, W.A. Tiller: J. Crystal Growth **152**, 79 (1995).
173 S. Uda, J. Kon, K. Shimamura, T. Fukuda: J. Crystal Growth **167**, 64 (1996).

174   S. Uda, J. Kon, K. Shimamura, J. Ichikawa, K. Inaba, T. Fukuda: J. Crystal Growth **182**, 403 (1997).

175   F.T. Wallenberger, N.E. Weston, S.A. Dunn: J. Non-Crystalline Solids **124**, 116 (1990).

176   F.T. Wallenberger, N.E. Weston, S.A. Dunn: J. Mater. Sci. **5**, 2682 (1990) F.T. Wallenberger, N.E. Weston, K. Motzfeldt, D.G. Swartzfager: J. Am. Ceram. Soc. **75**, 629 (1992).

178   J.P. Wilde, D.H. Jundt, L. Galambos, L. Hesselink: J. Crystal Growth **114**, 500 (1991).

179   J.K. Yamamoto, A.S. Bhalla: Mat. Res. Bull. **24**, 761 (1989).

180   J.K. Yamamoto, S.A. Markgraf, A.S. Rhalla: J. Crystal Growth **123**, 423 (1992).

181   S. Yin: Proc. SPIE **2969**, 586 (1996).

182   S. Yin, J. Zhang, F.T.S. Yu: Proc. SPIE **2844**, 153 (1996).

183   S. Yin, B.D. Guenther, F.T.S. Yu: Optical Memory & Neural Network **5**, 35 (1996).

184   S. Yin: Proc. IEEE **87**, 1962 (1999).

185   S. Yin: J. Nonlinear Optical Physics & Mat. **8**, 147 (1999).

186   S.Z. Yin: J. Nonlinear Optical Physics & Materials **8**, 147 (1999).

187   S.Z. Yin: Proc. IEEE **87**, 1962 (1999).

188   D.H. Yoon, T. Fukuda: J. Crystal Growth **144**, 201 (1994).

189   D.H. Yoon, T. Fukuda, J. Korean Ass. of Crystal Growth **4**, 405 (1994).

190   D.H. Yoon, M. Hashimoto, T. Fukuda: Jpn. J. Appl. Phys. **33**, 3510 (1994).

191   D.H. Yoon, P. Rudolph, T. Fukuda: J. Crystal Growth **144**, 207 (1994).

192   D.H. Yoon, I. Yonenaga, T. Fukuda: Cryst. Res. Technol. **29**, 1119 (1994).

193   D.H. Yoon, I. Yonenaga, T. Fukuda, N. Ohnishi: J. Crystal Growth **142,** 339 (1994).

194   A. Yoshikawa, K. Hasegawa, T. Fukuda, K. Suzuki: Ceramic Eng. Sci. **20,** 258 (1998).

195   A. Yoshikawa, B.M. Epelbaum, K. Hasegawa, S.D. Durbin, T. Fukuda: J. Crystal Growth **205**, 305 (1999).

196   A. Yoshikawa, B.M. Epelbaum, T. Fukuda, K. Suzuki, Y. Waku: Jpn. J. Appl. Phys. **38**, L55 (1999).

197   A. Yoshikawa, K. Hasegawa, J. H. Lee, S. D. Durbin, B. M. Epelbaum, D.H. Yoon, T. Fukuda, Y. Waku: J. Crystal Growth **218**, 67 (2000).

198   G.W. Young, J.A. Heminger: J. Crystal Growth **178**, 410 (1997).

199   F.T.S. Yu, S. Yin: Optical Memory & Neural Networks **1**, 289 (1992).

200   Y.M. Yu, V.I. Chani, K. Shimamura, T. Fukuda: J. Crystal Growth **171**, 463 (1997).

201   Y.M. Yu, V.I. Chani, K. Shimamura, T. Fukuda: J. Crystal Growth **177**, 74 (1997).

202   Z. Yu, R. Guo, A. S. Bhalla: J. Crystal Growth **233**, 460 (2001).

203   L. Zhang, C. Madej, C. Pedrini, B. Moine, C. Dujardin, A. Petrosyan, A.N. Belsky: Chem. Phys. Lett. **268**, 408 (1997).

204   H. Riemann, B. Hallmann, A. Lüdge in : Jahresbericht 1995, Institut für Kristallzüchtung (IKZ 1995) p. 17.

2 Fundamentals of Growth Dynamics of the μ-Pulling Down Method

Satoshi Uda

The growth dynamics of the μ-pulling down method are fully presented, focusing on the basic requirements for growth and on the solute distribution in the solid and melt during growth. The μ-PD method bears the differentiated solute transportation zones, i.e. the capillary and molten zones that characterize the effective partition coefficient with unity for the electrically neutral solute. A large temperature gradient near the interface is another attribute leading to an interface electric field that modifies the solute partitioning for ionic solutes in the melt.

2.1 Conservation of Mass and Heat and Meniscus Stability of the Molten Zone

The fundamental requirements for fiber growth are demonstrated by taking Si fiber growth as an example [1]. A Schematic illustration of the micro pulling down method is shown in Fig. 2.1. 5N-purity Si blocks were mounted in a cone-shaped crucible, 10 mm ϕ_{OD} and 6 mm ϕ_{ID}, made of graphite of high purity and density, and they were heated to melt print by the r.f. heating method. The capillary channel (0.5 mm diameter and 3 mm length) was made at the top part of the crucible cone and the melt flows down through it and comes out at its opening where the seeding and successive pull down growth process takes place. The growth direction was always <111>. The diameter of the grown Si fibers ranged from 0.2 to 1.0 mm while the length was up to 15 cm. The pulling rate was varied from 0.1 to 5.0 mm/min. To avoid oxidation of silicon, the growth chamber of the silica tube was flushed with high purity argon gas with a flow rate of 500 cc/min. The growth interface was monitored and kept uniform to keep the diameter constant.

One should maintain the molten zone volume and the zone shape constant in order to grow a stable fiber with constant diameter. This in turn requires three basic crystal growth principles: conservation of mass, conservation of energy, and mechanical stability of the system [2].

In practice, one should try to set up the growth system confine any perturbation to be as small as possible and thus enable growth conditions to be invariant during the entire growth process. This is attained by stable fiber pull down rates and heater output power plus symmetric configuration of the capillary nozzle to the crucible cone and uniform diameter of the capillary nozzle.

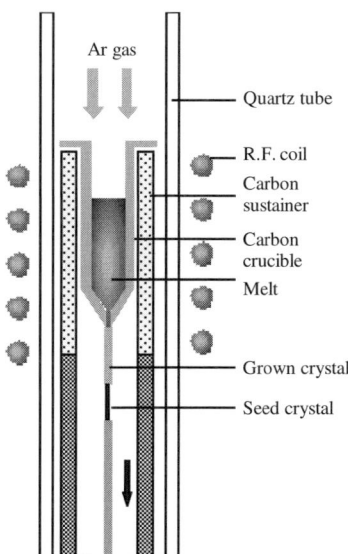

Ar gas

Quartz tube

R.F. coil

Carbon
sustainer

Carbon
crucible

Melt

Grown crystal

Seed crystal

Fig. 2.1. Schematic illustration of μ-PD growth apparatus.

Three basic parameters for fiber growth are as follows [2] and Si fiber crystal growth is taken as an example:

(a) Conservation of mass requires, when neglecting the density difference between the melt and crystal,

$$\left(d_{cry}\right)^2 v_{cry} = \left(d_{cap}\right)^2 \bar{v}_{cap}$$ (2.1)

where d_{cry} and d_{cap} are the diameters of the growing crystal and the capillary, respectively, while v_{cry} is the growth velocity and $\bar{v}_{cap}$ is the mean flow velocity in the capillary channel. Velocity perturbation easily breaks the mass conservation condition yielding fluctuations of the diameter; or intentional breaking of mass conservation leads to an interesting solute distribution which will be discussed in Sect. 2.2.1.3.

(b) Conservation of energy is described by

$$\rho_S \Delta H V_{cry} + K_L G_L = K_S G_S$$ (2.2)

where ρ_S is the density of the solid, ΔH is the latent heat, K_L and K_S are the thermal conductivity of the liquid and solid, respectively, and G_L and G_S are the temperature gradient in the liquid and solid. The first term on the left-hand side is the latent heat of crystallization and the second term is the heat flux from the melt toward the interface while the first term on right-hand side is the heat flux in the crystal away the growth interface. The first term in left-hand side is much smaller than other two terms in (2.2) in the fiber growth because of the very high temperature gradient at the interface. This enables much faster growth of single-crystal fiber than for a bulk crystal.

(c) Shape stability needs

$$\phi = \phi_0 \tag{2.3}$$

where ϕ is the angle between the meniscus and the growth axis (Fig. 2.2) and ϕ_0 is a material constant which is a function of crystallographic orientation but is independent of growth parameters such as fiber growth rate, diameter, and zone length. ϕ_0 is constrained by the surface tension relationship between solid, liquid and gas, which is described by (2.4) [2].

$$\cos(\phi_0) = (\gamma_{SG})^2 + (\gamma_{LG})^2 - \frac{(\gamma_{SL})^2}{2(\gamma_{SG})(\gamma_{LG})}, \tag{2.4}$$

where γ_{ij} is the interfacial energy between the solid-liquid, solid-gas, and liquid-gas interfaces. Measured values for ϕ_0 for Si are reported to be about $11 \pm 1°$ [3].

The fundamental parameters involved in the µ-PD method are the crucible radius at the bottom, R_1, the radius of the fiber, $R_2 \left(= \frac{1}{2} d_{cry} \right)$ and the molten zone height, H. The relationship between these three parameters is theoretically given by (2.5) [4],

$$H = R_2 \cos(\phi_0) \left[\cosh^{-1}\left(\frac{R_1}{R_2 \cos(\phi_0)} \right) - \cosh^{-1}\left(\frac{1}{\cos(\phi_0)} \right) \right]. \tag{2.5}$$

Figure 2.2 illustrates the trace of the solid-liquid-gas junction with increase of the crystal radius, R_2 and simulated curves by (2.5) with $\phi_0 = 9°$, $11°$ and $13°$ are also drawn. Triple points with various ratios of R_2/R_1 lie nearly on the $9°$ line and a nonlinear fitting technique ϕ_0 to be $9.2°$ which is appreciably less than $11 \pm 1°$ [3]. This may arise for the gravity effect, i.e. the vertex of the angle, ϕ, points upwards or downwards. One should keep the dimensional parameter relationship (2.5) constant during growth for stable fiber growth. The meniscus profile of the molten zone, $z(r)$, is also given by (2.6) [4],

$$z(r) = R_2 \cos(\phi_0) \left[\cosh^{-1}\left(\frac{R_1}{R_2 \cos(\phi_0)} \right) - \cosh^{-1}\left(\frac{1}{\cos(\phi_0)} \right) \right] \tag{2.6}$$

where r is the radial distance from the growth axis and $R_2 < r < R_1$. It should be noted that the stability of the meniscus is closely related to that of the molten zone height.

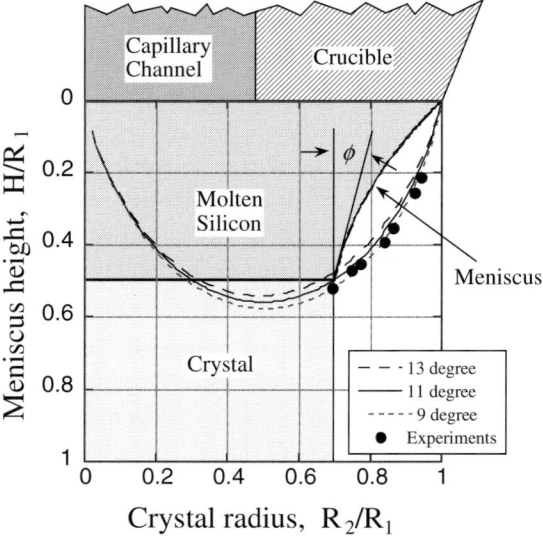

Fig. 2.2. Stability relationship between the crystal radius, crucible radius at the bottom and the molten zone height. ϕ is the angle between the meniscus and the growth axis.

2.2 Solute Transport in Melt and Solid during Growth via the μ-Pulling Down Method

2.2.1 Analytical Approach

One can gain an idea of solute transport in the melt and solid by investigating the impurity distribution in a growing crystal via the micro pulling down (μ-PD) technique (Fig. 2.1). This growth method has the great advantage of growing a single-crystal fiber with good compositional homogeneity [5–7]. Theoretical analysis of the crystallization process of this method has been attempted [7, 8].

Further elaboration of the analysis of the crystallization process via the μ-PD method was made [9] by taking as an example the fiber crystal growth of $Si_{1-x}Ge_x$ in such a way that the solute diffusion boundary volume was differentiated into the molten zone (zone I) next to the interface and the capillary zone (zone II) leading to the bulk melt (Fig. 2.3) and the steady-state solution was obtained for each zone.

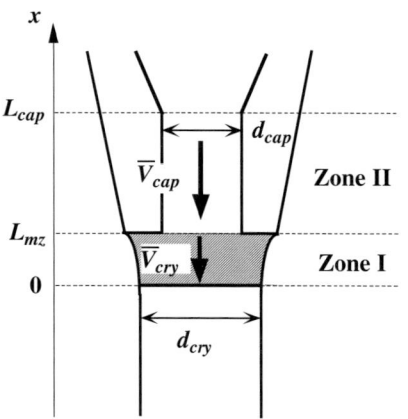

Fig. 2.3. Solute accumulation volume consisting of two zones and the coordinate system for solute concentration analysis.

The solute mass balance at the zone boundary between zones I and II is essential for steady-state growth and the instant breakage of the mass balance represents the onset of an intermediate transient. Additionally, the equilibrium partition coefficient, k_0 and the diffusion constant, D, for Ge in the $Si_{1-x}Ge_x$ melt ($x = 0.05$) were obtained for growth by the μ-PD method, which were compared with those obtained based on the Czochralski method, noting the difference between these two growth methods in the magnitude of the effect of the radial flow component on partitioning.

Two features of the μ-PD method for $Si_{1-x}Ge_x$ crystal growth should be noted; i.e. (1) the volume ratio of the bulk melt to the growing single-crystal fiber is very large so that the solute concentration change in the bulk melt is very small for a considerable length of the fiber; for instance, the solidified melt fraction, g, is only 0.03 for a 15 cm long crystal fiber of 500 μm diameter grown from a 2 g initially charged melt; and (2) for a certain growth velocity, the so-called solute diffusion boundary layer can be as long as desired by adjusting the length of the capillary nozzle which is free from convection.

There are well-known experimental methods for obtaining the equilibrium partition coefficient, k_0 and the effective partition coefficient, k, i.e., (i) via a BPS plot [10] and (ii) via the Scheil equation, represented by (2.7) and (2.8), respectively.

$$k = \frac{k_0}{k_0 + (1 - k_0)\exp\left(-\frac{V}{D}\delta_C\right)} \tag{2.7}$$

$$C_S = C_0 k (1 - g)^{k-1} \tag{2.8}$$

where V is the growth velocity, δ_C is the thickness of the solute diffusion boundary layer, C_s is the solute concentration in the solid while C_0 is the initial solute concentration in the liquid. One should note that (2.7) is valid only when the system is in a steady sate and k obtained from (2.8) is reliable only if the mixing is complete. However, these conditions are not met in practice for the normal Czochralski method except for these cases; i.e., the system could be in a pseudo-steady state to use (2.7) as long as the solidified melt fraction, g, is small, and one may be allowed to use (2.8) when k_0 is close to unity [11] or a vigorous melt flow removes the solute boundary layer.

2.2.1.1 Steady State Analysis

Since the consumption of the melt is very small during fiber growth by the μ-PD method because of the high volume ratio of the bulk melt to growing crystal, the pseudo-steady state is attainable even for the case where k_0 is appreciably less than unity. It is also possible at high growth rate that the solute enriches to be C_0/k_0 at the interface, leading k to be unity since the size of the liquid volume which is kept from convection can be large. Figure 2.3 illustrates that the solute diffusion volume consists of the molten zone (zone I) next to the interface and the capillary nozzle (zone II) through which the solute transforms from the bulk melt to the molten zone. It is assumed that these two zones are not significantly affected by melt convection and the solute flux is considered to be one-dimensional and is then described basically in terms of the gradient of the solute concentration and the growth velocity. The justification of this assumption will be discussed later. For steady-state growth, differential equations can be set up for each zone in a co-ordinate system tied to the moving interface:

<Zone I>

$$D_{Ge}\frac{\partial^2 C_L^I}{\partial x^2} + V_{cry}\frac{\partial^2 C_L^I}{\partial x} = 0 \qquad\qquad 0 \le x \le L_{mz} \qquad\qquad (2.9a)$$

and the boundary conditions are

$$V_{cry}(1-k_0)C_L^I + D_{Ge}\left(\frac{\partial^2 C_L^I}{\partial x}\right) = 0, \qquad\qquad \text{at } x = 0 \qquad\qquad (2.9b)$$

and

$$C_L^I = C_{Lmz} \qquad\qquad \text{at } x = L_{mz}, \qquad\qquad (2.9c)$$

for $r < d_{cap}/2$ where r is the radial distance from the axis of symmetry and d_{cap} is the capillary diameter.

<Zone II>

$$D_{Ge}\frac{\partial^2 C_L^{II}}{\partial x^2} + V_{cap}\frac{\partial^2 C_L^{II}}{\partial x} = 0 \qquad\qquad L_{mz} \le x \le L_{cap} \qquad\qquad (2.10a)$$

and the boundary conditions are

$$C_L^{II} = C_{Lmz} \qquad\qquad \text{at } x = L_{mz}, \qquad\qquad (2.10b)$$

and good mixing at $x = L_{cap}$ in the cone is assumed since the capillary is long enough for C_L^{II} to decrease to be close to C_0 near the outlet to the cone and such a low C_L^{II} easily becomes C_0 at $x = L_{cap}$ by convection in the cone. Thus,

$$C_L^{II} = C_0 \qquad\qquad \text{at } x = L_{cap}. \qquad\qquad (2.10c)$$

In (2.9) and (2.10), C_L is the solute concentration in zone j (j = I, II), and C_{Lmz} is the steady-state concentration at $x = L_{mz}$. L_{mz} is the position of the zone boundary betweens zones I and II while L_{cap}.is the location of the entrance of the capillary nozzle from the crucible. V_{cry} and V_{cap} are the velocities of the axial melt flow in zone I and zone II, respectively. It should be noted that V_{cry} is equal to the growth velocity when the density difference between the solid and liquid is neglected. However, V_{cap} in zone II has a parabolic profile [8] and the value for V_{cap} used in (2.10) is the average of $V_{cap} = 0$ at the wall to V_{cap} = maximum at the center. Thus, $\overline{V}_{cap}$ should be used in place of V_{cap}. Similarly, an average rate, $\overline{V}_{cry}$ should be used instead of V_{cry} since there is a change in flow cross-section at $x = L_{mz}$ which causes the inhomogeneity of V_{cry} near the zone boundary. Thus, (2.9a), (2.9b) and (2.10a) should be replaced with (2.9d), (2.9e) and (2.10d), respectively:

$$D_{Ge} \frac{\partial^2 C_L^I}{\partial x^2} + \overline{V}_{cry} \frac{\partial^2 C_L^I}{\partial x} = 0 , \qquad\qquad 0 \le x \le L_{mz} \qquad\qquad (2.9d)$$

$$\overline{V}_{cry}(1 - k_0)C_L^I + D_{Ge}\left(\frac{\partial C_L^I}{\partial x}\right) = 0 , \qquad \text{at } x = 0 \qquad\qquad (2.9e)$$

and

$$D_{Ge} \frac{\partial^2 C_L^{II}}{\partial x^2} + \overline{V}_{cap} \frac{\partial C_L^{II}}{\partial x} = 0 . \qquad\qquad L_{mz} \le x \le L_{cap} \qquad\qquad (2.10d)$$

Solute mass conservation should hold at the zone boundary:

$$\pi\left(\frac{d_{cry}}{2}\right)^2 \left(D_{Ge}\frac{\partial C_L^I}{\partial x} + \overline{V}_{cry}C_L^I\right) = \pi\left(\frac{d_{cap}}{2}\right)^2\left(D_{Ge}\frac{\partial C_L^{II}}{\partial x} + \overline{V}_{cap}C_L^{II}\right)$$

$$(2.11)$$

$$\text{at } x = L_{mz}$$

where d_{cry} is the diameter of the crystal. Solutions for zones I and II are given in (2.12a) and (2.12b), respectively:

$$C_{L(x)}^{I} = C_{Lmz} \frac{(k_0 - 1)\exp(-\overline{V}_{cry} / D_{Ge}) - k_0}{(k_0 - 1)\exp(-L_{mz}\overline{V}_{cry} / D_{Ge}) - k_0} \tag{2.12a}$$

and

$$C_{L(x)}^{II} = C_{Lmz} + (C_{Lmz} - C_0)\frac{\exp[(L_{cap} - x)\overline{V}_{cap} / D_{Ge}](-1 + \exp[(x - L_{mz})\overline{V}_{cap} / D_{Ge}])}{1 - \exp[(L_{cap} - L_{mz})\overline{V}_{cap} / D_{Ge}]}$$

$$\tag{2.12b}$$

where

$$C_{Lmz} = \frac{A}{B} \tag{2.12c}$$

and

$$A = C_0\overline{V}_{cap} \exp(L_{cap}\overline{V}_{cap} / D_{Ge})(1 - k_0 + k_0 \exp(L_{mz}\overline{V}_{cry} / D_{Ge})), \tag{2.12d}$$

$$B = \overline{V}_{cap} \exp(L_{mz}\overline{V}_{cap} / D_{Ge})(1 - k_0)$$

$$+ k_0 \exp[(\overline{V}_{cap} + \overline{V}_{cry})L_{mz} / D_{Ge}]\left(\overline{V}_{cap} - \left(\frac{d_{cry}}{d_{cap}}\right)^2 \overline{V}_{cry}\right)$$

$$+ k_0\overline{V}_{cry}\left(\frac{d_{cry}}{d_{cap}}\right)^2 \exp[(L_{cap}\overline{V}_{cap} + L_{mz}\overline{V}_{cry}) / D_{Ge}]. \tag{2.12e}$$

Fig. 2.4a illustrates the calculated Ge concentration distribution in zones I and II when $\overline{V}_{cry}$ is chosen to be 0.5, 2.0 and 8.0 cm/h, and d_{cry}/d_{cap} is fixed to be 1.6. Physical constants used for the calculation are listed in Table 2.1. Values of D and k_0 are those obtained by the initial transient analysis which will be shown in Sect. 2.2.2.2. It should be noted that there is a noticeable deflection at the zone boundary at $x = L_{mz}$ when $\overline{V}_{cry}$ is 2.0 cm/h and the Ge concentration at the interface is less than C_0/k_0, which is the interface concentration for steady-state growth with $k = 1$. Thus, strictly speaking, the growth of $\overline{V}_{cry} = 0.5$ or 2.0 cm/h is not truly steady state but pseudo-steady state since the large volume ratio of the bulk melt to fiber crystal retains the Ge concentration in the bulk melt almost constant.

However, the deflection is not obvious for $\bar{v}_{cry}$ = 0.5 cm/h and $\bar{V}_{cry}$ = 8.0 cm/h for different reasons; i.e., good mass transport communication holds very small growth velocity while a very high velocity requires steep concentration gradients with similar magnitudes for both zones at the zone boundary. Figure 2.4b demonstrates that the mass transport communication between zones I and II becomes poorer as the ratio of d_{cry}/d_{cap} is greater. Note that there is no deflection at the zone boundary when d_{cry}/d_{cap} = 1.0.

Table 2.1. Physical parameters used for the analysis of the Ge distribution.

C_0	5.0 at%
k_0	0.43
D_{Ge}	$5.6 \cdot 10^{-5}$ cm^2 s^{-1}
L_{mz}	0.03 cm
L_{cap}	0.33 cm

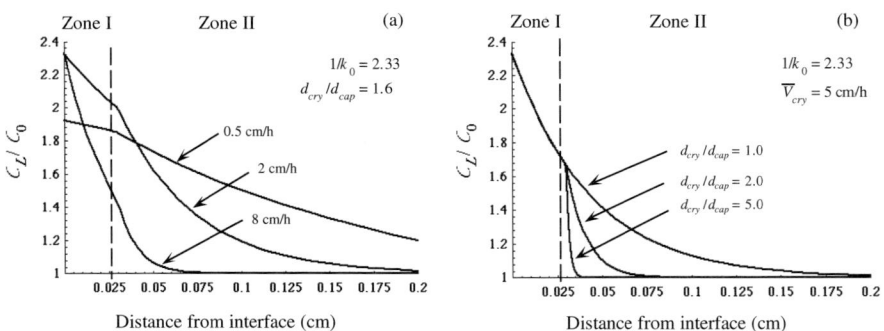

Fig. 2.4. Calculated curves of steady-state Ge distribution in zones I and II **(a)** with various growth rates for constant d_{cry}/d_{cap} (=1.6) and **(b)** various d_{cry}/d_{cap} for constant $\bar{V}_{cry}$ (= 5 cm/h). Note that there is no deflection at the zone boundary when d_{cry}/d_{cap} = 1.0.

2.2.1.2 Initial Transient Analysis

It is not possible to obtain the initial transient solution for the adjoined two zones in either analytical or numerical form since the boundary conditions are not symmetric and the solutions never converge. However, we may be allowed to use Smith et al.'s equation [12] for the analysis of the initial transient growth since the deflection point at the zone boundary is much less discernible when d_{cry}/d_{cap} is close to unity or V_{cry} becomes large enough to lead k to be unity. In this case, the Ge flux in the vicinity of the interface and primarily along the axis of symmetry should be considered rather than all over the zone. Thus, V_{cry} is used here rather than $\bar{V}_{cry}$:

$$\frac{C_S}{C_0} = \frac{1}{2}\left\{\begin{array}{l}\left[1 + \mathrm{erf}\left(\dfrac{\sqrt{\left(V_{\mathrm{cry}}/D_{\mathrm{Ge}}\right)}x}{2}\right) + (2k_0 - 1)\exp\left[\dfrac{-k_0(1-k_0)V_{\mathrm{cry}}}{D_{\mathrm{Ge}}}x\right]\right] \\ \cdot\,\mathrm{erfc}\left[\dfrac{(2k_0-1)}{2}\sqrt{\left(V_{\mathrm{cry}}/D_{\mathrm{Ge}}\right)}x\right]\end{array}\right\} \qquad (2.13\mathrm{a})$$

which is derived from the solution to the differential equation (2.13b) subject to the boundary conditions (2.13c), (2.13d) and (2.13e):

$$D_{\mathrm{Ge}}\frac{\partial^2 C_{\mathrm{L}}}{\partial x^2} + V_{\mathrm{cry}}\frac{\partial C_{\mathrm{L}}}{\partial x} = \frac{\partial C_{\mathrm{L}}}{\partial t}, \qquad\qquad (2.13\mathrm{b})$$

$$V_{\mathrm{cry}}(1 - k_0)C_{\mathrm{L}} + D_{\mathrm{Ge}}\left(\frac{\partial C_{\mathrm{L}}}{\partial x}\right) = 0 \qquad \text{at } x = 0 \text{ for all } t \geq 0, \qquad (2.13\mathrm{c})$$

$$C_{\mathrm{L}} = C_0 \qquad\qquad \text{at } t = 0 \text{ for all } x \geq 0 \qquad (2.13\mathrm{d})$$

and

$$C_{\mathrm{L}} = C_0 \qquad\qquad \text{at } x = \infty \text{ for all } t \geq 0. \qquad (2.13\mathrm{e})$$

It should be checked whether Equation (2.13b) approximately illustrates steady-state growth with large V_{cry} via the μ-PD method when $t = \infty$. Equation (2.9b) matches (2.13c) while (2.10c) is equivalent to (2.13e) because L_{cap} is large enough to be considered as $x = \infty$, which is verified in Fig. 2.4 for large growth velocity. However, (2.13a) does not take account of the effect of lateral melt convection on partitioning. This issue is very important and will be discussed later, combined with experimental results.

2.2.1.3 Intermediate Transient Analysis

An abrupt change of growth velocity during steady-state growth from V_{cry} to V' ($V_{\mathrm{cry}} < V'$) yields a transient rise in the concentration in the liquid in the vicinity of the interface [12] and consequently a concentration rise in the solid. Then after a while the system will be back to the same steady state as that before the velocity changes. Here V_{cry} is also used. This transient was given by Smith et al. [12] when time was measured from the instant at which the velocity change occurs. In a practical case, the sudden change from the small velocity in the pseudo-steady state to a velocity large enough for the steady state is much easier to achieve experimentally occur and this transient may be found by solving (2.13b) subject to the boundary conditions, (2.13c), (2.13e) and (2.14a):

$$C_L = C_0 \left\{ 1 + \frac{p}{k_0} \exp\left[\left(-V_{cry} / D_{Ge} \right) x \right] \right\} \qquad \text{at } t = 0 \text{ for all } x \geq 0 \qquad (2.14a)$$

which is a modification of Tiller et al.'s well-known concentration equation [13]. Instead of their $q \ (= 1 - k_0)$ [13], $p = \alpha - k_0$ is used and α is the degree of the solute saturation in the liquid at the interface, taking the range of $k_0 \leq \alpha \leq 1.0$:

$$\alpha = \frac{C_{L(0)}}{C_0 / k_0} . \qquad (2.14b)$$

It should be noted that (2.14a) is the true steady-state solution when $\alpha = 1.0$ while the system with $\alpha = k_0$ is under the control of complete mixing. The concentration in the solid for $\alpha < 1.0$ is found by replacing some of q of (43) in reference [12] with p:

$$\frac{C_S(x)}{C_0} = 1 - \frac{1}{2} \operatorname{erfc} \left[\frac{\sqrt{\left(V'_{cry} / D_{Ge} \right)} x}{2} \right]$$

$$+ p \frac{\frac{1}{2} - V_{cry} / V'_{cry}}{k_0 - V_{cry} / V'_{cry}} \exp\left[-\frac{V_{cry}}{V'_{cry}} \left(1 - \frac{V_{cry}}{V'_{cry}} \right) \frac{V'_{cry}}{D_{Ge}} x \right] \operatorname{erfc} \left[\left(\frac{V_{cry}}{V'_{cry}} - \frac{1}{2} \right) \sqrt{\left(V'_{cry} / D_{Ge} \right) x} \right]$$

$$+ \frac{2k_0 - 1}{2} \frac{p + k_0 - V_{cry} / V'_{cry}}{k_0 - V_{cry} / V'_{cry}} \exp\left[-k_0 q \left(V'_{cry} / D_{Ge} \right) x \right] \operatorname{erfc} \left[\left(k_0 - \frac{1}{2} \right) \sqrt{\left(V'_{cry} / D_{Ge} \right) x} \right].$$

$$(2.15)$$

Note that zone I + zone II was here again treated as one zone. Figure 2.5 illustrates the calculated intermediate transient for the case of $V_{cry} = 0.5$ cm/h and $V_{cry} / V'_{cry} = 0.1$, 0.2 and 0.5 where $\alpha = 0.85$ is obtained by using (2.12a). $d_{cry} / d_{cap} = 1.6$ and the other parameters used for the calculation are listed in Table 2.1.

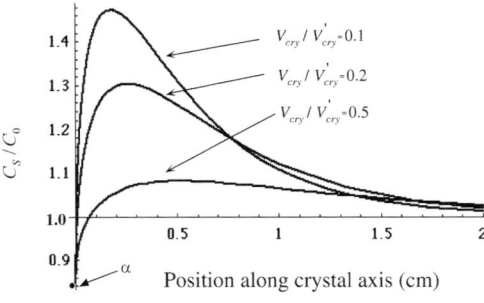

Fig. 2.5. Illustration of calculated intermediate transient for the case of V_{cry} = 0.5 cm/h and V_{cry} / V'_{cry} = 0.1, 0.2 and 0.5. α = 0.85 and d_{crv} / d_{cap} =1.6. The origin is arbitrarily positioned in the pseudo-steady-state region.

2.2.2 Example of Fiber Crystal Growth

$Si_{1-x}Ge_x$ of 2 g total weight with $x = 0.05$ was synthesized as a source material from the raw materials of 5N Si and 4N Ge. The crucible was made of graphite of high purity and density. The tip of the crucible has a flat-round cross-section of 1.1–2.0 mm diameter, which contains a capillary nozzle of 0.5 mm inner diameter and 3 mm length (Fig. 2.1). The experiment was performed under a flowing Ar gas atmosphere of 500 cm^3/min. The diameter of grown $Si_{1-x}Ge_x$ crystal fibers ranged from 0.8 to 1.9 mm and the pulling rate was varied from 0.5 to 10 cm/h. The grown samples were cut and polished along the axis of symmetry, and accurate concentrations were determined with an EPMA.

The gravity segregation of Ge was a serious concern for $Si_{1-x}Ge_x$ growth and experiments were performed to minimize its effect. After the Si and Ge raw materials were ground and well mixed, they were charged into the graphite crucible and instantly melted by the rf heating method. The melt on the outlet of the capillary nozzle was touched by a Si seed grown along the <111> direction and the initial transient was initiated immediately following the melting of the charge so that there was not enough time for gravity segregation to effectively take place before the initial transient started. Moreover, after the system seemed to reach the steady state, the concentration was constant for a fairly long region, which showed little accumulation effect due to gravity even during the growth, and the steady-state concentration of Ge measured with an EPMA along the central axis of the crystal was very close to the Ge content of the initial charge. Thus, it was assumed that the initial concentration in zones I, II and the cone was almost constant and nearly equal to C_0.

2.2.2.1 Steady State

The critical growth rate for differentiating pseudo-steady-state growth and steady-state growth is ~ 3 cm/h under the growth conditions in Table 2.1, which matches the expectation of (2.12a) and (2.12b).

It was difficult to completely quench down the furnace system by turning off the loaded power to freeze in the steady-state distribution of Ge in both the melt and solid across the interface since the thermal capacity of the bulk melt in the crucible was large, preventing the rapid quenching of the melt in the capillary and the molten zone. Thus, it is not appropriate to use (2.12a)–(2.12e) to deduce k_0 and D_{Ge}. Figure 2.6 represents the Ge distribution near the interface in the quenched sample. It should be noted that the Ge concentration distribution showed a steep positive gradient toward the molten zone starting at $x = 0$. This is because the quenching experiment drove a very large V'_{cry} leading to a large V'_{cry}/V_{cry} which consequently yielded the abrupt rise of the Ge concentration. The value of C_s/C_0 is about 2.3, which needs $V'_{cry}/V_{cry} > 10^3$ when $V_{cry} = 5$ cm/h. It should be noted that $C_s/C_0 \approx 2.3$ is very close to $1/k_0$.

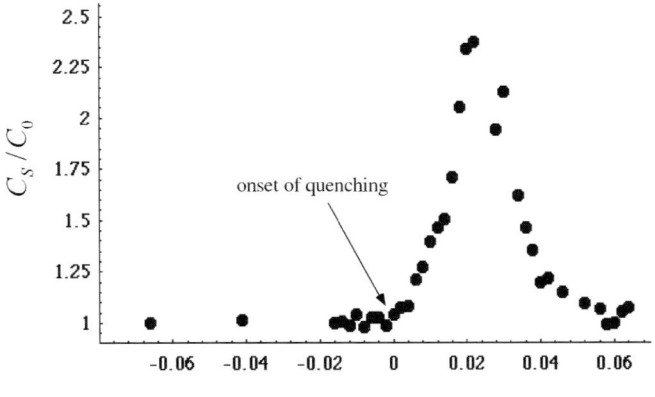

Position along crystal axis (cm)

Fig. 2.6. Ge distribution near the interface in the quenched sample. Note that the rapid growth occurred after quenching. V_{cry} = 5 cm/h and d_{cry}/d_{cap} =1.6.

2.2.2.2 Initial Transient

Figure 2.7 represents the Ge distribution in the initial transient to steady state. V_{cry} was 5 cm/h. By fitting the data in Fig. 2.7 with (2.13a) via nonlinear programming, the *polytope technique* [14], k_0 was found to be 0.43 and D_{Ge} was found to be $5.6 \cdot 10^{-5}$ cm^2 s^{-1}. The value of D_{Ge} is an order of magnitude smaller than those of various dopants in the Si melt which were previously reported [15, 16]. This difference will be discussed later and is associated with the effect of the radial flow component on the solute flux and partitioning.

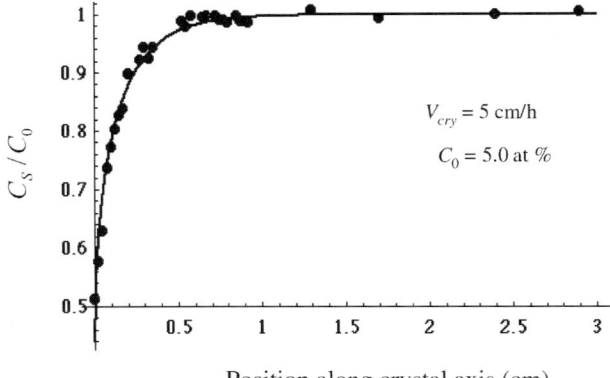

Position along crystal axis (cm)

Fig. 2.7. Ge distribution in the initial transient to steady state. V_{cry} = 5 cm/h and d_{cry}/d_{cap} = 1.6. The solid line was calculated from (2.13a). The origin is the junction between the seed and the grown crystal.

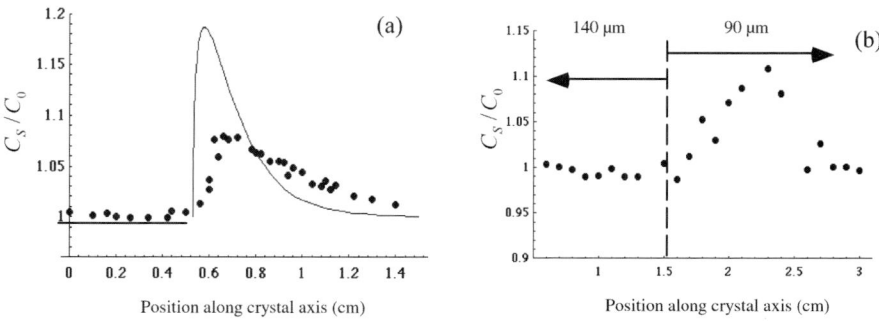

Fig. 2.8. Intermediate transient rise of Ge concentration due to (a) the sudden change of growth velocity, $V_{cry} = 3.0$ to 6.0 cm/h, while d_{cry}/d_{cap} is kept at 1.6, and (b) the sudden reduction of the molten zone thickness from 140 to 90 µm while V_{cry} is fixed at 3.0 cm/h. The solid line in (a) calculated from (2.15) is drawn over the plot for comparison. The origin is arbitrarily positioned in the steady-state region.

2.2.2.3 Intermediate Transient

Since the abrupt change of the growth velocity was not satisfactory in experiments, the concentration rise due to the sudden increase of the growth velocity from V_{cry} to V'_{cry} was much lower than expected by calculation. This is represented in Fig. 2.8a for the case of $V_{cry} = 3$ cm/h and $V'_{cry} = 6$ cm/h where very little change of the molten zone was affirmed. It should also be noted that the recovery to the steady state is slower than expected by calculation from (2.15) since $d_{cry} \neq d_{cap}$. The sudden volume reduction of the molten zone, keeping V_{cry} constant, gave a similar change of the Ge concentration which is illustrated in Fig. 2.8b. These intermediate transients are explained in terms of the solute accumulation diversity in the vicinity of the interface due to the sudden change of the growth conditions, which is, for convenience, measured by the magnitude of the imbalance of the Ge mass flow, ΔM, at the boundary between zones I and II, i.e.,

$$\Delta M = J_\mathrm{I} S_\mathrm{I} - J_\mathrm{II} S_\mathrm{II} \tag{2.16a}$$

$$= \pi \left(\frac{d_{cry}}{2}\right)^2 \left(-D_{Ge}\frac{\partial C}{\partial x} - V^\mathrm{I} C_\mathrm{L}^\mathrm{I}\right) - \pi \left(\frac{d_{cap}}{2}\right)^2 \left(-D_{Ge}\frac{\partial C}{\partial x} - V^\mathrm{II} C_\mathrm{L}^\mathrm{II}\right)$$

$$\text{at } x = L_{mz} \tag{2.16b}$$

where S is the cross section of zone j ($j = $ I, II). When the system is in steady state, $\Delta M = 0$, $V^\mathrm{I} = \overline{V}_{cry}$ and $V^\mathrm{II} = \overline{V}_{cap}$. In the intermediate transient, the magnitude of the nonzero ΔM directly reflects the extent of the concentration deviation from the steady state. The abrupt increase of the growth velocity from V_{cry} to V'_{cry} leads to - $J_\mathrm{I} > -J_\mathrm{II}$ since the flow velocity change, ΔV^j in accordance with the growth velocity change is larger than ΔV^{II}, and the difference is more distinct when the ratio of d_{cry}/d_{cap} is greater and the mass transport communication is poorer between zones I and II (see Fig. 2.4b). The sudden reduction of the molten zone size requires a smaller V^{II} leading to the same flux relationship, $-J_I > -J_{II}$. This yields the transitional increase of the Ge concentration in the melt in the vicinity of the interface introducing the transitional rise of Ge concentration in the solid. It is the ratio of d_{cry}/d_{cap} that determines how fast the system returns to the steady state; i.e., the large d_{cry}/d_{cap} retards recovery.

It is thus suggested that the desired impurity distribution in Si single-crystal fibers can be designed by the manipulation of the imbalance of the impurity mass flow at the zone boundary.

2.2.3 Effect of Flow in the Molten Zone

2.2.3.1 Effect of Marangoni Flow and Central Down Flow in the Molten Zone

The influence of melt convection in the molten zone on partitioning due to Marangoni flow and the central down flow from the capillary should be evaluated before trying to fit the initial transient region with (2.13a) [12] to calculate k_0 and D for Ge. This is because (2.13a) was deduced only for the no-mixing case. The most serious problem may be the effect of Marangoni flow in the molten zone on partitioning. Marangoni flow was clearly perceived in the μ-PD method though it looked less operative than that observed during the fiber growth of oxide crystals. The Marangoni number, Ma, was roughly estimated without taking account of the 5 at% Ge content and its value, $Ma \approx 70$, is small; however the Marangoni velocity is larger than the growth rates. Physical and geometrical constants for the pure Si melt [17–20] used for the calculation are listed in Table 2.2.

The effect of the disturbance by the central down flow from the capillary to the molten zone should also be discussed. The effect may be expressed in terms of the magnitude of scratching the solute diffusion layer which determines how far the solute accumulation is from the no-mixing case or complete mixing case. Figure 2.9 shows the radial concentration variations in crystals which were found for steady-state growth with different ratios of d_{cry}/d_{cap}. A large ratio such as $d_{cry}/d_{cap} > 3.8$ leads to a large $\overline{V}_{cap}$, yielding a large central down flow from the capillary to the molten zone scratching the solute boundary layer in the center and lowering the concentration (Fig. 2.9c). If the convection in zone I had been strong enough for complete mixing, no concentration variation along either the radial or axial direction would have been observed. On the contrary, for growth with $1.0 < d_{cry}/d_{cap} < 2.8$, the solute boundary layer is solid enough all over the interface to have the interface concentration saturation of C_0/k_0 in the liquid leading to an unity of the effective distribution coefficient, k_{eff}^{Ge}. This tells us that there was a well-developed solute accumulation layer with a gradient near the interface which is not much influenced by the convection when V_{cry} is large and $1.0 < d_{cry}/d_{cap} < 2.8$, giving k_{eff}^{Ge} the value unity. The above discussion suggests that the effect of convection in zone I was not so significant that there was a decay of Ge concentration from the interface toward the capillary zone and the Ge concentration at the interface is saturated in many cases. For those cases, we can use the no-mixing model for the analysis without large errors.

Table 2.2. Physical and geometrical parameters for calculation of the Marangoni number (*Ma*)

Surface tension gradient with temperature	$\partial\gamma/\partial T$ [a]	-0.08 dyn cm^{-1} K^{-1}
Temperature difference between the top of the Molten zone and the interface	ΔT	50 K
Molten zone length (= L_{mz})	H	0.03 cm
Viscosity	$\bar{\eta}$ [b]	7.2 x 10^{-3} dyn cm^{-2} s
Thermal conductivity	K	0.67 J cm^{-1} s^{-1} K^{-1}
Density	ρ [c]	2.53 g cm^{-3}
Specific heat	c_p	1.04 J g^{-1} K^{-1}

$Ma = -(\partial\gamma/\partial T)(\Delta T\, H/\bar{\eta}D_T)$ where D_T is the thermal diffusivity and $D_T = K/\rho c_p$

[a]Ref.[17], [b]Ref[18], [c]Ref[19, 20]

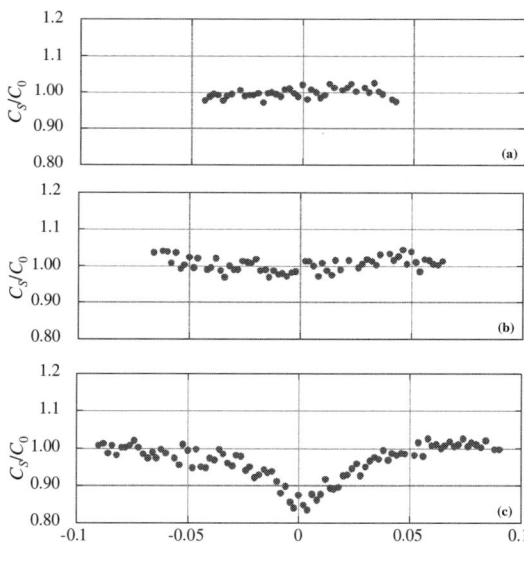

Fig. 2.9. Radial distribution of Ge in steady-state growth: (**a**) d_{cry}/d_{cap} = 1.8, (**b**) d_{cry}/d_{cap} = 2.8 and (**c**) d_{cry}/d_{cap} = 3.8. d_{cap} = 500 μm, L_{mz} = 300 μm and V_{cry} = 6.0 cm/h. The origin is the center of the crystal.

Radial distance from center (cm)

2.2.3.2 Effect of Radial Flow Component in the Molten Zone

For the past few years, the effect of the radial flow component of melt convection on partitioning has been vigorously studied [21–24] since people have realized that the BPS theory [10] does not take account of the influence of the radial flow component, which often underestimates the elimination of the solute pile up ($k_0 < 1.0$) in the vicinity of the interface. Here Yen and Tiller's theory [21] is employed to evaluate the degree of the influence of the radial flow component on partitioning during steady-state growth via the μ-PD method. The flux equation can be expressed by (2.17) rather than the one-dimensional analysis represented by (2.9a) or (2.10a):

$$D_{Ge}\nabla^2 C_L + V_{cry}\nabla C_L = 0. \tag{2.17}$$

However, they tried to incorporate the effect of the radial flow into the conventional one-dimensional differential equation as a modifier to the effect of the axial flow component [21]:

$$D_{Ge}\frac{\partial^2 C_L}{\partial x^2} + V_{cry}\frac{\partial C_L}{\partial x} - q = 0 \tag{2.18}$$

where q is a net convective loss term. Using (2.18), they reevaluated Kodera's analysis [15] and found that diffusion constants of various impurities in the Si melt, which were acquired based on BPS plots, lay in the range of $\sim 10^{-4} cm^2 s^-$. These values were nearly ten times larger than those for other liquid metals. Thus, they call them an effective diffusion constant which was larger than the true diffusion coefficient, D_L, and written in (2.19) [21]:

$$D_{eff} = D_L\left(1 - \frac{\int q dx}{D_L \dfrac{\partial C_L}{\partial x}}\right). \tag{2.19}$$

In this section, D is calculated to be $5.6\cdot 10^{-5} cm^2 s^{-1}$ which is consistent with the reevaluation of diffusion constants by Yen and Tiller [21]. Thus it is concluded that the effect of the radial flow component is very small and the initial transient analysis along the axial direction by using (2.13a) by Smith et al. [12] and the steady-state analysis by setting up (2.12) are reasonable and fully acceptable.

2.3 Influence of the Interface Electric Field on Solute Partitioning

2.3.1 Interface Electric Field

The interface electric field influences solute partitioning leading to an electric field-dependent effective solute partition coefficient [25, 26]. This is peculiar to growth by the micro pulling down (μ-PD) method or the laser-heated pedestal growth (LHPG) technique which accompanies a significantly high temperature gradient at the solid-liquid interface. The existence of such an electric field was experimentally reported on congruent $LiNbO_3$ by D'yakov et al. [27], where it was found that the electric field generated via the temperature gradient is associated with the thermoelectric power (Seebeck effect) while an additional electric field is associated with a charge separation effect which has a linear relationship with the growth velocity. The Seebeck effect was found at the solid-liquid interface for the crystal growth of InSb [28], which was used as a monitor of interface temperature fluctuations. This thermoelectric power contribution was achieved via the use of automatic poling during fiber growth [29] and was also found in the abnormal diameter-broadening of TiO_2 crystal growth via the LHPG method by Yen and Tiller [30]. They also showed strong uphill diffusion of Al and Si in TiO_2 crystals due to an electric field arising from the thermoelectric power. On the other hand, Feisst and Räuber [31] showed that k_{eff} varied over a very wide range when an electric current was applied to the interface via a seed and a Pt crucible and that this technique led to a periodic ferroelectric domain structure in $LiNbO_3$ to be used for nonlinear optical devices [32]. The importance of these field effects on solute partition coefficients based on the concept of a field-modified partition coefficient, k_{E_0}, and an effective growth velocity, V_{E_L}, has been documented by Uda and Tiller [25] by taking as an example the redistribution of Cr^{3+} ion during $LiNbO_3$ fiber growth. They also showed experimentally and analytically that the strong electric field operating on the liquid dragged the Cr^{3+} ions from the $LiNbO_3$ crystal to the melt through the interface, in which case the field-modified equilibrium partition coefficient, k_{E_0}, has a negative value [26].

Moreover, the interface electric field significantly influences the partitioning of the intrinsic $LiNbO_3$ melt ionic species, i.e. Li^+, OLi^-, $Nb_2O_4^{2+}$ and O^{2-} [33, 34] and the dynamic congruent melt composition for bulk crystal growth was discussed as a function of growth parameters, i.e. growth velocity, V, temperature gradient in the liquid, G_L, and solute boundary layer thickness, δ_c [35], which predominate the interface electric fields.

The μ-PD or LHPG methods make it possible to explore these field effects by controllably changing the growth velocity, V, and the temperature gradients in the liquid, G_L, and in the solid, G_S, near the interface, over a wide range. In order to clarify the effect of these electric fields, the μ-PD technique was used to grow a single-crystal fiber by consuming all the melt and then the solute redistribution for the whole crystallization process was analyzed. A schematic illustration of the μ-PD furnace for oxide crystals is illustrated in Fig. 2.10. The best features of the μ-PD technique for electrically neutral materials are discussed in Sect. 2.2 and they are (1) melt convections can be much reduced in the capillary pipe [9] and thus, (2) the effective partition coefficient is normally unity for electrically neutral materials so that a uniform solute concentration can be achieved, and (3) the crucible itself serves as a heater and suitable thermal symmetry is maintained during crystal growth.

During crystal fiber growth either by μ-PD or LHPG method, an interface electric field is mainly produced by thermoelectric power caused by a temperature gradient which becomes larger as the crystal diameter is reduced [25]. Uda and Tiller [25] measured the axial temperature gradient during $LiNbO_3$ fiber crystal growth by the LHPG method and they achieved a temperature gradient as high as

13000 °C/cm in the solid when the crystal diameter was 250 μm; the corresponding electric field was 9.9 V/cm. Luh et al. [29] showed the occurrence of ferroelectric polarization due to thin thermoelectric power during $LiNbO_3$ fiber crystal growth, resulting in the formation of a single-domain $LiNbO_3$ fiber crystal. A single-domain structure was also observed in a $LiNbO_3$ fiber crystal grown by the μ-PD method [5].

In this section, solute partitioning will be discussed in the steady state, initial transient and terminal transient under the presence of intrinsically occurring electric fields, taking as an example Mn-doped $LiNbO_3$ single-crystal fiber growth by the μ-PD method. In particular, the physical meaning of the negative signs of the field-modified effective growth velocity, V_{E_L}, and the equilibrium partition coefficient, k_{E_0} will be discussed. This happens when the growth velocity is smaller than the critical velocity, V^*, and each ionic species has its own specific V^* [36].

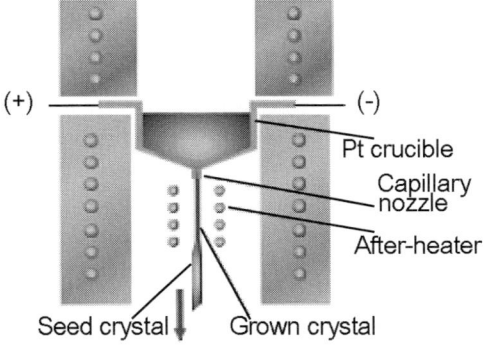

$(+)$ —————— —— $(-)$

Pt crucible

Capillary
nozzle

After-heater

Seed crystal   Grown crystal

Fig. 2.10. Schematic illustration of μ-PD furnace for the growth of oxide materials.

2.3.2 Experimental Setup

2.3.2.1 μ-PD Installation

The μ-PD system shown in Fig. 2.10 was used to grow round single-crystal fibers of Mn: $LiNbO_3$ in an air atmosphere. It comprises a platinum crucible, 3×4×12 mm^3, serving also as a heater by direct imposition of electric power. The capillary nozzle, 300–1300 μm of diameter and 2 mm long, is attached to the bottom of the crucible and single-crystal fibers are pulled down from the outlet of the capillary nozzle. An after-heater was equipped to lower the temperature gradients near the interface to avoid cracking during growth.

2.3.2.2 Preparing the Source Material

5N-grade powders of Li_2CO_3 and Nb_2O_5 were well mixed to given congruent melt composition, 48.38 mol% Li_2O [37] with 3 wt% of MnO_2 (5N grade). They were sintered in a platinum crucible at 1100 °C for 6 hours to give the source material.

2.3.2.3 Growth of Single-Crystal Fiber

There are several basic parameters involved in crystal growth by the μ-PD technique that are discussed in Sect. 2.2: (1) conservation of mass at the boundary between the capillary pipe and the molten zone, (2) conservation of energy represented by the heat of fusion, the temperature conductivity of the solid and liquid, and (3) mechanical stability of the molten zone shape represented by the interfacial wetting angle [38]. These parameters confine the stable molten zone volume to be very small and, consequently, the free surface area is very small so that the free surface height in most cases is less than 50 μm. This led to little effect of Marangoni flow.

The source materials were melted in a platinum crucible and the melt came down through the capillary nozzle to its outlet where the seed crystal was attached. Mn-doped single-crystal fibers of constant diameter were grown with the *a*-axis orientation with the crystal diameter ranging from 500 to 1300 μm. The Growth rate, V, was 8.33 μm/s. Crystals were grown consuming all the melt in the crucible so that grown crystals included the initial transient, steady state and the terminal transient regions. Some specimens were quenched very rapidly during growth by abruptly turning off the heater power in order to freeze the solute distribution in the melt. The grown samples were cut and polished along the axis of symmetry, and accurate concentrations were determined by electron probe microanalysis (EPMA).

2.3.3 Experimental Results

2.3.3.1 Equilibrium Partition Coefficient, k_0, for Mn

It is important to evaluate the size of the equilibrium partition coefficient, k_0, for Mn relative to unity since the influence of electric field on the partitioning will be discussed in this section. The specific interest is whether k_0 is less than unity. Mn distribution in the initial transient region is plotted in Fig. 2.11 for the bulk crystal grown by the Czochralski method under a temperature gradient of 120 °C/cm. The source material is the same as that used for fiber crystal growth. Although partitioned Mn in the crystal diffused more or less into the seed crystal which initially did not contain Mn, it is evident that k_0 is less than unity, which is consistent with the previously reported data [39].

2.3.3.2 Axial Distribution of Mn in the Grown Crystal Fiber

The Axial distributions of Mn in grown fiber crystals of 500 μm and 1300 μm $\varnothing_d$ are shown in Fig. 2.12 from the initial transient to the terminal transient region with the entire range of the melt solidified fraction, g. The well-defined steady-state region represents nearly no mixing freezing by the μ-PD technique. However, an interesting inconsistency was found: (1) the rise of Mn concentration in the initial transient indicated that the equilibrium partition coefficient, k_0, is less than unity, while the concentration decline in the terminal transient indicated that k_0 is greater than unity and (2) the steady-state concentration is higher than the bulk concentration, C_0, which means that the effective partition coefficient, k, is greater than unity despite the general idea that $k > 1.0$ cannot be deduced from $k_0 < 1.0$ in the BPS equation [10]. These cannot be explained from the normal freezing view, but should be discussed in association with the influence of the interface electric field on partitioning. It should also be noted that this effect was more remarkable for fiber with smaller diameter, 500 µm $\varnothing_d$, in such a way that the steady-state Mn concentration was higher, i.e., the field-modified effective partition coefficient, k_E, is larger, and the Mn concentration drop started at an earlier solidification stage, i.e. at smaller g.

The field-modified effective partition coefficient, k_E, was directly observed to be greater than unity by investigating the Mn concentration in the solid and liquid near the interface which was frozen in the rapidly quenched sample. This is shown in Fig. 2.13 and k_E was about 1.2 for the fiber crystal growth with 800 µm crystal diameter and 8.33 µm/s growth velocity. The value of k_E was strongly affected by the interface electric field and the diffusion constants for Mn in the solid and liquid. This will be discussed in the following.

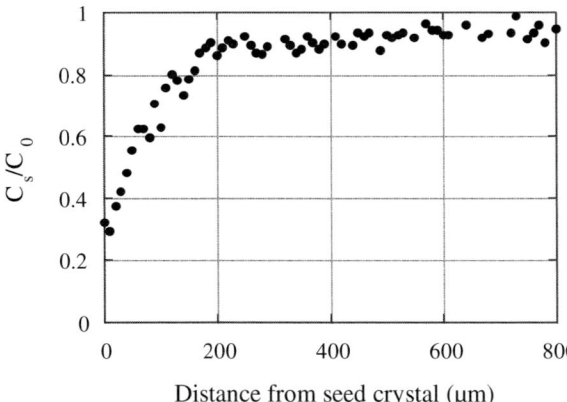

Fig. 2.11. Axial distribution of Mn in the initial transient region of the bulk crystal grown by the Czochralski method: $\varnothing_d = 15$ mm and $G_L = 120\ °C/cm$.

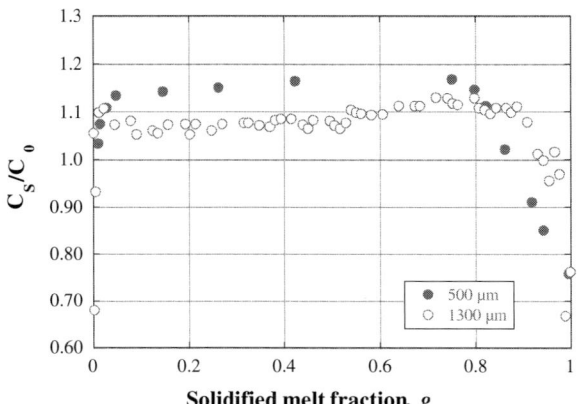

Fig. 2.12. Axial distributions of Mn in the whole length of fiber crystals grown by the µ-PD method. $V = 8.33$ µm/s.

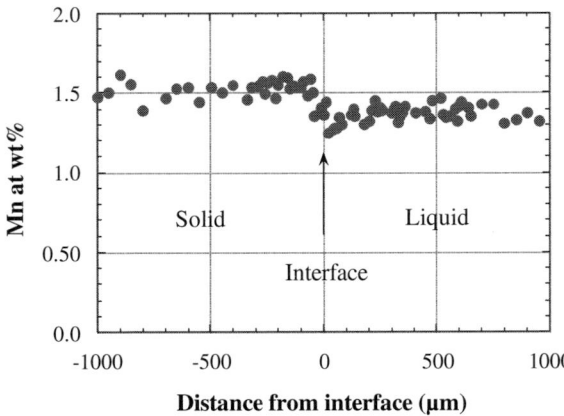

Fig. 2.13. Steady-state axial distribution of Mn in the solid and liquid near the interface. The field-modified effective partition coefficient, k_E, is about 1.2: $\varnothing_d = 800$ μm and $V = 8.33$ μm/s.

2.3.4 Analytical

2.3.4.1 Potential Difference and Electric Fields at the Interface

During crystallization of ionic melts, an electric field will be present in both the liquid and the solid. D'yakov et al. [27] pointed out that for congruent $LiNbO_3$ the thermoelectric potential difference, $\Delta\phi$, between a location at temperature T_1 in the solid and a location at temperature T_2 in the liquid is given by

$$\Delta\phi = \alpha_S (T_1 - T_i) + \alpha_L (T_i - T_2) + \alpha_i V \tag{2.20}$$

where T_i is the interface temperature, α_i is the crystallization-EMF coefficient, while α_S and α_L are the thermoelectric coefficients for the solid and the liquid, respectively. They [27] found $\alpha_S = 0.76 \pm 0.02$ mVK^{-1}, $\alpha_L = -0.4$ mVK^{-1} and $\alpha_i = 1.25 \pm 0.2$ mV sec μm^{-1}. Thus, the interface electric field operating on the liquid, E, is subdivided into two fields: the charge-separation-effect-related field, E_c, and the Seebeck-effect-related fields: E_t, i.e.

$$E = E_c + E_t. \tag{2.21a}$$

In particular, electric fields operating on the liquid near the interface, E_{cL} and E_{tL} are important and they are represented by

$$E_{cL} = \alpha_i V / 2\delta_c \tag{2.21b}$$

and

$$E_{tL} = -\alpha_L G_L \qquad (2.21c)$$

where δ_c is the thickness of the solute boundary layer.

2.3.4.2 Flux Equations and Interface Mass Conservation

The μ-PD method has a specific advantage for the analysis of solute transport that since the free surface area is very limited and the height of the free surface is 25–50 μm in many cases, there is little effect of Marangoni convection, Bénard convection and buoyancy-driven convection [9], as discussed in the previous section. Thus, it is reasonable to take a one-dimensional analysis for the Mn distribution. In a coordinate system tied to the moving interface, the flux J^j for the jth species in the liquid is given by [25]:

$$J^j = -D^j \frac{\partial C^j}{\partial z} - VC^j + \frac{D^j z^j e(E_c + E_t)C^j}{k_B T} \qquad (2.22a)$$

where D^j is the diffusion constant, z^j is the valence of the jth species, and k_B is the Boltzmann constant. Following the approach of [25], the flux equations of the interface for the jth species in the liquid and solid are given by

$$J_L^j = -D_L^j \left(\frac{\partial C_L^j}{\partial z} \right)_i - V_{E_L}^j C_{L(i)}^j \qquad (2.22b)$$

and

$$J_S^j = -D_S^j \left(\frac{\partial C_S^j}{\partial z} \right)_i - V_{E_S}^j k_0 C_{L(i)}^j. \qquad (2.22c)$$

V_{E_L} and V_{E_S} are the effective velocities for the liquid and the solid, respectively and are given by

$$V_{E_L}^j = V - \frac{D_L^j z_L^j e(E_{cL} + E_{tL})}{k_B T} = V - q_L^j; \qquad q_L^j = \frac{D_L^j z_L^j e(E_{cL} + E_{tL})}{k_B T} \qquad (2.23a)$$

$$V_{E_S}^j = V - \frac{D_S^j z_S^j e(E_{cS} + E_{tS})}{k_B T} = V - q_S^j; \qquad q_S^j = \frac{D_S^j z_S^j e(E_{cS} + E_{tS})}{k_B T} \qquad (2.23b)$$

where L and S denote liquid and solid, respectively, and q_β^j (β = L, S) is the field-driven flux term. The interface conservation condition is given by

$$V_{E_L}^j \left(1 - k_{E_0}^j\right)C_{L(i)}^j = -D_L^j \frac{\partial C_L^j}{\partial z} + D_S^j \frac{\partial C_S^j}{\partial z}$$ (2.24)

where k_{E_0} is the electric field-modified equilibrium partition coefficient:

$$k_{E_0}^j = k_0^j \frac{V_{E_S}^j}{V_{E_L}^j}.$$ (2.25a)

If D_S^j is small enough to be neglected, (2.25a) becomes

$$k_{E_0}^j = k_0^j \frac{V}{V_{E_L}^j}.$$ (2.25b)

It should be noted that $k_{E_0}^j$ goes to infinity when $V_{E_L}^j \to 0$; however, this is not a singularity of the physical problem, but an artifact arising from the introduction of the effective velocity, $V_{E_L}^j$ and the field-modified equilibrium partition coefficient, $k_{E_0}^j$. The field-modified effective partition coefficient, k_E^j, is given by replacing k_0^j by $k_{E_0}^j$ and V by $V_{E_L}^j$ in the BPS equation [10]:

$$k_E^j = \frac{k_{E_0}^j}{k_{E_0}^j + \left(1 - k_{E_0}^j\right)\exp\left[-V_{E_L}^j \delta_c / D_L^i\right]}.$$ (2.26)

Hereafter, Mn was chosen for the jth species.

2.3.4.3 Critical Growth Velocity, V*

Fig. 2.14 shows the k_{E_0} variation as a function of growth velocity, V, for different temperature gradients. For each k_{E_0}, there is a critical velocity, V^*, which is represented by the intercept of the V-axis with the asymptotic line. That is, the effective velocity, V_{E_L} becomes 0 when the growth velocity V is equal to the critical velocity, V^*. $V^*$is calculated from (2.23a):

$$V^* = \frac{2D_L \delta_C z_L e \alpha_L G_L}{D_L z_L e \alpha_i - 2\delta_C k_B T}.$$ (2.27)

V^* moves toward the $V = 0$ origin as G_L decreases, and is very small and practically negligible when the temperature gradient is low, which is the case with normal Czochralski growth. However, V^* becomes very important for crystal growth by the μ-PD technique which accompanies a very high temperature gradient at the interface. Note that k_{E_0} changes sign from negative to positive at $V = V^*$. It should also be noted that there is a specific V^* for each ionic solute as there is a specific k_{E_0} for them.

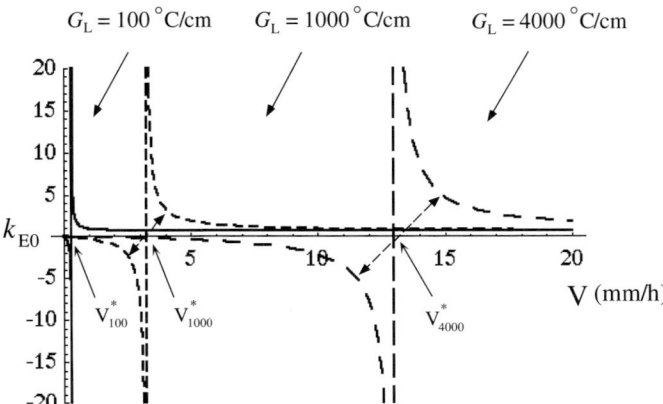

Fig. 2.14. k_{E_0} variation for Mn as a function of growth velocity, V, for different temperature gradients. The critical velocity V^* is represented by the intercept of the V-axis with the asymptotic line. The dotted arrows show the pair of positive and negative values for k_{E_0}. A parameter set of $D_L = 5.8 \times 10^{-6} \mathrm{cm}^2 \mathrm{s}^{-1}$, $D_S \approx 1 \times 10^{-8} \mathrm{cm}^2 \mathrm{s}^{-1}$, $\delta_C = 50 \, \mu\mathrm{m}$ and $k_0 = 0.5$ was used.

2.3.4.4 Negative Signs of Effective Velocity, V_{E_L}, and the Field-Modified Equilibrium Partition Coefficient, k_{E_0}

Equations (2.23a), and (2.25a) and (2.25b), tell us that the field-modified effective growth velocity, V_{E_L}, and the field-modified equilibrium partition coefficient, k_{E_0}, take a negative value when $V < V^*$. A negative V_{E_L} leads C_L to 0 at $z = 0$ as $t \to \infty$. This predicts that cations such as Mn^{4+} near the interface are driven away by an interface electric field toward the far bulk liquid while the growth system after the seed is touched is held without pulling down ($V = 0$). On the other hand, pulling down the crystal with growth velocity V, which is large enough for V_{E_L} to be positive, pushes Mn^{4+} ions toward the interface. This Mn migration in the liquid is calculated by using the concentration equation for the initial transient region which takes account of the field effect. Solute conservation in a coordinate system moving at velocity V, is generally given by

$$-\frac{\partial}{\partial z} J_\beta^j = \frac{\partial C_\beta^j}{\partial t}$$

$$= D_\beta^j \frac{\partial^2 C_\beta^j}{\partial z^2} + \frac{\partial}{\partial z}\left\{\left[V - \frac{D_\beta^j z^j eE}{k_B T} - u(z)\right]C_\beta^j\right\}; \qquad \beta = S, L, \qquad (2.28)$$

where $u(z)$ is a convection term which may be neglected in a simple no-mixing model which is the case with the μ-PD method. Setting up the differential equation (2.29),

$$\frac{\partial C_L}{\partial t} = D_L \frac{\partial^2 C_L}{\partial z^2} + V_{E_L} \frac{\partial C_L}{\partial z},$$

(2.29)

and combining with the appropriate boundary conditions,

$$C_L = C_0 \qquad\qquad \text{at } t = 0 \text{ for all } z \text{ and } z = \infty \text{ for all } t, \qquad (2.30a)$$

$$V_{E_L}\left(1 - k_{E_0}\right)C_L = -D_L \frac{\partial C_L}{\partial z} \qquad\qquad \text{at } z = 0 \text{ for all } t, \qquad (2.30b)$$

$$V = 0 \qquad\qquad \text{for } 0 < t, \qquad (2.30c)$$

the following solution is obtained:

$$C_L = C_0 + \frac{C_0}{2k_{E_0}}\left\{\begin{array}{l} -\left(k_{E_0} - 1\right)\exp\left(-\frac{V_{E_L}}{D_L}z\right)\text{erfc}\left(\frac{-V_{E_L}t + z}{2\sqrt{D_L t}}\right) - k_{E_0}\text{erfc}\left(\frac{V_{E_L}t + z}{2\sqrt{D_L t}}\right) \\[2mm] + \left(2k_{E_0} - 1\right)\exp\left[\frac{\left(k_{E_0} - 1\right)\left(V_{E_L}z + k_{E_0}V_{E_L}^2 t\right)}{D_L}\right]\text{erfc}\left[\frac{V_{E_L}\left(2k_{E_0} - 1\right)t + z}{2\sqrt{D_L t}}\right] \end{array}\right\}.$$

(2.31)

Inserting $V_{E_L} = -q_L$ ($V \to 0$) into (2.31), the migration of Mn is illustrated in Fig. 2.15 by taking the parameter set of $G_L = 4000$ °C/cm, $D_L = 5.8\times10^{-6}$ cm^{2}s^{-1}, $\delta_C = 50$ μm and $k_0 = 0.5$. D_S is neglected here. The value of D_L was estimated from Fig. 2.12 by evaluating k_E (2.26). The C_L value at the interface ($z = 0$) by (2.31) is equivalent to that given by Tiller and Sekerka [40]. It is clearly shown that Mn is being dragged away from the interface during the period of 600 seconds. In this case, $V_{E_L} = -2.8$ μm/s. The negative V_{E_L} shows that Mn^{4+} ions are being driven away toward the far bulk liquid by the interface electric field and thus, the Mn concentration at the interface gradually decreases and nearly reaches zero. When the holding period becomes longer, the interface completely depletes the Mn ions. The Mn distribution due to these negative-valued parameters was experimentally observed. Figure 2.16 shows that the melt held for 20 minutes after seed touch without the start of pulling down completely depleted Mn in the interface region. An abrupt rise of Mn concentration occurred in the molten zone probably because the amount of melt is finite and the magnitude of the field operating on the liquid rapidly decayed with distance from the interface while an infinite amount of melt and the constant field strength was assumed for the calculation for Fig. 2.15.

Then, after 600 s, the crystal suddenly starts to grow at the pulling-down rate, 8.33 μm/s leading to positive V_{E_L} = 3.7 μm/s, and Mn ions then flow back towards the interface. Thus, the Mn concentration in the melt at the interface gradually increases during the initial transient up to the steady state. This recovery process is numerically calculated in Fig. 2.17. Accordingly, the partitioned Mn in the solid progressively raises its concentration during the initial transient, which looks like the initial transient for $k_0 < 1.0$ (Fig. 2.12) though k_{E_0} = 1.13. On the other hand, anions such as O^{2-} ion can accumulate near the interface by the same electric field and the calculated concentration pile up is illustrated in Fig. 2.18. The accumulation of oxygen near the interface yields oxygen bubbles in the liquid [41].

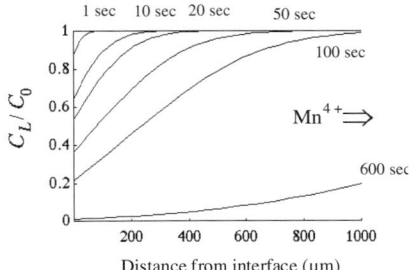

Fig. 2.15. Depletion of Mn near the interface by an interface electric field during period of 600 s (0 < t < 600). The parameter set of G_L = 4000 °C/cm, D_L = 5.8×10^{-6} cm^{2}s^{-1}, δ_C = 50 μm and k_0 = 0.5 was used.

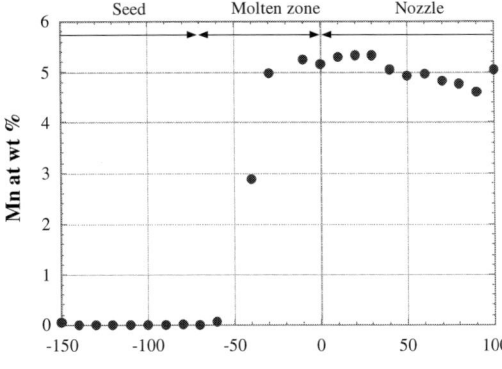

Fig. 2.16. Illustration of Mn depletion in the liquid near the interface observed 20 minutes after seed touch without pulling down. The dimension of the seed crystal is 900 × 500 μm.

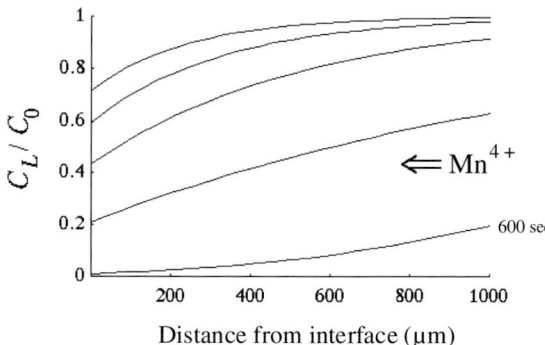

Fig. 2.17. Numerical representation of the recovery process of Mn flow toward the interface after the start of pulling down at the rate of 8.33 µm/s ($t > 600$).

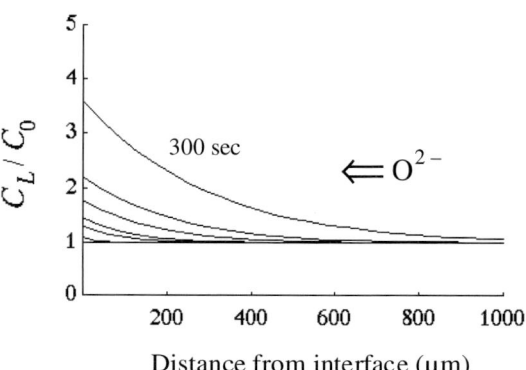

Fig. 2.18. Accumulation of oxygen near the interface by an interface electric field during the period of 300 s. $G_L = 4000$ °C/cm and $\delta_c = 50$ µm. For the oxygen ionic species, $D_L = 1 \times 10^{-5}$ cm^{2}s^{-1} and $k_0 = 0.48$ are used [42]. Calculated $V_{E_L} = 2.42$ µm/s.

2.3.4.5 Solute Distribution at the Critical Velocity, V*

It should be noted that there is a specific differential equation in place of (2.29) for the case of $V = V^*$ which is obtained by inserting 0 into V_{E_L} in (2.29) and replacing k_{E_0} by $k_0 V / V_{E_L}$ in (2.30b). The solution for this special case was given by Tiller and Sekerka [40]:

$$C_L = C_0 \left\{ 1 - \mathrm{erfc}\left(\frac{z}{2\sqrt{D_L t}} \right) + \exp\left[\frac{k_0 V}{D_L} (z + k_0 V t) \right] \times \mathrm{erfc}\left(\frac{z + k_0 V t}{2\sqrt{D_L t}} \right) \right\}. \tag{2.32}$$

By using (2.32) for the case of $V = V^*$, there is no discontinuity in the solute concentration profile in the liquid when the growth velocity, V, passes the critical value, V^*. Accordingly, the solute concentration in the solid is also continuous around V^*. When the crystal grows with negative $V_{E_L}^j$ and $k_{E_0}^j$, a positive ionic solute, j, such as Mn, is completely depleted in the melt near the interface when the system is in the steady sate (t $\rightarrow$ ∞). Thus, the concentration of the ionic solute partitioned into the solid, C_S, in the steady state is zero. Then, if the growth velocity, V, suddenly increases its $V > V^*$, C_S takes a nonzero value depending on V. By taking the alternate steady-state growth with $V < V^*$ and $V > V^*$, the alternate layers consisting of the layer having the ionic solute free portion and the layer with a certain amount of the solute could be formed.

Another interesting point related to the characteristics of k_{E_0} is that the k_{E_0} can basically take any value, i.e. $k_{E_0} < 1$, $k_{E_0} = 1$ and $k_{E_0} > 1$ (Fig. 2.14), which leads to a practically wider choice of the field-modified effective partition coefficient, k_E, despite the fact that the effective partition coefficient, k, is normally constrained in the region below unity ($k_0 < 1$) or above unity ($k_0 > 1$).

2.4 Solute Redistribution by an Interface Electric Field during Fiber Crystal Growth

2.4.1 Interface Electric Field along the Radial Direction

The influence of an interface electric field along the axis of symmetry was mainly discussed in the previous section. However, Imai et al. [43] investigated the compositional variation of potassium lithium niobate (KLN) fiber crystals grown by the μ-PD method and found that K was rich in the core. This may have been driven by the thermoelectric power operating on the ionic melt in a radial direction since such a radial solute distribution was not observed for the fiber growth from the neutral melt by the μ-PD technique [9].

In this section, the temperature distribution and associated thermoelectric power in the solid near the interface were measured along both the axial and radial directions during $LiNbO_3$ crystal fiber growth by the μ-PD method. Then the Mn concentration profile was analyzed along the radial direction in fiber crystals with different diameters by comparing with the axial analysis. A two-dimensional analysis was attempted concerning the radial effect of the interface electric field on solute partitioning [44]. Finally, one can view the influence of the interface electric fields on partitioning for crystal growth from the ionic melt via the μ-PD method.

2.4.2 Experimental Procedure

2.4.2.1 Measurement of Temperature Gradient and the Thermoelectric Power

The axial temperature gradient in the solid and melt near the interface was measured during crystal growth. Two thermocouples (R-type: Pt-PtRh 13%), which were attached to the seed crystal 150 μm apart from each other, were inserted within the capillary nozzle of 800 μm $\varnothing_d$ in such a way that one was at the center and the other was located in the periphery (Fig. 2.19). They were pulled down giving successive axial temperatures in the core and peripheral part of the melt in the nozzle, molten zone and of the solid as the seed crystal was pulled. The diameter of the crystal enclosing these thermocouples was 800 μm.

LiNbO$_3$ is ferroelectric at room temperature and has a Curie temperature at around 1150 °C [45, 46] where LiNbO$_3$ transforms from ferroelectric to paraelectric. The thermoelectric power caused by the temperature gradient in the solid near the solid-liquid interface was measured along both the axial and radial directions in the paraelectric temperature region (> 1150 °C) so that there was no polarization effect due to ferroelectric characteristics on the potential measurements. Figure 2.20 illustrates the thermoelectric power measurement. The axial thermoelectric power in the solid was surveyed by using a pair of platinum wire electrodes (50 μm ∅$_d$) which enclosed the fiber with 800 μm spacing and the upper electrode was located 800 μm below the interface (Fig. 2.20a). For the radial measurement, one platinum wire was buried in the crystal normal to the growth axis in such a way that its top was situated at the center of the crystal, and another wire was attached to the surface of the fiber crystal with the same height level as the first one (Fig. 2.20b).

The temperature distribution in the capillary nozzle, the molten zone and the solid near the interface is shown in Fig. 2.21. It should be noted that the melt temperature became higher toward the peripheral part in the capillary zone since the platinum crucible itself served as a heater while the temperature in the solid was higher in the center, which was consistent with the experimental observation that the interface was concave to the melt. Thus, there was an inversion of radial temperature gradient in the molten zone. This inversion plays an important role in relation to the magnitude of the radial interface electric field that will be discussed later. It should also be noted that the temperature gradient in the solid near the interface was nearly 4000 °C/cm for the fiber of 800 μm ∅$_d$.

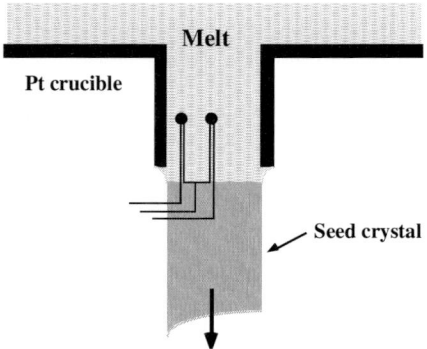

Fig. 2.19. Schematic illustration of the set-up of the temperature measurement using two thermocouples.

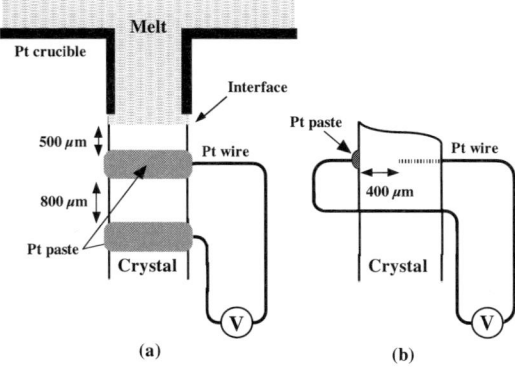

Fig. 2.20. Schematic illustration of measurements of thermoelectric power along (a) the axial direction and (b) the radial direction.

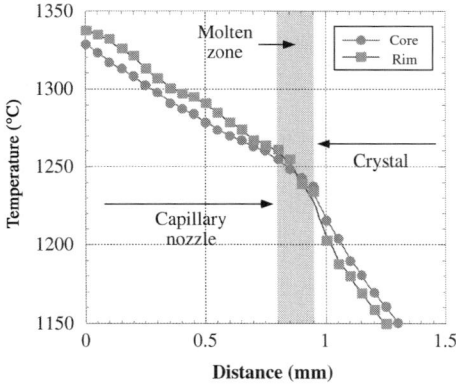

Fig. 2.21. Temperature distributions in the capillary nozzle, molten zone and the solid near the solid-liquid interface. The origin is the point where the measurement started.

Electric fields in the solid near the interface were obtained for crystal fibers with several diameters based on the thermoelectric potentials measured 800 μm below the interface. Figure 2.22 shows that the electric fields along the axial and radial directions have a similar magnitude and increased from 0.8 to 1.6 V/cm as the crystal diameter decreased from 1300 to 500 μm. It should be noted that there is no crystallographic dependence of thermoelectric power since the measured portion was in the paraelectric region. It should also be noted that the electric field obtained 800 μm below the interface showed a lower value than that expected at the interface, thus, the electric field at the interface should have been higher. This may be evaluated by using the potential equation (2.20) derived by D'yakov et al. [27]. Figure 2.21 shows that the temperature gradient at the interface was nearly 4000 °C/cm which led to an interface electric field as large as 3.0 V/cm. The temperature gradient at the point 800 μm below the interface was 1800 °C/cm which yields 1.4 V/cm, which is consistent with the experimental value.

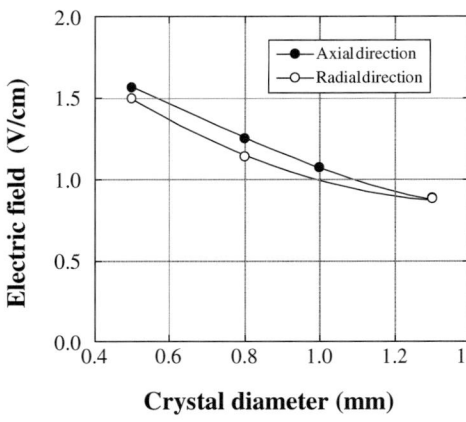

Fig. 2.22. Measured electric fields in the crystal near the interface as a function of crystal diameter. Note that the measurements were executed 800 μm below the interface.

2.4.2.2 Mn Distribution in the LiNbO₃ Fiber Crystal

The axial distribution of Mn in the grown fiber crystals of 500 and 1300 μm $\varnothing_d$ is shown in Fig. 2.12 in the previous section. One should note that the effect of the axial interface electric field was clearly observed even in the thick fiber of 1300 μm $\varnothing_d$. Figure 2.23 shows radial Mn distributions during the steady-state growth of crystal fibers with different diameters. Radial Mn distributions were almost homogeneous for fibers with 800 and 1300 μm $\varnothing_d$. There seemed to be a periodic concentration fluctuation in the fiber of 1300 μm $\varnothing_d$ in Fig. 2.23; however, this was not clear since the magnitude of the fluctuation was about 0.15 wt% which was comparable to or smaller than the detection limit of EPMA. On the other hand, there was a clear peak of Mn concentration at the center when the crystal diameter was 500 μm. One should remember that both the axial and radial electric fields in the solid near the interface were observed with similar magnitudes for fibers with 500–1300 μm $\varnothing_d$ and the axial effect was observed for the fiber of 1300 μm $\varnothing_d$. However, the radial influence was observed only for the fiber of 500 μm $\varnothing_d$. This difference will be discussed later in association with the role of the inversion of the radial temperature gradient in the molten zone.

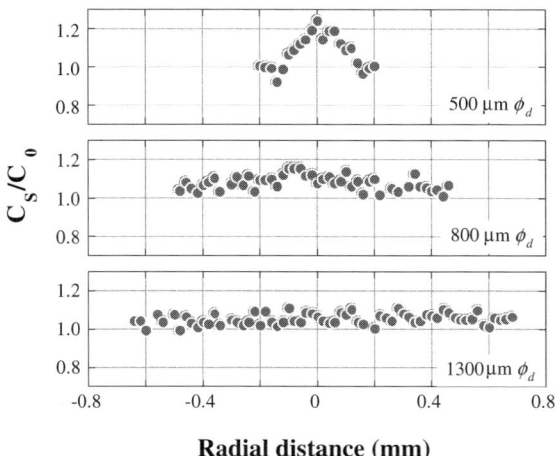

Fig.2.23. Radial distribution of Mn in LiNbO₃ fiber crystals with different diameters. Growth rate was 8.3 μm/s.

2.4.3 Analytical Investigation

The axial Mn distribution that is associated with an axial interface electric field was analyzed in the previous section by using the field-modified equilibrium partition coefficient, k_{E_0}, and the field-modified effective partition coefficient, k_E, combined with the effective velocity, V_E [36]. Here, the radial distribution of Mn will be analyzed at the same time in association with the effect of the radial interface electric field on partitioning. In a coordinate system tied to the moving interface at velocity V, the solute conservation condition under the influence of the interface electric field is generally given by

$$D_L \nabla^2 C + \varsigma_E \nabla C - u \nabla C = \frac{\partial C}{\partial t} \tag{2.33}$$

where D_L is the diffusion constant in the liquid, C is the solute concentration, V_E is the field-modified velocity and u is the flow velocity of the melt. Using a cylindrical coordinate system in which the z-axis is the pulling direction and the r-axis is the radial direction (Fig. 2.24), (2.33) for the steady state becomes

$$D_L \left[\frac{\partial^2 C}{\partial z^2} + \frac{1}{r} \frac{\partial}{\partial r} \left(r \frac{\partial C}{\partial r} \right) \right] + V_{E_z} \frac{\partial C}{\partial z} + V_{E_r} \frac{\partial C}{\partial r} + u_z \frac{\partial C}{\partial z} + u_r \frac{\partial C}{\partial r} = 0 \tag{2.34}$$

where u_z and u_r are the axial and radial components of the flow velocity, respectively, while V_{E_z} and V_{E_r} are the axial and radial components of the field-modified effective velocity, respectively. V_{E_z} is equivalent to V_{E_L} in (2.23a) in the previous section. A similar procedure as explained for (2.23a) for V_{E_z} was taken to deduce the radial field-modified effective velocity, V_{E_r} except that there is no radial growth velocity, i.e. $V_r = 0$. It is assumed that the charge distribution along the radial direction is constant so that the charge-separation-driven field along the radial direction, $(E_{c_L})_r = 0$. Since the temperature distribution satisfies $\dfrac{\partial T}{\partial r} = G_{Lr} = 0$ on the cylindrical axis, $r = 0$, it is also assumed that the Seebeck-effect-induced field along the radial direction, $(E_{t_L})_r$ is simply represented as a linear function of r with zero value at $r = 0$ and the maximum value $-\alpha_L G_{L_r}$ at the outermost boundary of the solute diffusion layer $(r = R^*)$. Thus, the equations for the radial electric field, E_r, and the field-modified effective velocity of the radial direction, V_{E_r}, are

$$E_r = \left(E_{c_L}\right)_r + \left(E_{t_L}\right)_r = \left(E_{t_L}\right)_r = \frac{r}{R^*}\alpha_L G_{L_r} , \tag{2.35a}$$

$$V_{E_r} = -\frac{D_L z E_r}{k_B T} = -\frac{r}{R^*}\frac{D_L z \alpha_L G_{L_r}}{k_B T} = \frac{r}{R^*} V_{E_r}^{R^*} \tag{2.35b}$$

and

$$V_{E_r}^{R^*} = \frac{D_L z \alpha_L G_{L_r}}{k_B T} , \tag{2.35c}$$

where R^* is the crystal radius and $V_{E_r}^{R^*}$ is the V_{E_r} for $r = R^*$. It should be noted that there is no variation of V_{E_r} along the axial direction since the magnitude of the electric field was assumed to be constant over the solute diffusion boundary layer.

In (2.34), the convection terms, u_z and u_r, can be omitted for the case of the μ-PD method which has little convection in the molten zone [9]. Diffusion in the radial direction is also neglected since the crystal diameter is larger than the diffusion layer thickness. Thus, combined with (2.35a), (2.35b) and (2.35c), Eqution (2.34) is reduced to

$$D_L \frac{\partial^2 C}{\partial z^2} + V_{E_z}\frac{\partial C}{\partial z} + V_{E_r}\frac{\partial C}{\partial r} = 0. \tag{2.36}$$

Using the dimensionless variables $Z = z / \delta c$, $R = r / R^*$ $\overline{C}$, $= C/C_\infty$ where $0 \le Z \le 1$ and $0 \le R \le 1$, and C_∞ is the solute concentration in the bulk liquid, (2.36) becomes

$$\frac{\partial^2 \tilde{C}}{\partial Z^2} + \alpha\frac{\partial \tilde{C}}{\partial Z} + \beta R\frac{\partial \tilde{C}}{\partial R} = 0 \tag{2.37}$$

where

$$\alpha = V_{E_z} \, \delta_C / D_L \,, \tag{2.38a}$$

and

$$\beta = V_{E_r}^{R^*} \, \delta_C^2 / D_L \, R^* \,. \tag{2.38b}$$

α and β are constants which should be determined by the growth conditions.

The boundary conditions for (2.37) are set up as follows. The solute conservation condition at the interface requires

$$\alpha(1 - k_{E_0})\tilde{C} + \frac{\partial \tilde{C}}{\partial Z} = 0 \qquad \text{at } Z = 0, \, 0 < R < 1; \tag{2.39a}$$

the solute concentration along the edge of the solute boundary layer ($Z = 1$) is equal to the bulk concentration, C_∞,

$$\tilde{C} = 1 \qquad \text{at } Z=1, \, 0 < R < 1, \tag{2.39b}$$

and since $\tilde{C} = C_s / C_\infty \approx C_s / C_0 \approx 1$ at $Z = 0$ and $R = 1$ during steady-state growth,

$$\tilde{C} = 1/ k_{E_0} \qquad \text{at } Z=0, \, R = 1. \tag{2.39c}$$

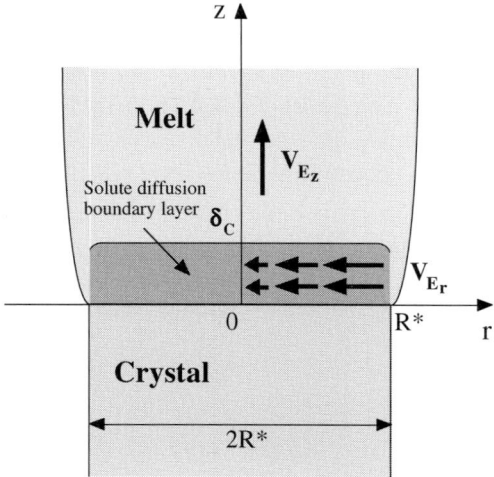

Fig. 2.24. The coordinate system for the solute conservation equation.

To solve (2.37), the method of Zuo and Guo [47] was employed, i.e., using the separation of variables technique, i.e.,

$$\tilde{C}(R, Z) = \Phi(R, Z) + \Psi(Z) \tag{2.40}$$

and

$$\Phi(R, Z) = \Gamma(Z)\Omega(R). \tag{2.41}$$

The complete solution is carried out in the appendix and gives the two-dimensional concentration, $\tilde{C}(R, Z)$,

$$\tilde{C}(R, Z) = \frac{k_{E_0} + (1 - k_{E_0})e^{-\alpha Z}}{k_{E_0} + (1 - k_{E_0})e^{-\alpha}}$$

$$+ \frac{e^{-\frac{1}{2}\left\{\sqrt{\alpha^2 + 4\gamma}(Z-2) + \alpha(Z+2)\right\}}\left\{e^{\sqrt{\alpha^2 + 4\gamma}(Z-1)} - 1\right\}\left(1 - \frac{1}{k_{E_0}}\right)R^{-\gamma/\beta}}{\left(e^{\sqrt{\alpha^2 + 4\gamma}} - 1\right)\left\{k_{E_0} + (1 - k_{E_0})e^{-\alpha}\right\}}, \tag{2.42}$$

where λ is the constant which is numerically determined for the given k_{E_0} and α (Eq. A12 in the appendix).

The qualitative solute concentration profile in the solute boundary layer is attempted by using (2.42) for growth of a fiber of 600 μm $\varnothing_d$ whose radial temperature gradients is assumed to be $G_{L_r} = 1500$ °C/cm, yielding a significant radial electric field effect on partitioning. Other growth conditions and physical constants for the calculation are $G_{L_z} = 5000$ °C/cm, $V = 10$ μm/s, $\delta_c = 40$ μm, $D_L = 5.8 \times 10^{-6} \text{cm}^2\text{s}^{-1}$ and $k_0 = 0.5$ which lead to a parameter set of $\alpha = 0.26$, $\beta = 9.7 \times 10^{-3}$, $\gamma = -1.68 \times 10^{-2}$ and $k_{E_0} = 1.34$. The calculated two-dimensional concentration is illustrated in Fig. 2.25a. The resultant solute concentration in the solid obtained by $k_{E_0} \times \tilde{C}(R, 0)$ is rich in the center (Fig. 2.25b), which is consistent with the experimental result for the case of 500 μm $\varnothing_d$ fiber growth. The solute accumulation in the center is more remarkable as the radial interface electric field is stronger.

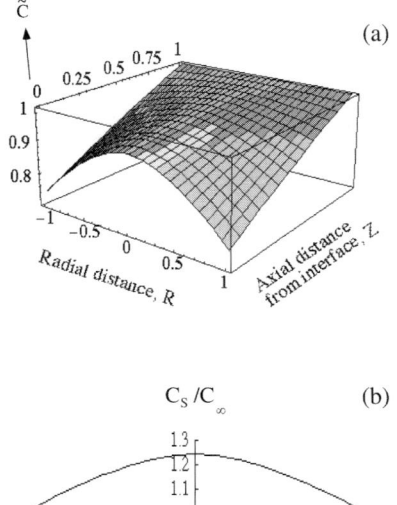

Fig. 2.25. (a) Illustration of solute distribution in the solute diffusion layer predominated by the interface electric field. $\alpha = 0.26$, $\beta = 9.7 \times 10^{-3}$, $\lambda = -1.68 \times 10^{-2}$ and $k_{E_0} = 1.34$. (b) Resultant solute distribution in the solid.

2.4.4 Influence of the Inverted Temperature Gradient on the Radial Electric Field

The interface electric field generated during crystal growth from the ionic melt by the μ-PD method was about 50 times larger than that of the normal Czochralski method. These electric fields operate axially as well as radially. The influence of the interface electric field operating along the axial direction was observed for the fiber of 1300 μm $\varnothing_d$ which was thickest and was expected to show the smallest electric field among fibers grown in the experiments. However, the effect of the electric field along the radial direction was observed to be significant in the fiber of 500 μm $\varnothing_d$, and in this case Mn was observed to be rich in the core of the fiber. The temperature in the solid was higher in the core, and the interface electric field in the solid operated in the direction from the center to the rim [34], which could not explain the Mn transportation toward the center in the solid. Therefore, this distribution in the solid was probably a consequence of the partitioning of the Mn distributed rich in the core in the liquid, which was caused by the electric field operating on the liquid from the periphery to the center of the fiber. However, the interface electric field in the liquid with the radial direction was not so strong unless the fiber diameter was as small as 500 μm $\varnothing_d$ though the electric field in the solid was significant even for the fiber of 1300 μm $\varnothing_d$ (Fig. 2.22). This may be related to the fact that there was an inversion of the radial temperature gradient somewhere in the molten zone. The inversion may have occurred in the upper molten zone in the case of the growth of the fiber of 500 μm $\varnothing_d$ because the heat release was more effective for the thinner fiber and the inversion easily occurred somewhere near the outlet of the nozzle (Fig. 2.26b). On the other hand, the inversion for the fiber growth of 1300 μm $\varnothing_d$ may have been shown up in the lower molten zone (Fig. 2.26a). A radial interface electric field, thus, could not be large for the thick fiber. In other words, it is possible to make the dopant distribution uniform by selecting a proper crystal diameter. In this case, the redistribution of the solute is determined by the thermoelectric power of the axial direction. The location of the inversion is schematically illustrated as a function of the fiber diameter in Fig. 2.27. On the other hand, there is no inversion of the radial temperature gradient in the LHPG method and a strong radial electric field as well as the axial field occurs at the solid-liquid interface. As a result the radial distribution of the dopant is not uniform in the crystal grown by the LHPG method.

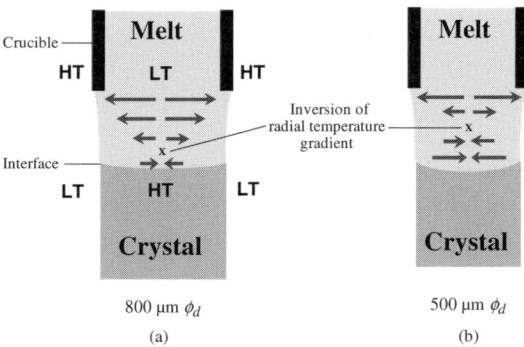

Fig. 2.26. Illustration of the radial electric field in the liquid and solid near the interface associated with the temperature gradient. (a) 800 μm $\varnothing_d$ and (b) 1300 μm $\varnothing_d$. HT and LT denote high temperature and low temperature, respectively.

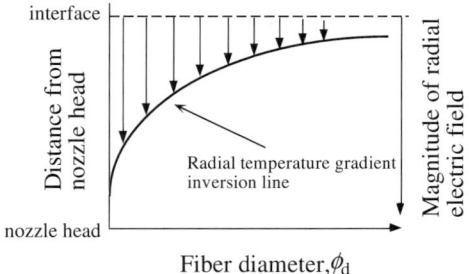

Fig. 2.27. Illustration of the location of the inversion of the radial temperature gradient as a function of the fiber diameter.

Acknowledgements

The author is indebted to Prof. Fukuda of Tohoku University and his research group with whom I had a great opportunity to study the significance of the μ-PD method. In particular, I thank Mr. J. Kon for his persistent endeavor to carry out the most difficult experiments.

Appendix

Solution for the partial differential equation, (2.37).

Inserting (2.40) into (2.37) gives (A1) and (A2),

$$\frac{\partial^2 \Psi(Z)}{\partial Z^2} + \alpha \frac{\partial \Psi(Z)}{\partial Z} = 0 \tag{A1}$$

and

$$\frac{\partial^2 \Phi(R,Z)}{\partial Z^2} + \alpha \frac{\partial \Phi(R,Z)}{\partial Z} + \beta R \frac{\partial \Phi(R,Z)}{\partial R} = 0. \tag{A2}$$

By combining the boundary condition, (2.39a) with (A1), $\Psi(Z)$ is solved:

$$Y(Z) = \frac{k_{E_0} + (1-k_{E_0})e^{-\alpha Z}}{k_{E_0} + (1-k_{E_0})e^{-\alpha}} \tag{A3}$$

which is identical to the solute distribution obtained by the BPS equation [10]. Equation (A2) changes into (A4) by replacing $\Phi(R,Z)$ with (2.41):

$$\frac{\partial \Omega(R)}{\partial R} + \frac{(\partial^2 \Gamma(Z)/\partial Z^2) + \alpha(\partial \Gamma(Z)/\partial Z)}{\beta R \Gamma(Z)} \Omega(R) = 0. \tag{A4}$$

In order to solve for $\Omega(R)$ and $\Gamma(Z)$, a constant λ is introduced to give (A5),

$$\frac{\partial \Omega(R)}{\partial R} + \frac{\lambda}{\beta R} \Omega(R) = 0 \tag{A5}$$

and

$$\lambda = \frac{(\partial^2 \Gamma(Z)/\partial Z^2) + \alpha(\partial \Gamma(Z)/\partial Z)}{\beta R \Gamma(Z)} \tag{A6}$$

Equation (A7) is deduced from (A6):

$$\frac{\partial^2 \Gamma(Z)}{\partial Z^2} + \alpha \frac{\partial \Gamma(Z)}{\partial Z} - \lambda G(Z) = 0. \tag{A7}$$

The solutions for (A6) and (A7) are

$$\Omega(R) = c_0 R^{-\lambda/\beta} \tag{A8}$$

and

$$\Gamma(Z) = c_1 e^{\left(-\alpha - \sqrt{\alpha^2 + 4\lambda}\right) Z/2} + c_2 e^{\left(-\alpha + \sqrt{\alpha^2 + 4\lambda}\right) Z/2}, \tag{A9}$$

where c_0, c_1 and c_2 are constants. The boundary conditions, (2.39b) and (2.39c) give c_1 and c_2,

$$c_1 = \frac{e^{\sqrt{\alpha^2 + 4\lambda}}\left(1 - \dfrac{1}{k_{E_0}}\right)}{c_0\left(e^{\sqrt{\alpha^2 + 4\lambda}} - 1\right)\left\{1 + k_{E_0}\left(e^{\alpha} - 1\right)\right\}} \tag{A10}$$

and

$$c_2 = \frac{1 - \dfrac{1}{k_{E_0}}}{c_0\left(e^{\sqrt{\alpha^2 + 4\lambda}} - 1\right)\left\{1 + k_{E_0}\left(e^{\alpha} - 1\right)\right\}}. \tag{A11}$$

Inserting (A10) and (A11) into (A9) and combining with (2.41), one finds that c_0 is canceled out. The boundary condition, (2.39a) leads to (A12) for λ,

$$\frac{\alpha^2\left\{e^{\sqrt{\alpha^2 + 4\lambda}} + k_{E_0}\left(1 - k_{E_0}\right)\left(e^{\sqrt{\alpha^2 + 4\lambda}} - 1\right)^2\right\}}{\left(1 + e^{\sqrt{\alpha^2 + 4\lambda}}\right)^2} + \lambda = 0, \tag{A12}$$

and λ is numerically obtained for the given α and k_{E_0}. For instance, a combination of $\alpha = 0.26$ and $k_{E_0} = 1.34$ gives $\lambda = -1.68 \times 10^{-2}$. Finally, the solution for $\tilde{C}(R, Z)$ is obtained:

$$\tilde{C}(R, Z) = \frac{k_{E_0} + (1 - k_{E_0})e^{-\alpha Z}}{k_{E_0} + (1 - k_{E_0})e^{-\alpha}}$$

$$+ \frac{e^{-\frac{1}{2}\left\{\sqrt{\alpha^2 + 4\gamma}(Z-2) + \alpha(Z+2)\right\}}\left\{e^{\sqrt{\alpha^2 + 4\gamma}(Z-1)} - 1\right\}\left(1 - \frac{1}{k_{E_0}}\right)R^{-\gamma/\beta}}{\left(e^{\sqrt{\alpha^2 + 4\gamma}} - 1\right)\left\{k_{E_0} + (1 - k_{E_0})e^{-\alpha}\right\}} \qquad \text{(A13)}$$

References

1 K. Shimamura, S. Uda, T. Yamada, S. Sakaguchi, T. Fukuda, *Jpn. J. Appl. Phys.* **35** (1996) L793.
2 R. S. Feigelson, *J. Cryst. Growth* **79** (1986) 669.
3 T. Surek, B. Chalmers, *J. Cryst. Growth* **29** (1975) 1.
4 V. A. Tatarchenko, *J. Cryst. Growth* **37** (1977) 272.
5 D. H. Yoon, I. Yonenaga, T. Fukuda, N. Ohnishi, *J. Crystal Growth* **142** (1994) 339.
6 D. H. Yoon, P. Rudolph, T. Fukuda, *J. Crystal Growth* **144** (1994) 207.
7 N. Schäfer, T. Yamada, K. Shimamura, H. J. Koh, T. Fukuda, *J. Crystal Growth* **166** (1996) 675.
8 H. J. Koh, N. Schäfer, K. Shimamura, T. Fukuda, *J. Crystal Growth* **167** (1996) 38.
9 S. Uda, J. Kon, K. Shimamura, T. Fukuda, *J. Cryst. Growth* **167**, (1996) 64.
10 J. A. Burton, R. C. Prim, W. P. Slichter, *J. Chem. Phys.* **21** (1953) 1987.
11 K. Shimamura, K. Sugiyama, S. Uda, T. Fukuda, *Jpn. J. Appl. Phys.* **34** (1995) 4894.
12 V. G. Smith, W. A. Tiller, J. W. Rutter, *Can. J. Phys.* **33** (1955) 723.
13 W. A. Tiller, K. A. Jackson, J. W. Rutter, B. Chalmers, *Acta Met.* **1** (1953) 428.
14 J. A. Nelder, R. Mead, *Computer J.* **7** (1965) 308.
15 H. Kodera, *Jpn. J. Appl. Phys.* **2** (1963) 212.
16 V. N. Romanenko, Y. M. Smirnov, *Inorg. Mater.* **6** (1970) 1527.
17 H. Sasaki, Y. Anzai, X. M. Huang, K. Terashima, S. Kimura, *Jpn. J. Appl. Phys.* **34** (1995) 414.
18 H. Sasaki, E. Tokizaki, X. M. Huang, T. Terashima, S. Kimura, *Jpn. J. Appl. Phys.* **34** (1995) 3432.
19 H. Sasaki, E. Tokizaki, T. Terashima, S. Kimura, *Jpn. J. Appl. Phys.* **33** (1994) 3803.
20 H. Sasaki, E. Tokizaki, T. Terashima, S. Kimura, *Jpn. J. Appl. Phys.* **33** (1994) 6078.
21 C. T. Yen, W. A. Tiller, *J. Crystal Growth* **118** (1992) 259.
22 A. G. Ostrogorsky, G. Müller, *J. Crystal Growth* **121** (1992) 587.
23 J. P. Garandet, J. J. Favier, D. Camel, *J. Crystal Growth* **130** (1993) 113.
24 J. P. Garandet, *J. Crystal Growth* **131** (1993) 431.
25 S. Uda, W. A. Tiller, *J. Crystal Growth* **121**, (1992) 93.
26 S. Uda, W. A. Tiller, *J. Crystal Growth* **126**, (1993) 396.
27 V. A. D'yakov, D. P. Shumov, L. N. Rashkovich, A. L. Aleksandrovskii, *Bulletin of the Academy of Sciences of the USSR. Physical Series* **49** (1986) 117.
28 J. R. Owen, E. A. D. White, *J. Crystal Growth* **42** (1977) 499.
29 Y. S. Luh, R. S. Feigelson, M. M. Fejer, R. L. Byer, *J. Crystal Growth* **78** (1986) 135.
30 C. T. Yen, D. O. Nason, W. A. Tiller, *J. Mater. Res.* **1** (1992) 980.

31   A. Feisst, A. Räuber, *J. Crystal Growth* **63** (1983) 337.
32   A. Feisst, P. Koidl, *Appl. Phys. Lett.* **47** (1985) 1125.
33   S. Uda, W. A. Tiller, *J. Crystal Growth* **121** (1992) 155.
34   W. A. Tiller, S. Uda, *J. Crystal Growth* **129** (1993) 341.
35   S. Uda, K. Shimamura, T. Fukuda, *J. Crystal Growth* **155** (1995) 229.
36   S. Uda, J. Kon, J. Ichikawa, K. Inaba, K. Shimamura, T. Fukuda, *J. Cryst. Growth* **179** (1997) 567.
37   P. F. Bordui, R. G. Norwood, C. D. Bird, G. D. Calvert, *J. Crystal Growth* **113**, (1991) 61.
38   R. S. Feigelson, W. L. Kway, R. K. Route, *Optical Engineering* **24**, (1985) 1102.
39   A. Räuber, in *Current Topics in Materials Science*, E. Kaldis, Ed. (North-Holland, Amsterdam, 1978), vol. 1, pp. 481.
40   W. A. Tiller, R. F. Sekerka, *J. Appl. Phys.* **35** (1964) 2726.
41   S. Uda, W. A. Tiller, *J. Crystal Growth* **152** (1995) 79.
42   W. A. Tiller, S. Uda, *J. Crystal Growth* **129** (1993) 328.
43   K. Imai, M. Imaeda, S. Uda, T. Taniuchi, T. Fukuda, *J. Cryst. Growth* **177** (1997) 79–87.
44   S. Uda, J. Kon, J. Ichikawa, K. Inaba, K. Shimamura, T. Fukuda, *J. Crystal Growth*, **182** (1997) 403.
45   P. K. Gallagher, J. H. M. O'Bryan, *J. Am. Ceram. Soc.* **68** (1985) 147.
46   H. M. O'Bryan, P. K. Gallagher, C. D. Brandle, *J. Am. Ceram. Soc.* **68** (1985) 493.
47   R. Zuo and Z. Guo, *J. Cryst. Growth* **158** (1996) 377.

3 Theoretical Analysis of the Micro-Pulling-Down Process

C.W. Lan

Theoretical analysis of the micro pulling down (μ–PD) process is presented using a finite-volume Newton method. The growth of Ge_xSi_{1-x} single-crystal fibers is used as an example to illustrate the role of transport phenomena in the diameter control, constitutional super-cooling, and solute segregation.

3.1 Introduction

The unique properties of near one-dimensional single-crystal fibers have attracted some attention for applications in optical and electronic devices [1–3]. Device-size as-grown fibers also reduce processing cost (for size reduction) significantly for device fabrication. There are several methods for growing fiber crystals [4], such as the edge-defined film-fed growth (EFG), floating zone (pedestal growth) methods, etc. Recently, the micro pulling down (μ-PD) process, a variant of the inverse EFG, developed by Fukuda's laboratory in Japan [5–9] has shown promis in producing single-crystal fibers with good diameter control and concentration uniformity. Several oxide [5–7] and semiconductor fibers [8, 9] have been grown, as well as high-temperature eutectic fibers [10]. The grown diameters range from 10 to 1500 μm for oxides and about 300 to 900 μm for semiconductors. Because of the high thermal gradients near the growth interface, it is also possible to use a very high growth rate without causing constitutional supercooling which leads to interface breakdown. In this process, some simple theoretical analyses for the operation limit [11] and solute distribution [12, 13] have been performed. However, no detailed modeling has been conducted. In particular, melt convection was ignored in the previous reports. In fact, even for EFG, detailed convective heat and mass transfer simulation coupled with capillary shaping and solidification has not been reported either. Since the system is small, the measurements of concentration and temperature are difficult. Furthermore, although buoyancy convection may be negligible due to the small physical dimension, thermocapillary convection (Marangoni flow) may still be important because of the high thermal gradients near the growth interface. Therefore, a detailed numerical simulation will allow crystal growers to better 'visualize' the process, and thus provide a foundation for process improvement and tuning.

In this chapter, a detailed numerical simulation is conducted in a self-consistent manner to analyze the μ–PD process for Ge_xSi_{1-x} fibers; $x = 0.05$. The melt flow, heat and mass transfer, the growth front and meniscus, and the grown fiber diameter for various process parameters are illustrated. The role of convection on solute transport will be emphasized. Some comparison with simple models for capillary shaping and solute diffusion as well as experimental observations is also performed. The model formulation and the solution scheme are described in the next section. Section 3.3 is devoted to the results and discussion, followed by conclusions in Section 3.4.

3.2 Model and Solution Scheme

The system to be modeled is sketched in Fig. 3.1. Since the melt height and the grown fiber length are much larger than the zone dimension (the area between the die and the growth front), to relieve computational load, while balancing model accuracy, the computational domain is chosen for the part near the melt zone only, as shown in Fig. 3.2a.

Since the melt height is much larger than the length of the capillary tube, a good approximation is to ignore the melt reservoir (the crucible), but include the melt height for meniscus calculation. Furthermore, it is believed that the melt flow in the reservoir does not affect the flow inside the channel much or the melt zone. In addition, the melt velocity at the top entrance (at $z = -L_h$) is assumed to be uniform. If a pseudo-steady state is achieved for the fiber pulling rate U_c, the melt velocity U_m at the upper boundary from the melt reservoir can be set to be $\rho_c U_c (R_c/R_h)^2/\rho_m$, where ρ_c and ρ_m are the crystal and melt densities, respectively.

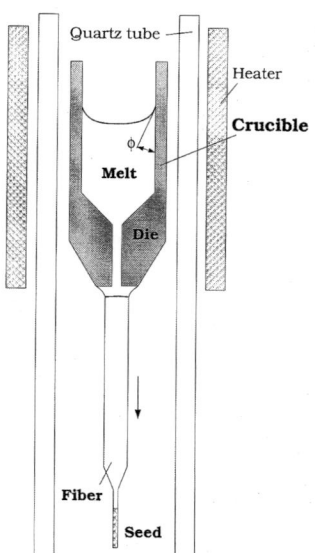

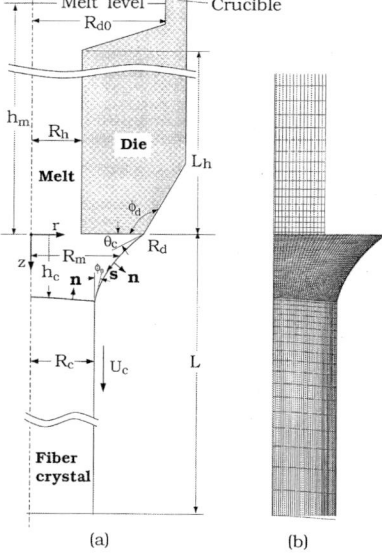

Fig. 3.1. Schematic diagram of μ–PD fiber growth process.

Fig. 3.2. (a) Schematic diagram of the computational domain; (b) a portion of a sample mesh for calculation.

Usually, the charge of the feed material is about 2 g, and the amount of the grown crystal is only 1 or 2% of the charge. Therefore, the change of the top melt level is very small. In fact, in practice, it is important to have enough melt for the fiber growth, and that can ensure the steadiness of the process, especially for capillary shaping; the effect of the melt height will be illustrated shortly. The top meniscus of the melt in the crucible is considered through the effective melt head h_{eff}, which will be explained shortly as well. The die temperature is assumed to be a linear distribution along the axial distance from $z = 0$. Furthermore, perfect attachment of the melt along the die edge is assumed.

Axisymmetry is further assumed. Hence, the melt velocity, temperature, solute concentration, and the free surface ($R_m(z)$), as well as the growth front ($h_c(r)$), are represented in the cylindrical coordinate system (r, z). It is also assumed that the solute is uniformly distributed in the melt reservoir and its concentration is C_0. Dimensionless variables are defined by the scaling length by R_d, velocity by α_m/R_d, temperature by the melting point T_m of silicon ($x = 0$), and concentration by C_0, where α_m is the thermal diffusivity of the melt. For convenience of representation, all the variables defined afterwards are dimensionless unless otherwise stated. The governing equations in the conservative-law form (or the so-called divergence form) for melt flow and heat and solute transfer can be described in terms of the stream function ψ, vorticity ω, temperature T, and concentration C as follows:

Equation of motion

$$\frac{\partial}{\partial r}\left(\frac{\omega}{r}\frac{\partial \psi}{\partial z}\right) - \frac{\partial}{\partial z}\left(\frac{\omega}{r}\frac{\partial \psi}{\partial r}\right)$$
$$+ \Pr\left[\frac{\partial}{\partial r}\left(\frac{1}{r}\frac{\partial}{\partial r}(r\omega)\right) + \frac{\partial}{\partial z}\left(\frac{1}{r}\frac{\partial}{\partial z}(r\omega)\right)\right] - \Pr Ra_T \frac{\partial T}{\partial r} + \Pr Ra_S \frac{\partial C}{\partial r} = 0. \qquad (3.1)$$

Stream equation

$$\frac{\partial}{\partial z}\left(\frac{1}{r}\frac{\partial \psi}{\partial z}\right) + \frac{\partial}{\partial r}\left(\frac{1}{r}\frac{\partial \psi}{\partial r}\right) + \omega = 0. \qquad (3.2)$$

Energy equation

$$-\frac{\partial}{\partial r}(ruT) - \frac{\partial}{\partial z}(rvT) + \frac{\partial}{\partial z}\left(r\alpha_i(T)\frac{\partial T}{\partial z}\right) + \frac{\partial}{\partial r}\left(r\alpha_i(T)\frac{\partial T}{\partial r}\right) = 0, \quad i = (m,c). \qquad (3.3)$$

Solute equation

$$-\frac{\partial}{\partial r}(ruC) - \frac{\partial}{\partial z}(rvC) + \frac{\Pr}{Sc}\left[\frac{\partial}{\partial z}\left(r\frac{\partial C}{\partial z}\right) + \frac{\partial}{\partial r}\left(r\frac{\partial C}{\partial r}\right)\right] = 0. \qquad (3.4)$$

Solute diffusion in the solid phase is neglected. In the above equations, Pr is the Prandtl number ($\Pr \equiv \alpha_m/v_m$), where v_m is the kinematic melt viscosity), Sc the Schmidt number ($Sc \equiv v_m/D$), and D the diffusivity of germanium in the melt. Also, α_i is the thermal diffusivity of phase i; $i = c$ for crystal and m for melt. Two important dimensionless variables, Ra_T and Ra_S, in the source term of the equation of motion are defined as follows:

$$\mathrm{Ra}_T \equiv \frac{g\beta_T T_m R_d^{\,3}}{\alpha_m \nu_m}\,; \qquad\qquad \mathrm{Ra}_S \equiv \frac{g\beta_S C_0 R_d^{\,3}}{\alpha_m \nu_m}\,,$$

where g is the gravitational acceleration and β_T and β_S are the thermal and solutal expansion coefficients, respectively. The stream function ψ and vorticity ω in the above equations are defined in terms of the radial (u) and axial (v) velocities as:

$$u = -\frac{1}{r}\frac{\partial \psi}{\partial z}, \quad v = \frac{1}{r}\frac{\partial \psi}{\partial r}, \tag{3.5}$$

$$\omega = \frac{\partial u}{\partial z} - \frac{\partial v}{\partial r}. \tag{3.6}$$

To solve the above governing equations, boundary conditions are also required. Most of the boundary conditions for the melt flow and heat and mass transfer can be found elsewhere [14]. Only some important boundary conditions are described here. The solute boundary conditions at the top entrance ($z = -L_h$) are set by the solute flux balance:

$$\boldsymbol{e}_z \cdot \nabla C = \left(\frac{\mathrm{Sc}}{\mathrm{Pr}}\right) \mathrm{Pe}_m (C - 1). \tag{3.7}$$

At the melt-crystal interface,

$$\boldsymbol{n} \cdot \nabla C = \left[\left(\frac{\rho_c}{\rho_m}\right) - K\right]\left(\frac{\mathrm{Sc}}{\mathrm{Pr}}\right) \mathrm{Pe}_c C(\boldsymbol{n} \cdot \boldsymbol{e}_z), \tag{3.8}$$

where $\boldsymbol{n}$ is the unit normal vector at the growth front pointing to the melt, $\mathrm{Pe}_m \equiv U_m R_d/\alpha_m$ and $\mathrm{Pe}_c \equiv U_c R_d/\alpha_m$ are the Peclect numbers of the melt and crystal, respectively, and K is the segregation coefficient according to the phase diagram. The density ratio (ρ_c/ρ_m) is considered, but its effect is small.

For the meniscus shape, the normal stress balance is used:

$$\boldsymbol{nn} : \boldsymbol{t} - \mathrm{Bo}(z + h_{\mathrm{eff}}) = 2H, \tag{3.9}$$

where $\boldsymbol{n}$ is the unit normal vector at the free surface pointing outward and $\boldsymbol{t}$ the shear stress tensor. $\mathrm{Bo} \equiv \rho_m g R_d^{\,2}/\gamma$ is the Bond number measuring the relative effects of gravity and surface tension, and h_{eff} the dimensionless effective melt height from $z = 0$; γ is the surface tension coefficient. If we know the static contact angle ϕ (see Fig. 3.1) and the dimensionless average melt height h_m, a simple approximation may be used to calculate h_{eff}:

$$h_{eff} = h_m - 2\mathrm{Bo}^{-1}\cos(\phi)R_d / R_{d0}, \tag{3.10}$$

where R_{d0} is the inner radius of the crucible. Clearly, if the crucible radius is small and the crucible can be wetted by the melt ($\phi \leq 90°$), h_{eff} can be smaller than the melt height h_m. In reality, it is not easy to accurately determine ϕ because the graphite reacts with Si and thus the wetting angle is changed. Therefore, we will use h_{eff} in the calculation and its effects can then be illustrated, which reflect the effects of the meniscus as well as the melt height.

In order to grow the fiber with a constant diameter, the growth angle constraint [15] needs to be specified:

$$\frac{dR_m}{dz} = -\tan(\phi_0), \tag{3.11}$$

where ϕ_0 is the growth angle. For example, $\phi_0 = 11°$ for Si in the <111> growth direction. However, the steadily grown fiber radius R_c is unknown a priori and needs to be calculated.

The tangential stress balance for the free surface is further required:

$$\boldsymbol{ns} : t = \mathrm{Ma}(\boldsymbol{s} \cdot \nabla T), \tag{3.12}$$

where $\boldsymbol{s}$ is the unit tangent vector at the free surface and the Marangoni number $\mathrm{Ma} \equiv (\partial\gamma/\partial T)R_d T_m/(\mu_m \alpha_m)$; $(\partial\gamma/\partial T)$ is the surface-tension-temperature coefficient and μ_m the melt viscosity.

The die temperature is assumed to be a linear function of the axial distance. At the surface of the material, heat transfer from the system to the ambient atmosphere is by both radiation and convection according to the energy balance along the material surface:

$$-\boldsymbol{n} \cdot \kappa_i \nabla T = \mathrm{Bi}(T - T_a) + \mathrm{Rad}_i(T^4 - T_a^4), \qquad i = (\mathrm{m,c}), \tag{3.13}$$

where $\boldsymbol{n}$ is the unit normal vector on the melt or crystal surface pointing outwards, κ_i the ratio of thermal conductivity of phase i to the melt, $\mathrm{Bi} \equiv hR_d/k_m$ the Biot number, and $\mathrm{Rad}_i \equiv \sigma\varepsilon_i T_m^3 R_d/k_m$ the radiation number; σ is the Stefan-Boltzmann constant, while ε_i is the surface emissivity of the melt or crystal. T_a is the effective ambient temperature, and is set to be a constant in this study.

At the end of the fiber in the domain ($z = L$), the zero flux boundary condition is used. The length L is made large enough so that the calculated molten zone length is insensitive to its value. L being 100 times R_d is found satisfactory. Furthermore, using the fixed-temperature boundary condition, i.e, $T = T_a$ at $z = L$, does not change the results much.

The governing equations with their associated boundary conditions are discretized by a finite volume method (FVM) [16]. A portion of a converged sample mesh for calculation is shown in Fig. 3.2b, where the meniscus shape, the growth interface, and the grown fiber radius are unknown *a priori*, and need to be solved with variables simultaneously. After the FVM approximation, the resulting nonlinear equations are then solved globally by Newton's method. Usually, five–seven iterations are enough for the solution to converge to an infinity norm of 10^{-6}. Details of the numerical scheme can be found elsewhere [16,17].

3.3 Results and Discussion

The physical properties of Ge_xSi_{1-x} and some input parameters used in the calculations are listed in Table 3.1. Since x is small (0.05), most of the physical properties of Si are adopted. The liquidus slope and the segregation coefficient are obtained from the phase diagram. In the present study, we are interested only in cases with $x = 0.05$, which was used in the previous experiments [13]. Also, the diffusivity values of Ge in Si from different reports [9, 13, 18] are quite different. However, according to a recent report [13], $D = 5.6 \times 10^{-5}$ cm^2/s seems to be more reasonable, and thus this value is used here. Before the results were presented, we also performed mesh refinement to ensure that the solution is not affected much by the mesh. Due to the large Sc value, the solute boundary layer is much thinner than the velocity one. Therefore, a fine grid spacing near the solid boundaries is necessary. Part of the sample mesh used in our calculations is shown in Fig. 3.2b. The number of unknowns resulted from this mesh is 16459, and one Newton iteration takes about 1 min in an HP9000/C180 workstation.

3.3.1 Meniscus Shape and Grown Fiber Size

The meniscus shape as well as the grown fiber diameter is mainly affected by the capillary statics and the melt height. The effect of melt convection is believed to be trivial, which will be shown shortly. Fig. 3.3a shows the calculated meniscus shapes for different h_{eff}'s; the grown fiber diameter is fixed and the growth angle ϕ_0 = 11°. An observed meniscus shape is also depicted for comparison. For $h_{eff} = 0$, the calculated shape is almost the same as that obtained from the low-Bo approximation (with an analytical solution), where the gravity effect is ignored [19]. This indicates that the effect of convection on the meniscus shape is trivial here. As compared with the observed shape, the agreement is reasonably good. In fact, the shape obtained by the CCD camera is not perfectly axisymmetric. More importantly, the wetting line at the melt/die junction is not clearly defined, which makes a better comparison difficult. The wetting angle with the die, i.e. θ_c (also see Fig. 3.2a, is slightly smaller than the Gibbs limit here; this limit is about 30° for the present case ($\phi_d = 120°$). Therefore, the dewetting is very likely.

Table 3.1. Physical properties of $Ge_{0.05}Si_{0.95}$ and some input parameters

Physical properties	Input parameters
$T_m = (1683.4 - 220x)$ K	$R_h = 0.0275$ cm
$\Delta H = 1803$ J g^{-1}	$R_d = 0.057$ cm
$k_c = 0.22\, T_m/T$ W cm^{-1} K^{-1}	$L_h = 0.33$ cm
$k_m = 0.64$ W cm^{-1} K^{-1}	$L = 3$ cm
$Cp_c = 1.038$ J g^{-1} K^{-1}	$T_d\vert_0 = 1673.4 - 1676.4$ K
$Cp_m = 1.059$ J g^{-1} K^{-1}	$T_d\vert_{L_h} = 1683.4$ K
$\partial\gamma/\partial T = -0.08$ dyn cm^{-1} K^{-1}	$T_a = 635.4$ K
$\gamma = 720 + (T - T_m)\, \partial\gamma/\partial T$ dyn cm^{-1}	$U_c = 1\sim6$ cm/h
$\mu_m = 0.007$ g cm^{-1} s^{-1}	$h = 4.4\times10^{-3}$ W cm^{-2} K^{-1}
$\beta_s = 0.013/(\text{at\% Ge})$	$C_0 = 5$ at%
$\beta_T = 1\times10^{-4}$ K^{-1}	
$\varepsilon_c = 0.7$	
$\varepsilon_m = 0.3$	
$\rho_c = 2.33$ g cm^{-3}	
$\rho_m = 2.55$ g cm^{-3}	
$K = 0.4$	
$\phi_0 = 11°$	

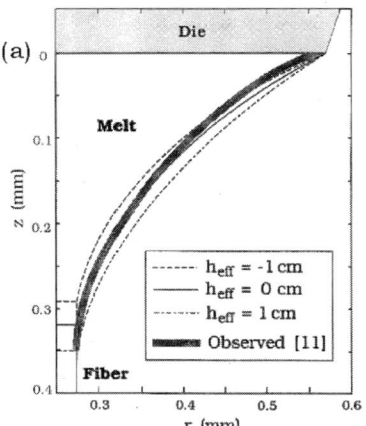
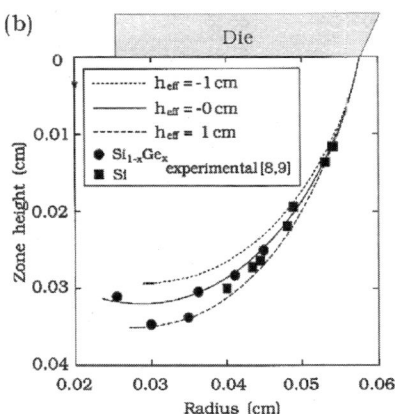

Fig. 3.3. (a) Meniscus shape for different h_{eff}'s; (b) relationship of zone height and grown radius for different h_{eff}'s.

The relationship of the zone height and the grown fiber size for different h_{eff}'s is summarized in Fig. 3.3b. This relationship is obtained by adjusting the die tip temperature $T_d\vert_0$. The experimental results are obtained from both the growth of Si [8] and $Ge_{0.5}Si_{0.95}$ [9]. As shown, the calculated results are in good agreement with the observed ones for the dimensional h_{eff} between 0 and 1 cm. Therefore, it is believed that the gravity force still plays a role here. Furthermore, for the same zone height, there is another solution branch with a smaller radius ($R_c < 0.03$ cm). However, this branch is not stable according to Surek's stability criterion [15], and should be avoided during growth.

3.3.2 Effect of Convection

Although the system is small, convection may still be important in such a small molten zone. To illustrate this, we have examined different convection modes and the calculated results are shown in Fig. 3.4. For this case, $T_d|_0 = 1676.4$ K, $D = 5.6 \times 10^{-5}$ cm^2/s, and $U_c = 3$ cm/h. The solute and flow fields are shown from Fig 3.4a to Fig 3.4c, while thermal fields from Fig 3.4d to Fig 3.4f. When the buoyancy and thermocapillary forces are turned off, the convection inside the melt is only due to the pulling of the fiber, as shown in Fig. 3.4a and Fig 3.4d. As the buoyancy force is considered in Fig. 3.4b and Fig 3.4e, two convection loops near the centerline and the free surface are induced. However, their intensities are very weak. The zero streamline is indicated by a dashed line with an asterisk. In fact, the excess Ge in front of the growth interface, being heavier at the bottom, stabilizes the system. As a result, the solute field is not affected much.

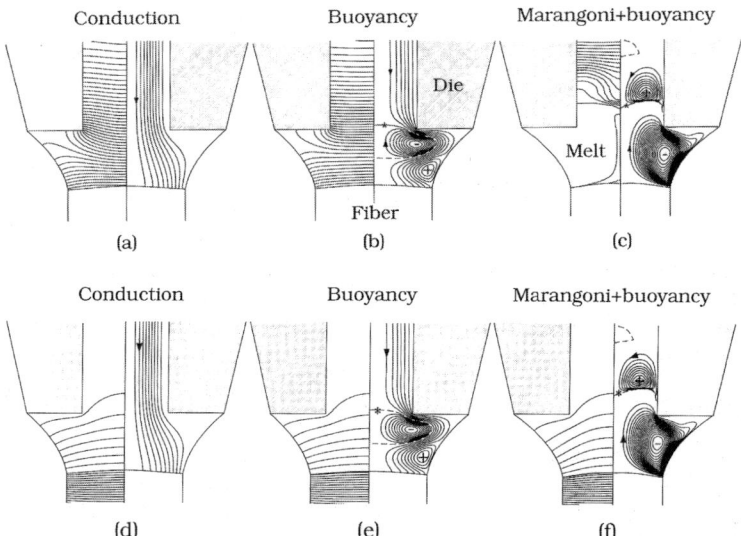

Fig. 3.4. Calculated fields for different convection modes: (**a**) and (**d**) for conduction only; (**b**) and (**e**) for buoyancy convection; (**c**) and (**f**) for both buoyancy and Marangoni convection. In the melt, $\Delta\psi = \psi_{max}/10$ for positive ψ, $\Delta\psi = \psi_{min}/20$ for negative ψ; $\Delta C = (C_{max} - C_{min})/50$, and $\Delta T = (T_{max} - T_{min})/10$, $\psi_{min} = 0$, $\psi_{max} = 4.203\times10^{-7}$ cm^3/s for (a), -3.965×10^{-7}, 6.475×10^{-7} cm^3/s for (b), and -0.001796, 5.146×10^{-6} cm^3/s for (c). $C_{min}/C_0 = 1$, $C_{max}/C_0 = 2.5061$ for (a) 1, 2.5024 for (b) and 1, 2.5524 for (c). $T_d|_0 = 1676.4$ K. $T_{max} = 1683.4$, $T_{min} = 1655.83$ K for (d), 1683.4, 1655.87 K for (e), 1683.4, 1655.32 K for (f).

However, as the thermocapillary force is considered, the effect becomes significant. As shown in Fig. 3.4c, the flow prevails in the melt zone. Because of the high temperature gradients near the growth interface, the induced melt speed is as high as 19.6 cm/s. In fact, such a high melt speed is not uncommon. In the calculations by Yeckel et al. [20] for floating-zone Si sheet growth, the same order of the melt speed was reported. As a result, the solute is well mixed in the molten zone. Outside the main flow loop, a small secondary cell is induced, and the solute fields are also affected there. However, further inside the hole, the solute mixing is poor, where the diffusion is dominant. The thermal field is also affected by the thermocapillary flow as well, as shown in Fig. 3.4f. Heat is delivered more effectively from the die to the fiber at the surface, which can be seen from the greater distortion of the isotherms near the meniscus. As a result, the melt zone at the surface becomes longer and thus the grown fiber diameter is reduced.

It should be pointed out that as mentioned previously for Fig. 3.3b, the solution obtained in Fig. 3.4 is not unique. A set of solutions with a much smaller fiber diameter is also obtained, and the multiplicity is mainly due to the nonlinear capillary statics [9]. According to Surek's stability criterion, this solution is not stable.

3.3.3 Effect of Pulling Rate

Different pulling rates are considered and their effects on the radial solute distribution in the melt are illustrated in Fig. 3.5. The flow and solute fields are also shown in the same figure. As shown in the solute fields, with a higher pulling rate, the solute gradients in the hole are higher. Because of the larger solute gradients, as expected, the radial solute segregation increases with the increasing growth rate. At $U_c = 6$ cm/h, the excess of Ge near the centerline also causes the interface depression there due to the lower liquidus temperature. In addition to the segregation, the increase of the pulling rate also reduces the grown fiber diameter, which can be seen from the field plots. Clearly, a greater release of the heat of fusion results in a longer zone, and hence a smaller fiber diameter. Due to the greater necking of the meniscus with increasing pulling rate, the Marangoni convection can penetrate further into the hole region, which can be seen by the location of the zero streamline and the diminishing of the secondary cell.

In practice, a major concern for the high growth rate in crystal growth is the onset of constitutional supercooling. To illustrate the possibility of supercooling, the actual and liquidus temperature profiles at the centerline from the growth front for different growth rates are plotted in Fig. 3.6. As shown, although we do not observe any supercooling for these three growth rates, the possibility of supercooling becomes higher with increasing growth rate, which can be seen from the angle between the two profiles at the growth front. Furthermore, the thermal gradients at the growth front are as high as 700 K/cm. Because of the lower crystal thermal conductivity, the thermal gradients in the crystal side are up to 2000 K/cm. Because of the high thermal gradients, a high growth rate is usually allowed for fiber growth.

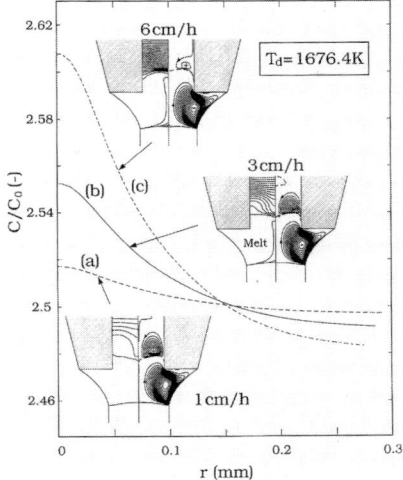
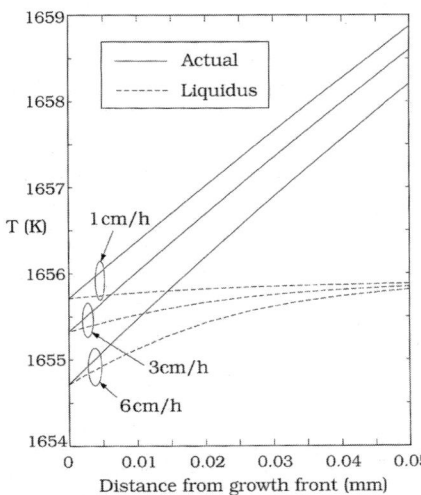

Fig. 3.5. Radial germanium distributions for different growth rates.

Fig. 3.6. Actual and liquidus temperature distributions in the melt at the centerline for different growth rates.

3.3.4 Effect of Die Temperature

During crystal growth experiments, it was observed that the control of the fiber diameter through the heater power (die temperature) is very effective. From Fig. 3.3b, it is obvious that the shorter the zone (the less power), the larger the grown fiber diameter. Therefore, calculations for $T_d = 1675.4$ K and 1673.4 K are also conducted, and the calculated results as well as that for $T_d = 1676.4$ K are summarized in Fig. 3.7 for comparison. From the field plots, the zone length decreases with decreasing die temperature leading to a larger fiber diameter. As the zone length decreases, the Marangoni convection becomes less likely to penetrate into the zone center. Instead, a secondary cell is induced. Due to the secondary flow cell, the solute segregation behavior is greatly changed. As shown, the radial solute distribution is reversed when the die temperature is reduced to 1673.4 K. This inversion of the segregation and the depletion of Ge at the centerline were also observed in growth experiments [13] when the fiber diameter was increased by reducing the die temperature.

We have also conducted calculations for smaller $|\partial\gamma/\partial T|$. Interestingly, the secondary cell is much harder to induce and, therefore, no inversion of the segregation is observed.

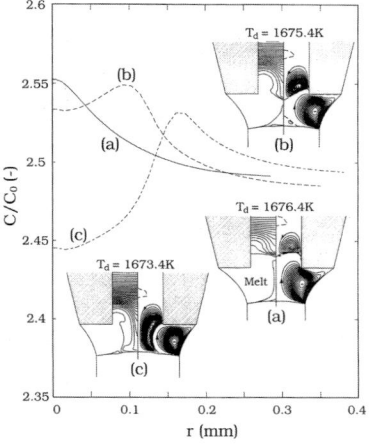 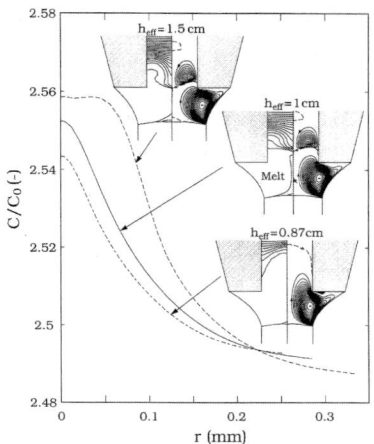

Fig. 3.7. Radial germanium distributions for different die temperatures.

Fig. 3.8. Radial germanium distributions for different h_{eff}'s.

3.3.5 Effect of Melt Height

As mentioned previously, the grown fiber amount is usually about 1 or 2% of the initial charge to the crucible. As a result, the melt height during growth does not change much. However, its value can affect the meniscus shape and the fiber size as shown in Fig. 3.3. Because of the change of the meniscus shape and the fiber diameter by the melt height, heat transfer can be changed as well. To further illustrate the effect of the effective melt height h_{eff} for the same die temperature and the growth rate (T_d = 1676.4 K and U_c = 3 cm/h), additional calculations are performed for h_{eff} = 0.87 and 1.5 cm, and their calculated results as well as that for h_{eff} = 1 cm are illustrated in Fig. 3.8. As shown from the field plots, when h_{eff} is reduced slightly to 0.87cm from 1 cm, the grown fiber diameter is significantly reduced. Because of the too small fiber diameter, this case is believed to be unstable. According to the analysis by Tatarchenko [19], the maximum stability of the meniscus is expected at $R_c \approx 0.8R_d$, and also from Fig. 3.3b, the fiber radius is very sensitive to the melt height when $R_c \approx 0.5R_d$. Furthermore, based on Surek's stability criterion [14] for capillary shaping, this case is not stable due to $dh_c(1)/dR_c < 0$. On the other hand, the increase rate of R_c per unit change of h_{eff} to 1.5 cm is smaller than that to 0.87 cm. Clearly, keeping R_c away from the point where $dh_c(1)/dR_c = 0$ makes the grown fiber radius less sensitive to the melt height. A similar observation was made during crystal growth experiments. However, with increasing fiber diameter, the radial solute segregation increases.

3.4 Conclusion

We have developed a numerical model to analyze the µ–PD growth of Ge_xSi_{1-x} fibers. The steady-state heat and mass transfer, melt flow, interface shapes, as well as the grown fiber diameter, are computed simultaneously using a robust finite-volume Newton method. The present analysis provides a direct view of the melt flow and solute transport in such a small system as well as a detailed understanding of the process. The calculated results can also be used directly the inverse EFG process.

The calculated fiber grown radius as a function of the zone length is in good agreement with measurements. Reasonable agreement for the meniscus shape is obtained as well. For the case with zero effective melt height, the calculated meniscus shape is almost identical with the analytical solution using the low-Bo approximation, which indicates that melt convection has very little effect on the meniscus shape. The melt flow in the hole is not affected much by the convection in the zone region. The calculated axial solute distribution in the zone region and the hole for the case without buoyancy and thermocapillary convections is in good agreement with that calculated using a plug flow approximation. In the zone region, the buoyancy convection is extremely weak due to the small physical dimension and the damping effect by the excess and heavier germanium segregated in the melt. However, thermocapillary convection is still strong in the melt zone leading to solute mixing there. Unfortunately, such mixing often causes larger segregation. Furthermore, the segregation increases with increasing growth rate. More interestingly, decreasing the die temperature or increasing the fiber radius could result in an inversion of the radial solute segregation as well as the depletion of germanium due to the formation of a secondary cell induced by thermocapillary flow. This inversion was also observed in experiments. Finally, we have also shown that the grown fiber radius is affected by the melt height. This indicates that different initial charges can grow fibers with different diameters. If the grown radius is less than half the die radius, dynamic instability occurs. For the same growth parameters, the grown fiber radius increases with increasing melt height.

References

1   L. Hesselink, S. Redfield, *Opt. Lett.* **13** (1988) 877.
2   S. Sudo, I. Yokohama, A. Cordova-Plaza, M. M. Fejer, R.L. Byer, *Appl. Phys. Lett.* **56** (1990) 1931.
3   H. Yoshinaga, K. Kitayama, *Appl. Phys. Lett.* **56** (1990) 1728.
4   R.S. Feigelson, Growth of fiber Crystals, in: E. Kaldis, Crystal Growth of Electronic Materials (Edited by E. Kaldis), North Holland, Amsterdam (1985) 127.
5   D.H. Yoon, I. Yonenaga, T. Fukuda, N. Ohnishi, *J. Crystal Growth* **142** (1994) 339.
6   Y.M. Yu, V.I. Chani, K. Shimamura, K. Inaba, T. Fukuda, *J. Crystal Growth* **77** (1997) 74.
7   B.M. Epelbaum, K. Inaba, S. Uda, T. Fukuda, *J. Crystal Growth* **179** (1997) 559.

8   K. Shimamura, S. Uda, T. Yamada, S. Sakaguchi, T. Fukuda, *Jpn. J. Appl. Phys.* **35** (1996) 703.

9   N. Schäfer, T. Yamada, K. Shimamura, H.J. Koh, T. Fukuda, *J. Crystal Growth* **166** (1996) 675.

10   A. Yoshikawa, K. Hasegawa, J.H. Lee, S.D. Durbin, B.M. Epelbaum, D.H. Yoon, T. Fukuda, Y. Waku, *J. Crystal Growth* **218** (2000) 67.

11   B.M. Epelbaum, K. Shimamura, S. Uda, H.J. Kon, T. Fukuda, *Cryst. Res. Technol.* **31** (1996) 1077.

12   H.J. Koh, N. Schäfer, K. Shimamura, T. Fukuda, *J. Crystal Growth* **167** (1996) 38.

13   S. Uda, H.J. Kon, K. Shimamura, T. Fukuda, *J. Crystal Growth* **67** (1996) 64.

14   C.W. Lan, *J. Crystal Growth* **169** (1996) 269.

15   T. Surek, B. Chalmers, *J. Crystal Growth,* **29** (1975) 1.

16   C.W. Lan, *Int. J. Numerical Methods in Fluids* **19** (1994) 19.

17   M.C. Liang, C.W. Lan, *J. Comp. Phys.* **127** (1996) 330.

18   H. Kodera, *Jpn. J. Appl. Phys.* **2** (1963) 212.

19   V.A. Tatarchenko, E.A. Brener, *J. Crystal Growth* **50** (1980) 33.

20   A. Yeckel, A. G. Salinger, J.J. Derby, *J. Crystal Growth* **152** (1995) 51.

4 Practice of Micro Pulling Down Growth

Boris M. Epelbaum

In this chapter we will focus on practical aspects of the micro pulling down growth method. Basic prerequisites for successful use of monocrystalline optical fibers are that they have to be grown with specific and homogeneous composition and smooth cylindrical geometry. In the first part of the chapter the problems of longitudinal homogeneity and surface quality of μ-PD fibers are addressed in detail. Direct implementation of these considerations is that significant improvement of fiber quality can be achieved only under precise control of the meniscus height. In the second part of the chapter the reverse situation is discussed: the intentional use of growth parameter variations during μ-PD for different growth studies. Here μ-PD appears as supporting instrumentality for bulk melt growth technologies. Because of the excellent stability and simplicity, the method has been proven to be a very useful research tool.

4.1 Growth of Homogeneous Fibers

As outlined in Chap. 1, growth of small-sized crystal fibers is motivated by their application to various optical devices. One key requirement of all applications is that fibers are homogeneously doped in the longitudinal direction [1, 9]. The attractive point of the method is the possibility to influence segregation during pulling using the adaptable geometry of μ-PD crucible [3] or pulling rate variations [4]. Nonuniformity of radial dopant in $LiNbO_3$ and Si-Ge fibers grown under nominally stable conditions was reported in [5, 6], but longitudinal distribution of dopants was always believed to be homogeneous, corresponding to $k_{eff} = 1$, since back-diffusion of melt species through the narrow capillary nozzle is negligibly small. Because of this intuitive assumption, the longitudinal homogeneity of μ-PD fibers and compositional transients have not been extensively investigated. Only in [7] does Chani point out an uneven distribution of Nd along YAG:Nd fibers grown with the use of a high-temperature version of the μ-PD method. However, current experimental observations during growth of lead tungstate $PbWO_4$ fiber crystals and $LiNbO_3$ of the stoichiometric composition from the Li-rich melt show that growth conditions and especially the meniscus height may have a pronounced effect on dopant segregation the in axial direction.

In this chapter the influence of convection in the meniscus on dopant segregation is discussed. Since the meniscus area is very small, the buoyancy convection can be ignored. On the other hand, thermocapillary (Marangoni) convection is very strong because of the high thermal gradient along the liquid free surface reaching 10^3–10^4 K/cm [2].

4.1.1 Experimental Observations

A standard μ-PD arrangement was used to grow $LiNbO_3$ fibers from the congruent and Li-rich melt, containing 54–58 mol% Li_2O. Fibers of $PbWO_4$ were grown using as a charge the crash of Czochralski material under the conditions described in detail in the following sect. 4.3.2. Growth was observed directly by a binocular telescope. Because of the steep temperature gradient near the interface, the growth process is very stable. Temperature deviations in the range of 10–25° lead to only moderate changes in meniscus height and very small changes in fiber diameter. Stoichiometric fibers of $LiNbO_3$ demonstrate a peculiar faceted appearance, whereas congruent fibers are rounded and nearly cylindrical (see details in sect. 4.2). Our observations on axial inhomogeneity during fiber growth can be summarized as follows:

- Transient areas until the fiber reaches steady-state growth mode are as long as 5–20 mm. For example, the first 7-10 millimeters of $LiNbO_3$ fiber, pulled from the Li-rich melt, are always nonfaceted, and only afterward can growth of a stoichiometric characteristically faceted fiber be established.
- A sharp decrease in meniscus height during steady state growth leads to capturing of impurities, bubbles, etc. The length of the defect part of a fiber is about 2–5 times the fiber diameter. Fig. 4.1 shows an example of this behavior during $PbWO_4$ growth, leading to a collapse of the crystal. The higher the meniscus height, the more pronounced is the degradation of crystal quality.
- An increase of the meniscus height normally leads to an improvement of fiber quality: a smaller amount of bubbles and inclusions, but also only within the transient area of 5–10 mm in length.

Direct observation of the meniscus during fiber pulling reveals that complicated and unsteady flow patterns are often present in the meniscus liquid pool, if its height-to-diameter ratio exceeds 0.2–0.3. The melt of $PbWO_4$ is transparent, and therefore the movement of small foreign particles sometimes present inside the meniscus can be easily detected. Also solid particles can be seen on the surface of the non-transparent $LiNbO_3$ melt, both in the meniscus and in the crucible. Only a rough estimation of their velocity was possible, but it is definitely in the order of centimeters per second.

4.1.2 Conditions for Growth of Axially Homogeneous Fibers

Let us now estimate the characteristics of melt flow in our system. As Marangoni convection is by far the dominating species transport mechanism, flow by buoyancy is not taken into account (this can be deduced from a comparison of the relevant dimensionless numbers, the Marangoni and Grashof numbers). Okano et al. [10, 11] have conducted a detailed study of melt convection in a floating zone arrangement using comparable sample geometry and using liquids with Prandtl numbers similar to $LiNbO_3$. According to their results the flow velocity u during μ-PD growth can be determined as following:

$$u = 0.106 \left| \frac{\partial \sigma}{\partial T} \right| \frac{dT}{dz} \frac{h}{\nu \rho} \tag{4.1}$$

for Reynolds numbers Re < 1, and

$$u = 0.974 \left[\frac{\left| \frac{\partial \sigma}{\partial T} \right| \alpha^{\frac{1}{2}} \frac{dT}{dz} h^{\frac{1}{2}}}{\nu \rho} \right]^{\frac{2}{3}} \tag{4.2}$$

for Reynolds numbers Re>1.

Here $\partial \sigma / \partial T$ is the derivative of the melt surface tension, α is the thermal diffusivity, ν is the kinematic viscosity, ρ is the melt density, dT/dz is the axial temperature gradient and h is the meniscus height.

Table 4.1 presents the material data for LiNbO$_3$ used for our analysis (available data for PWO are incomplete). When calculated according (4.1) and (4.2), flow velocities are in the range 0.2–30 cm/s for h changing from 0.001 cm to 0.03 cm and agree well with our experimental observations for the respective meniscus geometry. Considering the small size of the μ-PD meniscus, these are indeed very high values. Illustrative comparison: if it were possible to "switch off" thermo-capillary convection, we would have to rotate the fiber crystal at about 1000 min^{-1} in order to achieve the same mixing effect!

Table 4.1. Physico-chemical material data of congruent LiNbO$_3$ melt [12, 13]

ρ[g/cm^3]	ν[cm^2/s]	α[cm^2/s]	$\frac{\partial \sigma}{\partial T}$ [g/s^{2}K]	D[g/cm^3]
3.55–3.70	0.115	3.93·10^{-3}	−8.4·10^{-2}	5·10^{-4}

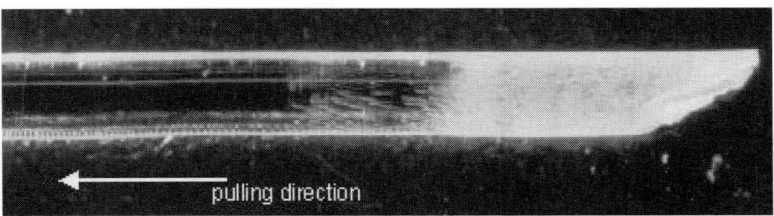

pulling direction

Fig. 4.1. Abrupt change in meniscus height during growth of a PbWO$_4$ fiber leads to capture of numerous W-rich inclusions and finally to crystal collapse. Note that the change in fiber diameter is very small.

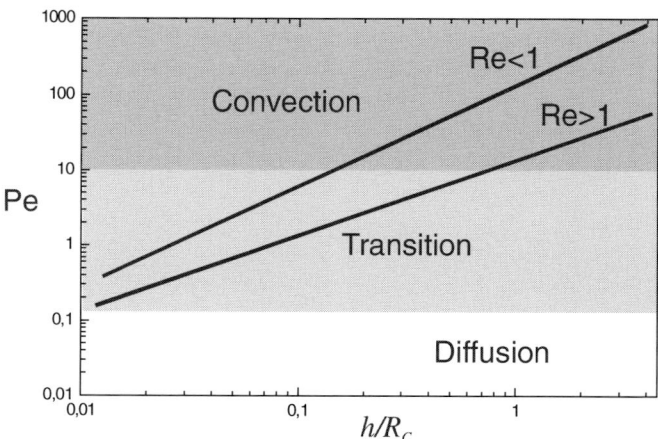

Fig. 4.2. Dependence of the Peclet number on the aspect value of meniscus height to diameter.

With the determined flow velocities we are able to evaluate the Peclet number $Pe = vh/D$ (D is the diffusion coefficient of the solute). This dimensionless Pe number scales the flow velocity and the rate of diffusive mass transfer. For $Pe \gg 1$ convection is the dominating transport mechanism, for $Pe \ll 1$ species transport by diffusion prevails. Figure 4.2 shows the dependence of the calculated Pe numbers on the aspect ratio h/R_C (R_C id the fiber radius). From this graph it is evident that the influence of the meniscus height on segregation is considerable. At small aspect ratios μ-PD growth occurs in a transition region in regard to dopant segregation. Both diffusion and convection are operative. At higher h/R_C ratios fluid mixing by Marangoni convection exclusively drives the incorporation of the dopant. From the graphical representation the boundary for the aspect ratio can be estimated at which the meniscus height h is sufficient to establish a segregation behavior typical for total mixing, i.e. k_{eff} differs considerably from unity. The respective heights for our fiber geometry are in the order of 0.005–0.01cm.

In Fig.4.3, a real scale drawing of a typical resistively heated μ-PD crucible is given, together with an indication of Marangoni induced flow patterns. The system comprises (1) the main crucible reservoir with the melt stirred effectively by Marangoni convection (the crucible center close to the nozzle is always hotter than the crucible legs attached to the cooled electrical feedthroughs), a nozzle (2) with steady Poiseuille flow and the meniscus (3) where the melt is once more subjected to Marangoni convection (as shown above, the flow velocity here depends strongly on the meniscus height). In Fig. 4.4 compositional solute profiles in the fiber and melt are shown for the case of low meniscus and weak Marangoni convection (Fig. 4.4a), and for high meniscus and intensive convection (Fig 4.4b). Both sketches represent steady-state fiber growth, and in both cases $k_{eff} = 1$. However, in the second case the rejected component (we are considering the dominating situation of $k_0 < 1$) is accumulated within the whole volume of the meniscus, since it diffuses in the direction of the liquid bulk only from the nozzle exit. The accumulation of components with $k_0 < 1$ is considerably higher in case b and amounts to $C_0 V/k_1$, where V is the meniscus volume and k_1 is the interfacial distribution coefficient, close to the equilibrium coefficient k_0.

Evidently, initial transients in μ-PD fibers can be described by the classical relationship of Pfann for zone melting. But in our case the liquid zone length not only affects the meniscus itself, but also the capillary channel, where melt flow is influenced by convection in the meniscus. This means that it is relatively wide, and reaches 0.5–1.0 of the fiber diameter, see Fig. 4.3. The transient length must consequently be 5–15 mm, depending on k_0. Figure 4.4 illustrates clearly the effect of meniscus height variations on longitudinal homogeneity of the fibers. When the meniscus decreases from its originally high position, all components, accumulated in the stirred meniscus pool (Fig. 4.4a) are captured. Since their concentration C_0/k_1 significantly exceeds the solubility limit, the formation of bubbles and inclusions is very likely. On the other hand, the increase of meniscus height from the initially low position enhances Marangoni convection and causes growth of a transient part of better crystal quality, until the meniscus pool is saturated.

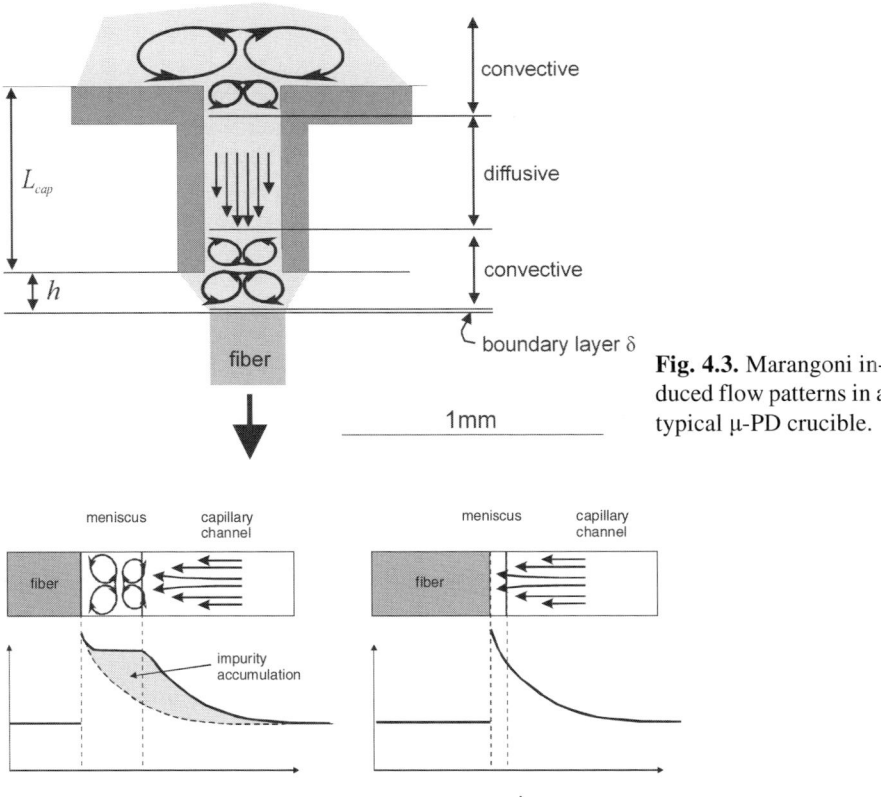

Fig. 4.3. Marangoni induced flow patterns in a typical μ-PD crucible.

Fig. 4.4. Compositional solute profiles in the fiber and in the meniscus generated by a high meniscus (**a**) and low meniscus (**b**).

Direct implementation of this consideration is that axially homogeneous fibers can be grown by the μ-PD method only under precise control of the meniscus height. Process stability during μ-PD should not irritate crystal growers. It is also reasonable to keep the meniscus height h as low as possible in order to avoid accumulation of constituents with $k_0 < 1$ in the meniscus pool and improve axial homogeneity. However, an intentional abrupt change in the meniscus height may be very useful during evaluation of new multicomponent compositions, since it helps to recognize the deviation from the congruent composition in a very simple and effective way. This will be discussed in the following sections.

4.2 Faceting of μ-PD Fibers and Improvement of Surface Quality

The prerequisite for optical application of μ-PD fibers is their production with specific orientations and smooth cylindrical geometry, since any diameter deviation leads to optical losses in devices. The formation of singular facets on the solid-liquid interface during growth results in a polygonal fiber morphology and peculiar surface irregularities (growth ridges) and should be avoided where possible. The crystalline quality of the μ-PD grown fibers is usually excellent, but observed fiber shapes are often different from being cylindrical due to the faceting effect.

In bulk crystal growth from the melt, high temperature gradients are normally considered to be useful for suppression of faceting, since the facet size r_f may be estimated to be inversely proportional to the square root of grad$T$ [14]. But in the case the fiber growth, despite of very high grad$T$, pronounced faceting was often found, sometimes leading to curious shapes with growth ridges being as large as the fiber itself [15].

The Formation of facets and ridges during CZ growth of lithium niobate was discussed in terms of capillarity by Reiche et al. [16] and was attributed to different wetting behavior of the melt on flat (faceted) and rough interfaces. They assume complete wetting of rough surfaces and only very limited wetting of the facets (estimated wetting angles are in the range 65–110°). The depression of the growth front near the facet is supposed to be dynamically unstable, so that upon slight disturbances the melt floods over the facet. In this model the value of the wetting angle on the facet is definitely overestimated. The model of [6] also fails to explain the periodicity of growth ridges. Detailed theoretical analysis of ridge formation was made by Voronkov [17]. According to [17] the melt slides along the facet after the actual growth angle ϕ, being different from the equilibrium angle ϕ_0, reaches some threshold value. The value $\phi - \phi_0$ for silicon crystals was calculated to be about 2°. Corresponding saw-like surface structures were estimated to be less than one micrometer high. This was termed 'microrelief' by Voronkov. However, growth ridges observed in μ-PD fibers are often much bigger. In this part some experimental results of fiber faceting during μ-PD growth are presented together with the capillary model to explain the size and periodicity of growth ridges.

4.2.1 Experimental Observations

In this study a standard μ-PD arrangement was used to grow $LiNbO_3$ fibers 500 μm in diameter from the congruent melt and from the 'stoichiometric' melt, containing 54–58 mol% Li_2O. To grow thin $LiNbO_3$ fibers less than 100 μm in diameter a modified version of μ-PD was employed (see detailed description in the following chapter). A thin platinum wire insert 50 μm in diameter was suspended inside the nozzle with the wire holder, so that the slightly sharpened insert end was below the nozzle opening. Final adjustment of the insert position was made after bringing the crucible with the insert to the working temperature. The insert was slowly pressed into the crucible by the platinum wire until the excessive length was disposed. With the use of a modified crucible much thinner fibers 30–50 μm in diameter can be grown.

Stable growth of fibers 500 μm in diameter and up to 150 mm in length was easily achieved along the c- and a-axis at pulling rates ranging from 0.5 to 2.5 mm/min. Thin fibers were grown at 0.3–0.5 mm/min. Crystals grown from the congruent melt were only slightly faceted, a typical example is shown in Fig. 4.5a, but fibers grown from the stoichiometric melt always demonstrated periodic growth ridges, Fig. 4.5b. Thin congruent fibers grown with the wire insert usually have continuous ridges nearly of the same size as the fiber itself (Fig. 4.6). The diameter of the faceted fiber was usually bigger than the nozzle diameter.

4.2.2 Discussion on Fiber Faceting

Since the method of micro pulling down is a kind of capillary controlled process the meniscus shape should be analyzed to gain a better insight into the faceting problem. In general, the shape of the meniscus is dictated by the balance of stresses caused by surface tension, hydrostatic pressure, dynamic pressure and viscosity. The conventional assumption is that capillary pressure dominates over hydrodynamic stresses in small-scale capillary growth systems [19, 20]. In our system this assumption is definitely valid for menisci of standard size, but it is not so in the case of thin fiber growth as discussed in [18]. The equations governing static meniscus shape are presented in detail in [19], and here already-known solutions relevant to our study will be used.

Figure 4.7 shows a section through the growth interface of a faceted fiber being pulled by the μ-PD technique. At the left side of the section the meniscus meets the nonfaceted surface, but at the right side an inclined facet exists on the periphery of the interface (see also Fig. 4.3a, a view of the interface from above). The meniscus at the right side is fixed on the sharp edge made by the facet with the lateral surface of the fiber. The phenomenon of meniscus fixation on facet edges was always directly observed by us during pulling of stoichiometric $LiNbO_3$ fibers. It is worth noting, that fixation of the meniscus is nothing but a consequence of the strong surface curvature on the edge (see the enlarged portion of Fig. 4.7) and in a strict physical sense the wetting condition is always valid here.

The facet as in Fig. 4.7 is positioned in a supercooled region relative to the rest of nonfaceted interface. This causes the facet to grow quicker than the main part of the fiber, and its size steadily increases. With the increase of facet radius r_F the angle $\alpha = \alpha_1 - \alpha_{hkl}$ between the meniscus surface and the facet plane decreases. At the moment when $\alpha = 0$, the meniscus slides along the facet to the initial position, and the process of facet growth repeats again providing growth ridge periodicity.

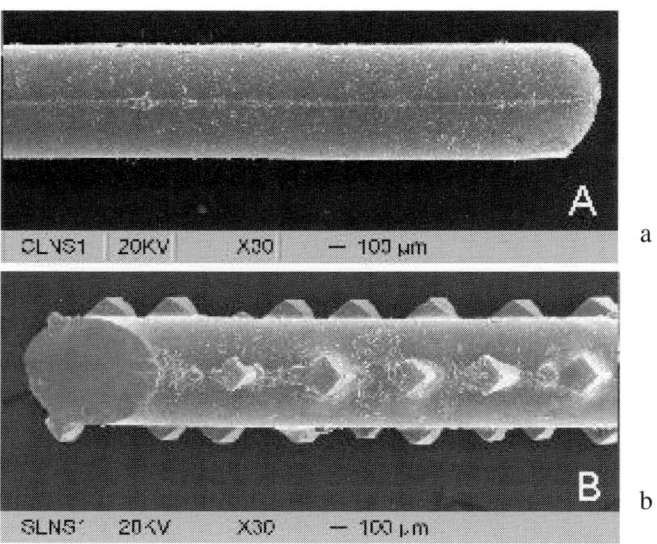

a

b

Fig. 4.5. SEM photomicrographs of c-oriented LiNbO₃ fibers: (**a**) fiber of congruent composition; (**b**) fiber of stoichiometric composition.

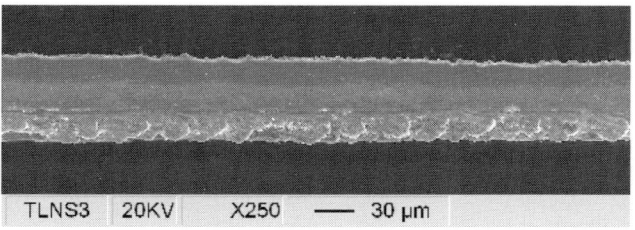

Fig. 4.6. SEM photomicrograph of a-oriented LiNbO₃ fiber. The fiber itself has a diameter of only about 20 μm, but growth ridges make it considerably bigger.

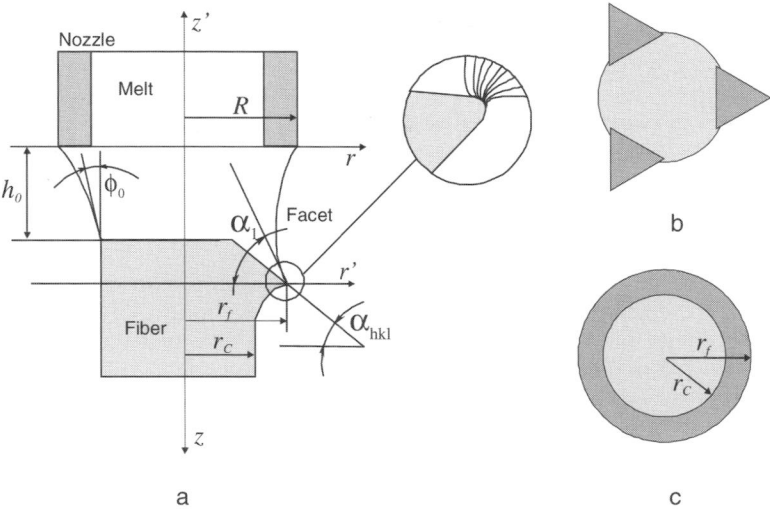

Fig. 4.7. (a) Section through a pulled fiber; **(b)** view of a growth interface from above; **(c)** model structure of growth interface used for calculation.

Let us now estimate the critical size of a facet. As a simplified axisymmetrical geometry we will consider a conical facet such as that depicted in Fig. 4.3c (naturally, such a facet cannot exist). According to the model presented above, the scheme of calculation should be as follows:

- First, for a given nozzle radius R, the meniscus height h_0 and growth angle ϕ_0 corresponding to the cylindrical crystal radius r_c have to be calculated. We will consider the meniscus height h_0 as a given parameter, since it is determined by controllable thermal conditions in our system. Note that h_0, ϕ_0 and the corresponding r_c belong to the nonfaceted part of the growth interface (the left side of Fig. 4.3a).
- The critical facet radius r_{FC} will be calculated from the condition of sliding $\alpha_1 = \alpha_{hkl}$ for the meniscus shown at the right side of Fig. 4.3a, but using the cylindrical geometry of Fig. 4.3c.

It is convenient to apply the generalized dimensionless form of the Young–Laplace equation as proposed in [19]:

$$z''r + z'(1 + z'^2)=0 \tag{4.3}$$

where σ is the surface tension of the melt, ρ is the density of the melt, and the capillary constant $\alpha = \sqrt{\dfrac{2\sigma}{\rho\gamma}}$ is used as the reference length. The coordinate system $z(r)$ is based on the lowest point of nozzle opening, see Fig. 4.7a. Using the boundary conditions $z'|_{r=R} = 0$ i.e. meniscus fixation on the sharp nozzle end and

$z'\big|_{r=r_C} = -\mathrm{tg}(\pi/2-\phi_0)$ i.e. the condition of the equilibrium growth angle ϕ_0, the analytical solution of (4.3) can be obtained in the form:

$$h_0 = r_C \sin(\pi/2-\phi_0)(\mathrm{arch}\frac{R}{r_C\sin(\pi/2-\phi_0)} - \mathrm{arch}\frac{1}{\sin(\pi/2-\phi_0)}). \qquad (4.4)$$

In Fig. 4.8 the fiber radius r_C dependence on the meniscus height h_0 is plotted for different growth angles ϕ_0. As discussed in [19, 20] the part of the curve $r_C = f(h_0)$, where $dr_C/dh_0 > 0$ corresponds to an unstable growth mode. Figure 4.8 represents the capillary problem during fiber pulling in a very illustrative way. Note the weak dependence of r_C on h_0 in the range $0.1R < h_0 < 0.3R$, which is the most important for practical growth .

For solving (4.3) for the faceted area (right side of Fig. 4.7) we should use the condition of meniscus fixation for both ends of the capillary curve. It is convenient here to change the coordinate system to $z'(r')$, see Fig. 4.7, since afterward we can get the solution in exactly the same form as (4.4).

By this means boundary conditions for (4.3) will be: $z\big|_{r=r_f} = 0$ and $z\big|_{r=R} = h_f$.

Taking into account that the point (r_f, h_f) belongs to the facet with α_{hkl}, i.e. $h_f = h_0 + (r_f - r_C(h_0))\sin\alpha_{hkl}$ we get for r_f

$$h_0 + (r - r_C(h_0))\sin\alpha_{hkl} = r_f \sin\alpha_1(\mathrm{arch}\frac{r_f}{R\sin\alpha_1} - \mathrm{arch}\frac{1}{\sin\alpha_1}). \qquad (4.5)$$

The value of $r_C(h_0)$ can be derived from (4.4) for certain h_0. During facet growth the angle α_1 decreases steadily approaching α_{hkl}. The melt slides along the facet at the moment when $\alpha_1 = \alpha_{hkl}$, leaving a mirror-like facet surface, see Fig. 4.5. The critical facet radius (sliding radius) r_{fc} can be therefore found from the simultaneous solution of (4.4) and (4.6).

$$h_0 + (r - r_C(h_0))\sin\alpha_{hkl} = r_{fc} \sin\alpha_{hkl}(\mathrm{arch}\frac{r_{fc}}{R\sin\alpha_{hkl}} - \mathrm{arch}\frac{1}{\sin\alpha_{hkl}}). \qquad (4.6)$$

In Fig. 4.9 the dependence of the critical facet radius r_{fc} on the meniscus height h_0 is plotted for different values of α_{hkl} as a result of numerical solution of the system of equations (4.4) and (4.6). Note that the critical facet size can be as big as twice the fiber diameter for facets sufficiently inclined with respect to the growth axis (for example $\alpha_{hkl} = 30°$). This should be considered as an upper limit estimation, because of the simplified geometry used for our calculations. Growth ridges on lithium niobate fiber shown in Fig. 4.5 B are built by the {01.2} pyramidal planes, having $\alpha_{hkl} = 57°$. Our calculations predict corresponding growth ridges $r_f/r_c \cong 1.2$, what is in a good agreement with experimental data.

The proposed model is indeed valid only under the assumption that undercooling of the order $\Delta T = \mathrm{grad}T(h_{fc} - h_0)$ is acceptable for growth of a certain face. In our experimental conditions $\mathrm{grad}T$ is about 0.5–0.6 K/μm, the difference h_{fc}-h_0 is in the range 100–150 μm, and the corresponding ΔT reaches 100 K. This explains the contrast in the appearance of fibers shown in Figs. 4.5a and 4.5b, since for fibers grown from noncongruent melt (actually solution growth) undercooling up to 100 K is possible. Rest droplets often observed on facets of stoichiometric fibers are of $LiNbO_3$- Li_3NbO_4 eutectic composition. In the case of growth from the congruent melt, only $\Delta T = 5$–10 K is possible, and the corresponding facet size is about 10 μm. This is still much bigger than predicted by the model of [17], since the condition of meniscus fixation on the sharp facet edge is present here. In that case the meniscus breaks free from the facet edge spontaneously because of fiber pulling mechanism vibrations. Small growth ridges on congruent fibers are for this reason less regular than big ones on stoichiometric fibers. During growth of very thin fibers the faceting effect has the same nature as described above, but it is even more pronounced since the meniscus is not fixed in upper part, but wets the pin surface. The mechanism of thin fiber faceting will be discussed later.

The Figure 4.9 clearly suggests the actions one can undertake to minimize the faceting effect during fiber pulling: (i) use of the smallest meniscus height still allowing defect-free growth and (ii) use of seed orientations where closely packed faces have α_{hkl} close to 90°. However, high temperature gradients peculiar to fiber growth methods are not useful during growth from noncongruent melt, until $\mathrm{grad}T$ is less than $(T_C - T_E)/(h_{fc} - h_0)$, where $T_C - T_E$ can be found from the corresponding phase diagram.

Listed modifications can significantly reduce the faceted area in single-crystal fibers grown by μ-PD, however growth ridges always remain, especially in fibers grown from incongruent melt.

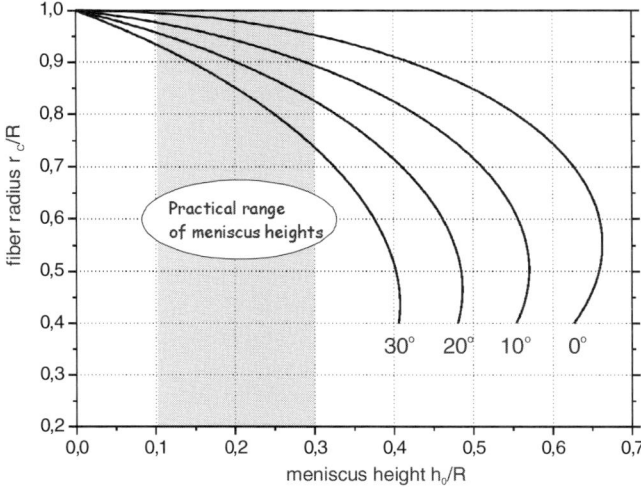

Fig. 4.8. Fiber radius as a function of the meniscus height for different growth angles.

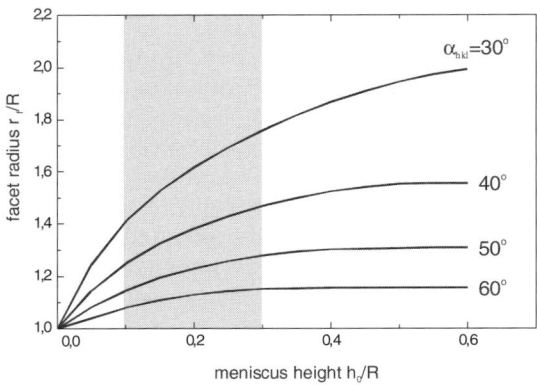

Fig. 4.9. Facet radius as a function of the meniscus height.

4.3 Crystal Growth Studies with the Micro Pulling Down Method

As discussed in Chap. 5, the μ-PD method is a fast way to grow high-quality crystal for property evaluation. Another important application of the method is the use of μ-PD in material science studies concerning defect formation, crystallization fields, and metastability. This growth technology can be successfully used as a research tool for bulk melt growth, prior to full-scale Czochralski or Bridgman experiments. In the following section we will illustrate this application of μ-PD by the examples of 'difficult' crystals such as terbium aluminum garnet and lead tungstate.

4.3.1 Terbium Aluminum Garnet – Phase Diagram and Crystal Growth

Terbium aluminum garnet $Tb_3Al_5O_{12}$ (TAG) is one of the most promising materials for Faraday isolators in the visible and near infrared region. Concerning the basic physical properties, i.e. Verdet constant and optical transparency, TAG outdoes the most commonly used materials in compact isolator devices.

The first successful growth of TAG single crystals was reported by Rubinstein et al [21]. The crystals were grown from an oxyfluoride flux and where rather small in size but large enough to measure their fundamental properties. Years later a horizontal Bridgman technique was applied by Kuzanyan et al. [22]. In their experiments they met with serious problems connected with the concurrent crystallization of the perovskite $TbAlO_3$ (TAP). Attempts to grow TAG from the stoichiometric melt by the Czochralski method led to crystals of millimeter scale only.

The results of DTA measurements within the binary system Al_2O_3-Tb_2O_3 can be summarized as follows: TAG melts incongruently at 1840 °C. TAP is the most stable binary phase in this system [23]. The region for primary crystallization of

TAP covers the chemical composition of TAG and suppresses the crystallization of TAG.

4.3.1.1 Fiber Growth Experiments

Growth experiments were performed with a high-temperature modification of a μ-PD apparatus [24]. A small iridium crucible rated for 1–2 cm^3 melt volume and an iridium afterheater were directly coupled into the field of a 10 kW RF generator, Fig. 4.10. The conical crucible bottom had a central capillary orifice 0.55 mm in diameter and 0.7 mm in length. The outside diameter of the crucible bottom was 1.0 mm. Platinum reflectors and an additional ceramic setup in the lower part of the heating assembly were added to reduce the axial temperature gradient. A $\langle 111 \rangle$ YAG seed in a ceramic holder was attached to an X-Y manipulator for final alignment with the orifice. The meniscus and growing crystal were viewed directly through the transparent protective quartz tube with a CCD camera. The growth process was controlled manually by cooperatively interacting adjustments of melt temperature and pulling rate at the initial stage of growth process and after each change of the pulling rate. The temperature was always adjusted to result in a fiber diameter about 10 % smaller than the outside diameter of the crucible bottom (see the enlarged view in Fig. 4.10), i.e. the meniscus height was held as low as possible.

Experiments were carried out for several melt compositions $(1 - x)$ Al$_2$O$_3$ + x Tb$_2$O$_3$ ranging from the stoichiometric composition of TAG $(x^{TAG} = 0.375)$ down to $x = 0.295$. The starting materials were 5N Al$_2$O$_3$ and 4N5 Tb$_4$O$_7$, and the growth atmosphere was pure argon. Since no remarkable evaporation of material from the crucible was observed the subsequent modifications of the melt composition were achieved by adding aluminum oxide directly into the crucible. After seeding and finding the proper melt temperature the pulling rate was varied between 0.15 mm/min and 4.0 mm/min.

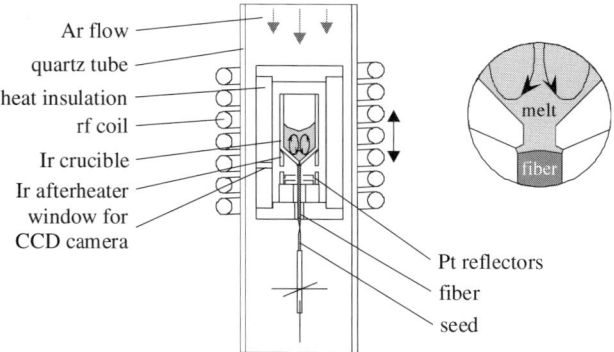

Ar flow
quartz tube
heat insulation
rf coil
Ir crucible
Ir afterheater
window for
CCD camera

melt
fiber

Pt reflectors
fiber
seed

Fig. 4.10. Sketch of the μ-PD apparatus and experimental setup.

4.3.1.2 Bulk Crystal Growth Experiments

According to the phase diagram of the binary system Al_2O_3-Tb_2O_3 the region of primary crystallization of TAG is $0.24 \leq x \leq 0.34$. When a melt of a composition within this range is cooled down it will be in equilibrium with TAG when its liquidus temperature is reached. Further cooling will lead to crystallization of TAG.

In the growth experiments one has to take into account strong segregation effects caused by the large difference in the chemical compositions of the melt ($x <$ 0.34) and the growing crystal ($x^{TAG} = 0.375$). As the melt composition will be shifted due to the higher consumption of Tb_2O_3 upon the growth of TAG, the maximum achievable yield is limited and can be calculated according to the lever rule:

$$g_{max} = \frac{x - x^{eut}}{x^{TAG} - x^{eut}} \tag{4.7}$$

where x^{TAG}, x and x^{eut} denote the concentrations of Tb_2O_3 in TAG, the starting melt and the eutectic, respectively. For starting melt compositions near the peritectic triple point ($x^{perit} \approx 0.33$) the maximum yield reaches approximately 50 %.

The Czochralski technique with rf heating was employed in the bulk growth experiments. Starting melts of different chemical compositions were used:

(i) $x = 0.32$ near the peritectic triple point and
(ii) $x = 0.29$.

The binary oxides (5N Al_2O_3 and 4N5 Tb_4O_7) were mixed and heated up very slowly in the growth crucible (iridium) under the growth atmosphere (flowing nitrogen) to allow for the complete reduction of Tb^{4+} contained the Tb_4O_7 and material synthesis. Then melting growth was carried out from an iridium wire with a pulling rate of 1.5 mm/h and crystal rotation of 15 rpm.

4.3.1.3 Results and Discussion

The grown TAG fibers were about 0.7 mm in diameter and typically 100 mm in length (see Fig. 4.11). X-ray powder analysis revealed that fibers grown from melts with $x > 0.32$ contained considerable amounts of TAP. Figure 4.12 shows a scanning electron microscope image (back-scattering electron mode) of a TAG fiber grown from the stoichiometric melt ($x = 0.375$). Obviously, the fiber is a mixture of two well-separated phases. By EDX measurements, the "light" phase was identified as TAG, the "dark" as TAP. The narrow black regions in the image are due to microcracks.

In contrast, fibers grown from alumina-rich melts ($x < 0.32$) consisted of pure TAG. The apparent perfection of the fibers (e.g. their transparency and development of crystallographic faces) became worse with decreasing Tb_2O_3 concentration in the melt. The best fibers were grown from a melt with $x = 0.317$ (Fig. 4.11). Independently of the melt composition, the highest quality was obtained at a pulling rate of 0.15 mm/min.

Fig. 4.11. TAG fiber grown from a melt with $x = 0.317$ (small division = 0.5 mm).

By the Czochralski technique polycrystalline samples of typically 5 g (accordingly 25 % of the input) were obtained. Crystal grown from a melt with $x = 0.32$ appeared semitransparent of white or slightly brownish color with small needles on the surface. X-ray powder analysis revealed the presence of TAP in the samples. In contrast, the crystals grown from the alumina-rich melt ($x = 0.29$) were mainly transparent. Crystallographic faces developed in part on the surface. Figure 4.13 shows an ultramicroscope image of a polished slice cut perpendicularly to the growth direction exhibiting a transparent matrix with enclosed milky-white regions. By X-ray analysis, corundum (a-Al_2O_3) was detected in the crystals. An explanation for this observation can be readily derived from the phase diagram: The crystallization of TAG consumes relatively more Tb_2O_3 than contained in the melt. Therefore the melt composition is shifted towards higher Al_2O_3 concentrations and consequently lower liquidus temperatures. If the mixing in the melt is poor and the pulling rate high the melt composition reaches (locally) the eutectic Al_2O_3/TAG which may be incorporated into the growing crystal.

In summary, TAG fibers have been grown using the micro pulling down method. In the experiments crystallization of TAP could be successfully avoided by growing from an alumina-rich melt (less than 32 mol% Tb_2O_3). The rapid growth and steep temperature gradient associated with the micro pulling down technique permits the crystallization of TAG even if the melt composition lies inside the region of primary crystallization of TAP [25].

In the Czochralski growth experiments crystallization of TAP was suppressed by starting from a melt with very low Tb_2O_3 concentration ($x = 0.29$). Crystals grown from melts with $x = 0.32$ still contained TAP.

The crystallization of TAG from melts with $x > x^{perit} \approx 0.33$ takes place under strong non-equilibrium conditions. Nonequilibrium crystallization is favored by the steep thermal gradients and by the high velocity of the solid–liquid interface that is typical for the micro pulling down technique. Deviations from thermodynamic equilibrium are smaller in the Czochralski technique. Accordingly, this method results in TAG crystallization only for melts with $x^{eut} < x < x^{perit}$; this is the region for primary crystallization.

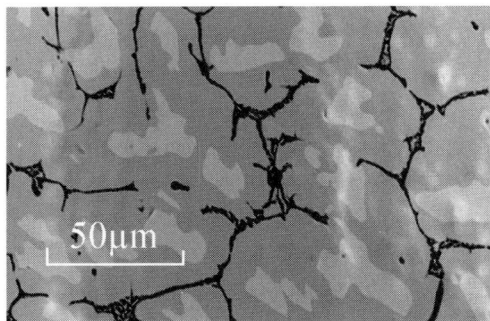

Fig. 4.12. BSE image of a TAG fiber grown from a stoichiometric melt (x = 0.375).

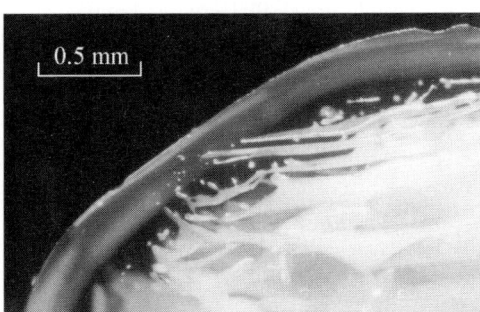

Fig. 4.13. Part of a TAG crystal grown by the Czochralski technique (x = 0.29).

4.3.2 Lead Tungstate – Problem of Incongruent Melt Vaporization During Growth

The optical and electrical properties of scheelite-structured lead tungstate $PbWO_4$ (PWO) crystals have been widely investigated in the 1970s as it promised to be a good material for acousto-optic modulators and ionic conductors [26–28]. Since the early work of Kroger [29] the luminescence properties of PWO have been considered exceptional compared to those of the other molybdates and tungstates. The systematic investigation of PWO emission was undertaken by Van Loo [30] and later by Groenink [31–32]. In nearly all cases the measurements were performed on small (0.5-1.0 cm in diameter, 2–3 cm in length) monocrystals grown by the Czochralski method.

Recently, the application of this material for precise electromagnetic calorimetry in high-energy physics has promoted the need for growth of large, optically clear single crystals [33]. Since PWO has been chosen as the scintillating medium for a crystal calorimeter of the new generation of the Large Hadron Collider (LHC) project at CERN [9] the problems associated with the production of high-quality large PWO crystals have received much attention. The LHC project requires the stirring amount about 100 tonnes of PWO crystals in relatively large monocrystalline items $3 \times 3 \times 23$ cm^3.

It should be noted that published data for absorbtion, emission and scintillation properties of PWO have been marked by significant diversity. Properties detrimental for LHC application, such as the relative amount of green emission component in radioluminiscence at room temperature which were found to be coupled with slow scintillation decay on the milliseconds time scale deviate significantly in various reports [35–36]. It would appear reasonable to suggest that the observation of green emission centres should be strong by connected with the structural perfection of the samples under investigation. Very recently Nitsch [37] reported on the growth of high structural quality PWO that showed the suppressed green emission component and consequently had especially suitable parameters for the LHC application.

The evident criteria of diversity in the characteristics of PWO samples is the difference in color. Much has been said in the literature about the nature of yellow coloration noticed in most parts of crystals. Some authors ascribe the coloration to deviation from stoichiometry, others to the presence of impurities, but chemical analysis has not definitely identified which impurity is responsible for the color. There are a lot of citations dealing with the growth practice of lead molibdate (PMO), in which significant vaporization of the melt was observed. It was found not to be detrimental to the growth process since the sublimate of PMO consisted of equal amounts of PbO and MoO_3 [38]. Using a mass spectrometer for high-temperature thermodynamic investigations the authors of [39] also found that in the mass spectrum of $PbWO_4$ there is a complete absence of lines corresponding to the dimer forms of lead and tungsten oxides, i.e. that $PbWO_4$ is vaporized without decomposition.

In this section, observations of incongruent lead oxide vaporization of PWO made with the use of the μ-PD method are discussed. Vaporization generates changes in Pb-W stoichiometry; consequently the fiber crystal formed from the continuously changing melt system becomes increasingly yellow. We suggest that μ-PD based analysis of growth problems can provide effective guidelines for technological improvements to achieve economically reasonable manufacture of PWO crystals.

4.3.2.1 Growth Experiments

The crucible structure is shown schematically in Fig. 4.14. The resistive heating of the platinum crucible was achieved by immediately passing the current through it. The crucible size was $8 \times 3 \times 2$ mm^3 (rated for the charge of 200–250 mg of row material). The nozzle was designed to ensure the effective partition coefficient of all melt constituent species to be equal to unity. It was made of platinum tube 0.7 mm in outer diameter and the nozzle length to capillary channel diameter ratio was 1:5, instead of the usually applied 1:2~3. The temperature of the nozzle was controlled to a high degree of accuracy by the power of the platinum wire after-heater, whereas the temperature of the crucible was adjusted to be $1150° \pm 10°$ C (which meant 20–40° higher than the PWO melting point). The alignment of the seed and the nozzle was controlled by the micro X-Y stage. Visual monitoring of the process was performed using an optical microscopic tube with the observation angle about 15–20°.

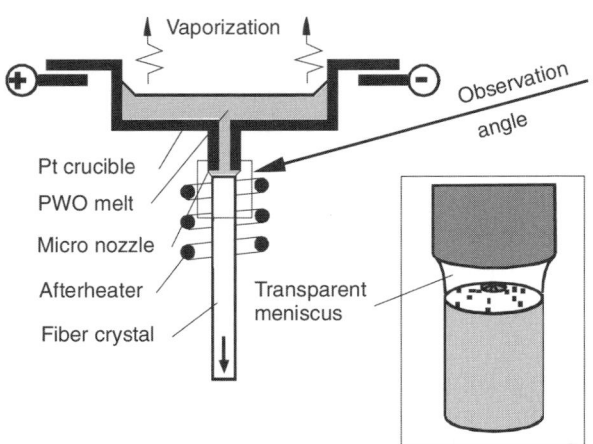

Fig. 4.14. Schematic of the μ-PD process showing the possibility of direct in situ observation of the growth interface.

PWO crystals 0.7 mm in diameter and 65–80 mm in length were grown in an air atmosphere along the c-axis. The pulling rate was varied in the range 0.15–0.45 mm/min. In all cases the initial charge was consumed completely (solidified fraction $g = 1.0$). Detectable evaporation of the melt was encountered. After growth the upper part of the furnace was covered with a white sublimate. The experiment was designed to observe the behavior of an exactly congruent starting material of highest purity. Because of this we used as a charge small pieces of CZ prepulled crystal which was grown by a conventional CZ method from 99.99% pure oxides mixed in stoichiometric proportion using a 40 mm Pt crucible. The weight of CZ crystal was approximately 45 g, so that only 15% of initial charge was used ($g = 0.15$). It was transparent and near colorless (a very minor yellow color could be observed using a bright white background).

The second phase identification was made by X-ray powder diffraction analysis using a Philips diffractometer (System PW-1700, CuKα radiation over the 2Θ range of 15° to 75°). The axial changes in defect structure were analyzed using crystal segments cut along the growth axis from the whole length of the crystal molded into a plastic holder by optical microscopy and by electron probe microanalysis (EPMA).

4.3.2.2 Discussion

All crystals grown in this study demonstrated the same type of defect structure along the length: the initial part ($g = 0$–0.15~0.35) was colorless and transparent without visible defects, the second ($g = 0.15$–0.85) became increasingly yellow colored, and multiple cracks appeared from $g = 0.6$–0.7. The last part ($g = 0.80$–1.0) was polycrystalline and bright-yellow in color. The dependence of the length of the transparent part on the pulling speed and a photograph of the initial part of the as-grown crystal ($g = 0.12$) is shown in Fig. 4.15. In the case of PWO, the μ-PD growth method provides excellent conditions for in-situ observation of the growth interface through a fully transparent meniscus.

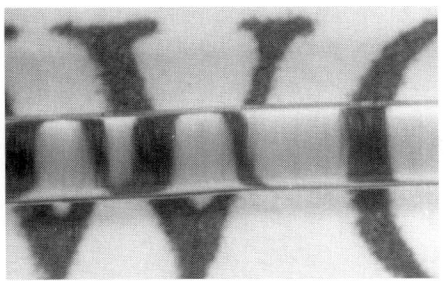

Fig. 4.15. As-grown c-oriented PWO crystal with a diameter of 700 μm.

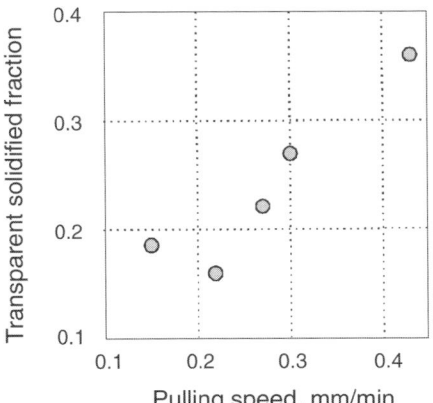

Fig. 4.16. Solidified fraction corresponding to the transparent colorless part of the PWO fiber crystal vs. pulling speed.

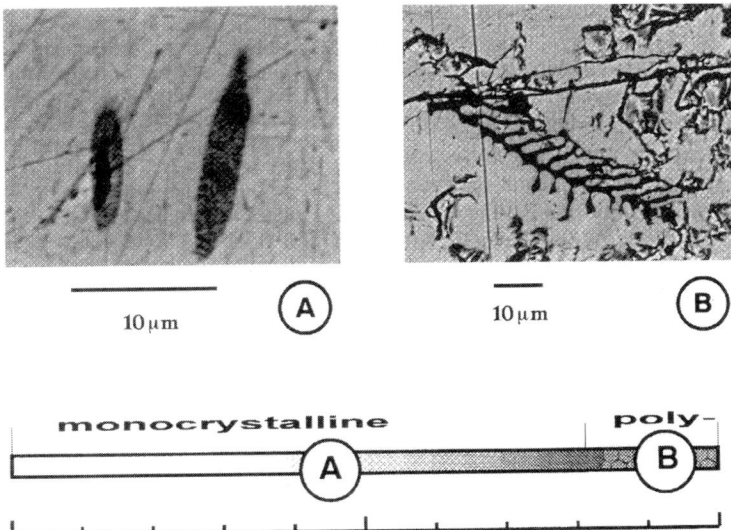

Fig. 4.17. Bubble-like eutectic (**A**) and lamellar (**B**) inclusions in a PWO microcrystal.

The crystallization front was directly visible to the operators eye: in the course of transparent growth it was convex upwards (toward the melt), smooth and without any defects. On the second stage specific defects look like the element of faceting formed in the middle part of the crystal, and the generation of second phase particles or bubbles, which were sometimes pushed away from interface, were observed. This second step is illustrated schematically in the right part of Fig. 4.14. In the last stage of polycrystalline growth the interface became completely irregular.

Stable growth of the crystal with constant diameter about 700 μm required the meniscus height to be established less than 50 μm. An attempt to increase the meniscus height always resulted in an unusual type of periodic instability of the diameter due to the high melt density. This phenomenon will be discussed elsewhere; here we would like only to make the point that shaped growth of PWO seems to be severely complicated by combination of high melt density with low surface tension [40].

The initial and final parts of the fiber PWO crystal were tested by X-ray analysis immediately after the growth and 10 days after being kept under room conditions. No changes in the X-ray pattern were found. The final polycrystalline part was revealed to be a binary mixture of tetragonal stolzite $PbWO_4$ (ICDD card 19-708) and an orthorhombic polymorph of WO_3 (ICDD card 20-1324). EPMA analysis detected the presence of elongated bubble-like inclusions in the colored monocrystalline part. They were composed of a mixture of $PbWO_4$ and WO_3 in a nearly eutectic composition. The length of inclusions was 2–10 μm. The eutectic nature of excessive tungsten oxide crystallization was verified with the observation of characteristic lamellar structures up to 150 μm in size of WO_3 plates incorporated inside the PWO matrix, see Fig. 4.16.

We find that the PWO melt evaporates incongruently under real growth conditions and the growing crystal is formed from the continuously changing melt system. It should be mentioned that the ratio of the free surface area to the melt volume is noticeably higher in μ-PD growth in comparison with the CZ method; however, CZ processing time is far longer. For example, the high-quality crystal described in Ref. [37] corresponds to $g = 0.25$–0.30 only.

The range of solid solution in $PbWO_4$-WO_3 is very limited. A high-temperature orthorombic polymorph of WO_3 was first reported to be quenchable to room temperature after stabilization by the addition of 2 mole percent of Nb_2O_5 [41]. The case of the $PbWO_4$-WO_3 system seems to be similar.

In summary, fiber crystals of $PbWO_4$ have been grown by the μ-PD method consuming the charge of high-purity raw material completely. Incongruent vaporization generates changes in stoichiometry and causes difficulties for PWO crystal growth from a tungsten excess melt. The yellow coloration of PWO crystals is attributable to excess WO_3. When the melt composition deviates severely from stoichiometry the excessive WO_3 produces inclusions of the eutectic mixture PWO(stolzite)-WO_3(orthorhombic polymorph). For effective growth of PWO from the melt, special care must be taken in keeping the melt composition constant during growth. The μ-PD method is shown to be a useful and versatile tool for crystal growth studies enabling us to investigate the growth process for the full range of solidification fraction from zero to unity.

4.3.3 Lead Tungstate – Search for Preferential Growth Direction

In the previous section the problems of charge preparation and incongruent evaporation of the PWO melt were discussed [37, 42, 8]. Incongruent evaporation was
found to generate changes in melt stoichiometry and to cause difficulties for PWO
growth from the tungsten-excessive melt. Nevertheless, up to now no data have
been published about the application of these findings to the CZ process. The second problem to be treated is crystal cracking during growth and post-growth mechanical treatment [43, 44, 45]. Weak cleavage planes were found to be (001),
(101) and (112). A reasonable choice of seed orientation can resolve the problem,
and is known to significantly affect crystal growth from a near-stoichiometric melt
[46]. However all reported crystals were grown along the c-axis, and only one
publication [45] reported the <104> direction as preferable over <001>.

In our view, the growth details for PWO have not yet been comprehensively
described in the available literature and as a consequence the economic aspects of
PWO growth remain unclear. For instance, the utilization of the starting material
in the CZ process (actual solidified fraction g of high-grade material) was not indicated in the cited references.

The approach of our investigation was to search for the optimal growth direction for PWO using a simple empirical procedure based on the micro pulling down
method with application of the results to CZ growth.

4.3.3.1 Experimental

The distinguishing features of the arrangement (see Fig. 4.18) used in the present
research were: (i) crucible height:diameter ratio equal to 1:1 instead of the previously used 1:3–1:4 and (ii) the absence of a capillary nozzle. The opening of 0.7
mm in the diameter was pierced by a sapphire pin in the crucible bottom, which
was 4×4 mm^2 in size. The edges of the opening were sharp enough to fix the meniscus (see the enlarged view in Fig. 4.18), but the growth interface was thereby
located close to the bulk melt, which was effectively mixed by natural convection,
thus allowing back-diffusion of rejected components. In this manner crucible proportions were adapted to simulate the usual CZ geometry and eliminate both enhanced melt evaporation and luck of melt partitioning, in contrast to the set-up described in the part 4.3.2.

The initial charge of 250–300 mg was taken from the shoulder part of a high-
purity pre-pulled CZ crystal of stoichiometric composition, and was completely
replaced in each growth run. The pulling rate was 0.5 mm/min in all experiments.
The fiber length was 70–90 mm, and the consumed melt fraction was 90–95%.

The seed holder was mounted on an X-Y micro stage, which permitted the seed
inclination to be varied from the vertical position up to 12°. The seed-to-orifice
alignment and the growth process were visually monitored with the use of a telescope.

The first c- and a-oriented seeds were cut from a CZ boule, then for subsequent
growth experiments the initial part of a previously grown fiber was used as seed.
In this way, in each sequence of growth experiments a fiber was withdrawn first

exactly along the *c*- or *a*-axis, and then 12, 18, 24, 30 and 36° from the original direction (see notation in Fig. 4.19).

4.3.3.2 Results and Discussion

Fiber crystals grown in this study featured a sharp transition from a colorless transparent part without any visible defects to a bright-yellow, highly defective part. The length of the transition region, marked by traces of bubble-like eutectic inclusions, was 2–3 fiber diameters. Evidently the proper growth process was disrupted within 3–4 minutes. The defective part of the fibers had multiple cracks, and usually separated into small pieces immediately after growth. Changes in the growth mode were easily detectable by visual inspection of the growth interface through the transparent meniscus. High-quality growth was accompanied by a smooth, convex interface (in the case of *a*-oriented seeds two diametrically opposed faces, presumably {101} or {103}, were detectable). From the moment of capturing the first inclusions, the interface became unstable, and subsequently completely irregular. These observations support the statement made above that µ-PD simulation of the CZ process is achievable by proper choice of the µ-PD crucible geometry. In the course of CZ processing of PWO the transition to highly defective growth also happens very quickly and is usually accompanied by a sudden change in diameter (a vivid example is given in Ref. [43], Fig. 1).

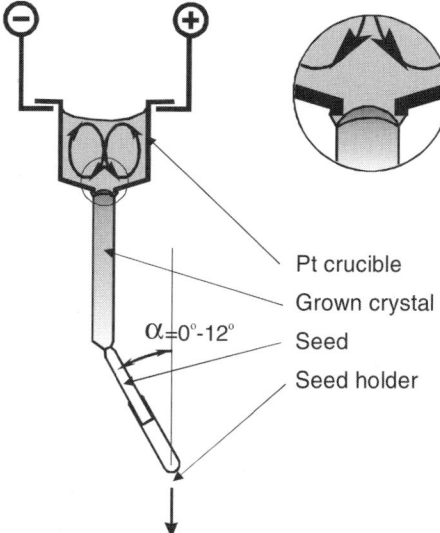

Fig. 4.18. Schematic diagram of the µ-PD process for the study of preferential seed orientation.

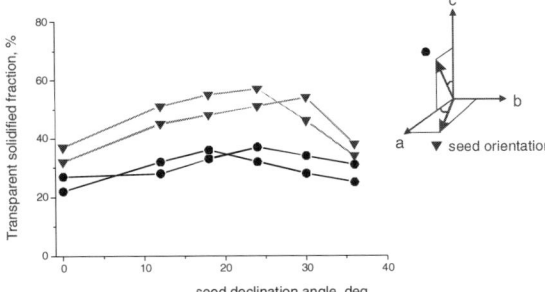

Fig. 4.19. Transparent solidified fraction of PWO fiber vs. seed orientation for the initial *a*-axis (triangles) and the c-axis (circles) growth sequences.

The variation in transparent fraction yield versus seed orientation is plotted in Fig. 4.19 for four sequences of growth experiments, with both c-axis and a-axis initial seeds. It can be seen that a- is obviously preferable over the *c*-orientation, and interfaces that correspond to axis orientation inclined 10–30° from the pulling direction were more resistant against changes in melt composition caused by incongruent evaporation. Similar findings in the context of subgrain generation were reported in [47] for CZ growth of lead molybdate.

The results of μ-PD growth studies were confirmed with the CZ technique. CZ crystals were grown at the Institute of Solid State Physics using the RUMO automatic diameter control crystal puller [48] equipped with a Pt crucible 80 mm in diameter and 80 mm in height. The crucible and afterheater were heated by an induction unit. The starting powders were PbO and WO$_3$ of 4N purity in the molar ratio of 49 to 51. The seed was cut 24° off the *a*-axis within the *a-b* plane. Crystal boules 32 mm in diameter and up to 170 mm in length were pulled at 2.5 mm/h. The high-quality, transparent part of the boule without any inclusions was, however, no longer than 120 mm, which corresponds to $g = 0.40$–0.45 (see Fig. 4.20). We believe that this can be only slightly increased (up to $g \approx 0.50$) by better adjustment of the lead oxide excess in the starting material.

In summary, a simple empirical procedure based on the micro pulling down method was devised to determine the optimal pulling direction for PbWO$_4$ crystals. Maximal yield of transparent material was obtained using seeds inclined 20–30° from the *a*-axis. CZ growth of a PWO starting from PbO-excessive melt confirmed our conjecture that the corresponding growth interface is more resistant to changes in melt composition caused by incongruent evaporation. Progress was made toward effective use of about half of the starting material.

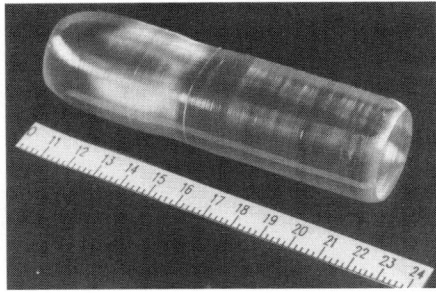

Fig. 4.20. As-grown PWO CZ crystal (defective end part removed).

Acknowledgments

The author is most grateful for the help and advice of his colleagues who have contributed in the development of these studies. In particular he would like to acknowledge Prof. T. Fukuda of IMR Tohoku University and Dr. D. Hofmann of the Institute for Materials Science 6 of the University of Erlangen-Nürnberg for many useful discussions. Also the author is indebted to Dr. A. Yoshikawa of IMR Tohoku University, Dr. S. Ganschow of IKZ Berlin and Dr. Gurjiyants of ISSP RAN Chernogolovka for valuable cooperation.

References

1   R.S. Feigelson, Materials Science and Engineering B1 (1988) 67.
2   P. Rudolph, T. Fukuda, Cryst. Res. Technol., 34 (1999) 3.
3   B.M. Epelbaum, K. Inaba, S. Uda, K. Shimamura, M. Imaeda, V.V. Kochurikhin, T. Fukuda, J. of Crystal Growth 176 (1997) 559.
4   H.J. Koh, N. Schaefer, K. Shimamura, T. Fukuda, J. Cryst. Growth, 167 (1996) 38.
5   S. Uda, J. Kon, K. Shimamura, T. Fukuda, J. Cryst. Growth 167 (1996) 64.
6   S. Uda, J. Kon, J. Ichikawa, K. Inaba, K. Shimamura, T. Fukuda, J. Cryst. Growth, 179 (1997) 567.
7   V.I. Chani, A. Yoshikawa, Y. Kuwano, K. Hasegawa, T. Fukuda, J. Cryst. Growth 204 (1999) 155.
8   B.M. Epelbaum, K. Inaba, S. Uda, T. Fukuda, J. Cryst. Growth 178 (1997) 426.
9   R.S. Feigelson, J. Cryst. Growth 79 (1986) 669.
10   Y. Okano, M. Ito, A. Hirata, J. Chem. Eng. Japan, 22 (1989) 275.
11   Y. Okano, M. Ito, A. Hirata, J. Chem. Eng. Japan, 22 (1989) 385.
12   J. Trauth, B.C. Grabmeier, J. Cryst. Growth 112 (1991) 451.
13   X. Chen, Q. Wang, X. Wu, K. Lu, J. Cryst. Growth 204 (1999) 163.
14   J. Brice, Cryst. Growth 6 (1970) 205.
15   N.Ohnishi, T.Yao, Jpn. J. Appl. Phys. 28 (1989) L278.
16   J. Reiche, B. Bohm, J. Hermoneit, et. al. Cryst. Growth 108 (1991) 759.
17   Voronkov, Izvestiya Akad. Nauk SSSR, Ser. Fiz. 52 (10) (1988) 1874.
18   A. Yoshikawa, B.M. Epelbaum, K. Hasegawa, S.D. Durbin, T. Fukuda, J. Cryst. Growth 205 (1999) 305.
19   A.V. Tatarchenko, Shaped crystal growth, Kluwer, Dordrecht, 1993.
20   J. Surek, Appl. Phys., 47 (1976) 4384.
21   L.G. Rubinstein, W.H. Van Uitert, J. Grodkiewicz, Appl. Phys. 35, 3069 (1964).
22   Kuzanyan, K.L. Ovanesyan, A.G. Petrosyan, G.O. Shirinyan, Dokl. Akad. Nauk Armyansk. SSR 74, 42 (1982) (in Russian).
23   Ganschow, D. Klimm, P. Reiche, R. Uecker, Cryst. Res. Technol., 34 (1999) 615–619.
24   B.M.Epelbaum, A.Yoshikawa, K.Shimamura, T. Fukuda, K. Suzuki, Y. Waku, J. Cryst. Growth, 198-199 (1999) 471.
25   S. Ganschow, B.M. Epelbaum, D. Klimm, A. Yoshikawa, T. Fukuda, Proc. SPIE 3724 (1999) 52–55.
26   Maksakov, A.M. Morozov et. al., Opt. Spectrosc. 14 (1963), 166.
27   van Loo and D.J. Wolterink, Phys. Lett. A47 (1974), 83.
28   Azarbayejani, J. Appl. Phys. 43 (1972), 3880.

29   Kroger, Nature (London) 159 (1947), 674.
30   van Loo, J. Solid State Chem. 14 (1975), 359.
31   Groenink and H. Binsma, J. Solid State Chem. 29 (1979), 227.
32   Groenink and G. Blasse, J. Solid State Chem. 32 (1980), 9.
33   A. Fyodorov, Korzhik et. al., Nuclear Sci. Symposium and Medical Imaging Confer-
      ence. 1994 IEEE Conference Record (Cat. No.94CH35762), 1 (1995), 114.
34   Lecoq, I. Dafinei, E. Auffray et. al., Nucl. Instrum. Methods A 365 (1995) 291.
35   Kobayashi, M. Ishii, Y. Usuki and H. Yahagi, Nucl. Instrum. Methods A 333 (1993)
      429.
36   Nikl, K. Polak, K. Nitsch et. al., Proc. SCINT95, Aug 28-Sept 1, 1995, Delft, The
      Netherlands.
37   Nitsch, M. Nikl, S. Ganschow et.al., J. Cryst. Growth 165 (1996), 163.
38   Bonner and G.j. Zudzik, J. Cryst. Growth 7 (1970), 65.
39   Nalivaiko, I.A. Rat'kovskii, Izv. Akad. Nauk SSSR Neorg. Mater., 17 (1981), No. 6,
      1132.
40   Swartz, T. Surek and B. Chalmers, J Electron Mater. 4 (1975) 255.
41   Roth and J.L. Waring, J. Research National Bureau of Standards – Physics and Chem-
      istry, 70A (1966), No. 4, 281.
42   S.C. Sabharwal, Sangeeta, D.G. Desai, S.C. Karandikar, A.K. Chauhan, A.K. Sangiri,
      K.S. Keshwani, M.N. Ahuja, J. Cryst. Growth 169 (1996) 304.
43   N. Senguttuvan, P. Mohan, S.M. Babu, C. Subramanian, J. Cryst. Growth 183 (1998)
      391.
44   N. Senguttuvan, P. Mohan, S.M. Babu, P. Ramasamy, J. Cryst. Growth 191 (1998)
      130.
45   S. Burachas, V. Martinov, V. Ryzhikov, G. Tamulaitis, H.H. Gutbrod, V.I Manko, J.
      Cryst. Growth 186 (1998) 175.
46   A.Yu. Bunkin, J. Cryst. Growth 123 (1992) 459.
47   Takano, S., Esashi, S., Mori, K., Namikata, T., J. Cryst. Growth 24/25 (1974) 437.
48   V.N. Kurlov, S.N. Rossolenko, J. Cryst. Growth 173 (1997) 417.

5 Crystal-Chemistry and Fiber Crystal Growth of Optical Oxide Materials

Valery I. Chani

Details of fiber crystal growth of known and novel oxide fiber crystals by the micro pulling down method are described. Application of the method for advanced materials research and optimization of crystal growth parameters is also outlined.

5.1 Introduction

Fundamental research and industrial applications of oxide single crystal materials in the semiconductor industry, optical communications, and medicine has shown enormous growth during recent years. Hundreds of institutions and commercial companies representing all economically developed countries participate in research projects related with the prediction, modification, and mass production of new crystalline materials for a wide range of applications. This represents a considerable increase in the number of participants of this process and the number of participating nations following fundamental discoveries in physics (lasers, novel optical effects and materials, etc.) and the invention of basic crystal growth techniques. Fiber crystals have also become the subject of intense study in recent years because of the extraordinary characteristics of thin elongated (quasi-one-dimensional) [1–3] crystals. Therefore an advanced growth technique, the micro pulling down (μ-PD) method, was developed recently to fabricate fibers of oxides, semiconductors and ceramic matrix composites with a high melting temperature [3–7].

Single-crystal fiber materials are of particular interest because of their unique characteristics, such as chemical uniformity, compact size, and, in some cases, high level of doping concentration. Although the majority of important oxide crystals, grown by the Czochralski (CZ) and other conventional melt growth methods, can be used in optical systems, fiber crystals appear to be more attractive, since growth conditions that can be established in the μ-PD system can result new unusual properties. These materials will most probably find applications in the fabrication of miniature laser sources and components of optical systems.

A number of very important materials are not available in single-crystal form because of difficulties related with the chemical and physical properties of the target solid-state materials and especially incongruent melting. As a result, much effort, time, and financial resources are necessary to develop crystal growth procedures resulting high-quality single-crystal ingots of these materials. Commercial production of the majority of optical oxide crystals is primarily based on solidification from the liquid phase.

Commonly, such methods can be simply classified into melt and flux growth techniques depending on the chemical properties of the target material, i.e. congruent or incongruent melting. In the former case, melt and crystal compositions are almost equal. Therefore, in general, the segregation phenomenon is weak. Thus, congruently melting materials can be produced using growth techniques based on simple solidification of isocompositional liquids. Generally, the μ-PD method is applicable for the growth of congruently melting compounds because high growth rates (1–10 mm/min) suppress segregation of constituents between the liquid and solid phases. However, the growth of incongruently melting compounds is also possible from the melts where the composition is slightly (5–10 %) different from that of the crystal. In this case the segregation phenomenon is detectable and has to be studied precisely, because the main problems related with such processes are determined by the rates of incorporation of all cations into the crystal. It is evident that the reproducibility of flux grown fibers is lower than those produced from melts of the same composition. Thus, the applicability of the μ-PD technique for the growth of any particular crystal should be first analysed based on the phase diagram of the corresponding mixture (congruency, volatility, etc.).

5.2 Apparatus and Procedures

The schematic diagram of the μ-PD apparatus is moderately simple (Fig. 5.1) and was described previously in Refs. [3, 4, 8–10]. The melts are placed in crucibles made of metals stable at high temperatures, corresponding to the melting point of the target crystalline material. Platinum, platinum/rhodium, iridium, and rhenium are the most popular crucible materials used in μ-PD systems. Normally, the crucibles are heated using a radio-frequency (RF) generator or through resistive heating of the crucible material.

At the very first stage the seed fiber crystal produced from a previous experiment or cut the from bulk crystal of the corresponding material is immersed into the crucible orifice. Thereafter the seed is pulled down, together with the growing fiber, using a precise pulling mechanism. Normally surface forces do not allow the melt to separate from the solid fiber material. Therefore the melt is passed through the orifice (sometimes called the nozzle) made in the bottom of the crucible. Application of the afteheater allows adjustment of the appropriate temperature gradient under the crucible and therefore regulation of the position of the solid/liquid interface in the vicinity of the crucible tip. The shape and location of the growth interface is one of the most important parameters determining the quality and uniformity of the resulting crystal, independently of the crystal growth technique applied, and the μ-PD method is not an exception. Therefore special attention is normally paid to monitor and detect the spatial distribution of the temperature gradient in the vicinity of the phase boundary. Two main techniques are mainly used to observe and to control the interface. In the case of relatively low temperatures (1000–1500 °C) simple optical microscopes are considered to be suitable tools to view the interface. However visual observation of the meniscus region, the solid/liquid interface, and crystals in high-temperature systems (1800–2000 °C and above) is normally made by a CCD camera and monitor. Similar to the CZ process, the μ-PD technique allows the so-called necking procedure (Fig. 5.2) used for decreasing the number of defects in the produced material.

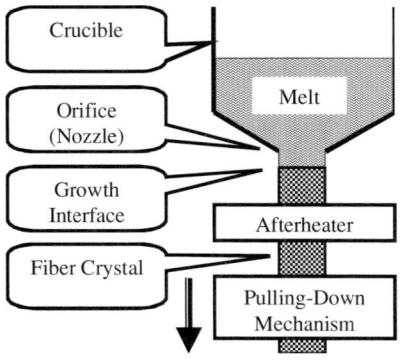

Fig. 5.1. Schematic of the micro pulling down (μ-PD) process.

5.2.1 μ-PD Systems with Resistive Heating

A schematic of the μ-PD apparatus with resistive heating is given in Fig. 5.3. Electric power is applied directly to the crucible container made of platinum or platinum/rhodium foil and pipe. A view of the crucible used in this process is given in Fig. 5.4. The crucible is heated due to resistive heating of the crucible material. The crystals are normally grown under air atmosphere with the pulling down rate ranged from 0.1 to 1.0 mm/min. The position and size/shape of the solid/liquid interface is adjusted manually by selection of a suitable magnitude of the electric currents flowing through the crucible and the afterheater. The pulling down rate is an additional variable allowing control and correction of the crystal size/shape and quality.

As a rule the crystal/nozzle diameter ratio can be adjusted in the range of 0.2–1.0 for the system shown in Fig. 5.3. Typically the growth processes is stopped after observation of any kind of crystal imperfection. Thereafter the crystal is separated from the molten zone, pulled down with the rate corresponding to the cooling rate of about 10–50 °C/min, and removed from the holder.

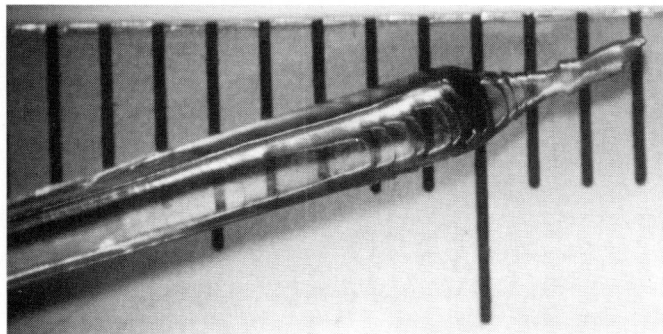

Fig. 5.2. KNbO$_3$ fiber crystal produced using the "necking" procedure (scale in mm) [11].

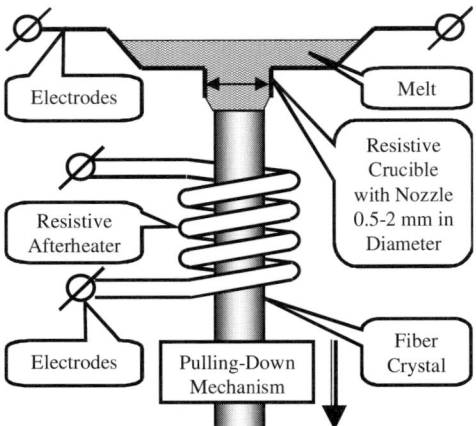

Fig. 5.3. Schematic of μ-PD system with resistive heating.

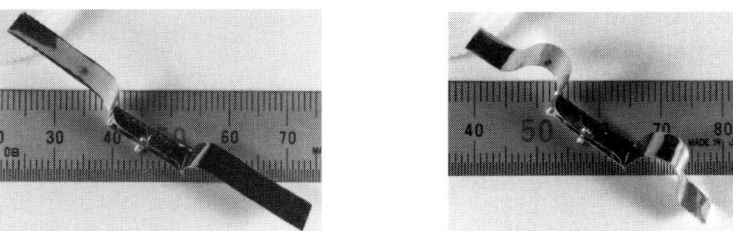

Fig. 5.4. Pt μ-PD crucibles used for crystal growth using apparatus based on direct resistive heating of the crucible (Fig. 5.3).

In contrast to bulk crystals, very slow cooling of the fiber materials after growth is generally not necessary because the difference between the temperatures in the core and peripheral parts of the fiber is low due to the small diameter. Heating of the seeds before growth is also made with a relatively high rate. At a second stage of some experiments the remaining melt is removed from the crucible using the same seed with a high pulling rate of 0.50–2.00 mm/min. The fast pulling rate suppresses segregation of constituents between the remaining melt and the solid. This way only a negligible part of the melt remains in the crucible after the growth run, and the cleaning of the crucibles for further experiments can be made much easier and faster.

It is noted that the effectiveness of the heating and temperature distribution in the vicinity of the liquid/solid interface also depends on the shape of the crucible. Therefore there is the additional possibility of controlling crystal growth conditions by modification of the crucible shape (Fig. 5.4.).

5.2.2 μ-PD Systems with RF Heating

A schematic presentation of a μ-PD crucible and the growth procedure for the system with radio-frequency (RF) heating is given in Figs. 5.5 and 5.6. The heating system shown in Figs. 5.5 and 5.7 was specially designed for the fabrication of fiber crystals and composites at high temperature (above 1800 °C).

The crucible of about 30–50 mm in height by 14–16 mm in diameter is placed on an alumina pedestal in a vertical quartz tube and is heated using an RF generator. The calibrated orifice of about 0.3–1.0 mm in diameter is made in the crucible bottom to allow the melt to flow in the direction of the solid/liquid interface. A view of the Ir crucibles used for the growth of the fiber crystals discussed below is given in Fig. 5.8.

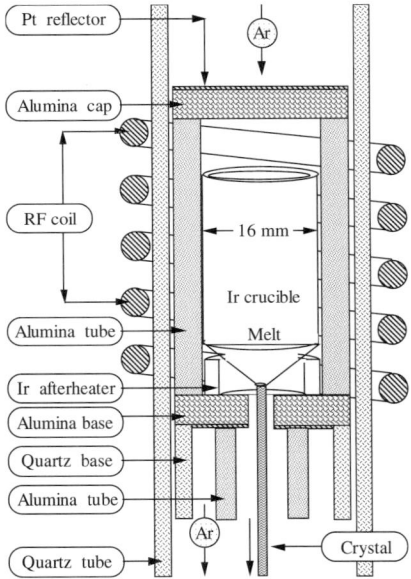

Fig. 5.5. Schematic of μ-PD system with RF heating.

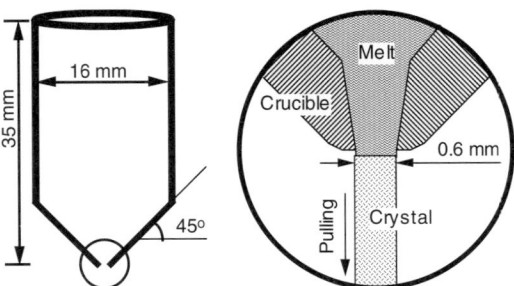

Fig. 5.6. Schematic representation of the μ-PD crucible and solid/liquid interface.

Fig. 5.7. Views of the μ-PD apparatus with RF heating (left) and the hot zone (right).

Fig. 5.8. μ-PD crucibles and the afterheater (below right crucible) used in the apparatus with RF heating (scale in mm).

In the system used in [4, 5, 9, 10, etc.] the crucible temperature was controlled by the power of the RF coil which was about 90 mm long with seven windings (Figs. 5.5 and 5.7). The crucibles (Fig. 5.8) were charged for up to 100 vol% regarding the powder of the starting materials or about 5–20 vol% regarding the melt. A high-density and high-purity (99.7%) alumina ceramic was used to surround the crucible for the thermal insulation. The crystals were grown in an Ar atmosphere (gas flow 2 l/min) to avoid oxidation of the crucible. Visual observation of the meniscus region, the solid/liquid interface, and the crystals was made by a CCD camera and monitor (Fig. 5.7). The spatial resolution of the observations was in the range of 50–100 μm.

5.2.3 Wetting Properties of the Melt

The main problem of fiber crystal growth from melts that are wettable with respect to the crucible material is related with the overflow of the melt due to good wetting properties. Evolution of the liquid behavior in μ-PD systems used for crystal growth from nonwettable ($Y_3Al_5O_{12}$ (YAG) garnet fibers in [5, 9]) and wettable ($Tb_3Ga_5O_{12}$ (TGG) crystals in [10, 12]) melts is given in Fig. 5.9. In the case of a nonwettable melt of YAG and other YAG related materials [4] the adjustment of the desired fiber diameter was easily made by the control of the RF power supply. Generally an increase of the crucible temperature resulted in a decrease of the crystal diameter similar to that known for the conventional CZ technique.

In contrast, the TGG melt was observed to wet the Ir crucible [10]. Thus, the higher the RF power applied to the crucible, the larger the surfaces of the crucible/liquid and crystal/liquid interfaces, as also shown in Fig. 5.9. Therefore the control of the size of the solid/liquid interfaces was quite different from that of the nonwetting melts and difficult at the starting stage of the experiments on TGG fiber growth. Overheating of the melt was accompanied by its overflow through the outer surface of the crucible. This way the cross-section of the molten zone was increased with a corresponding increase of the crystal diameter. As a result the crystal diameter and heat transfer from the melt were also increased, and fast solidification was observed due to rapid displacement of the crystal/liquid interface. To prevent this phenomenon the fibers were grown at minimal possible overheating of the crucible which allows minimizing the width of the liquid films positioned between the crystal and the crucible, as shown in Fig. 5.9. This way the overflow was suppressed by minimization of the channel thickness (the liquid film between the crystal and the crucible) allowing melt transport to the outer wall of the crucibles.

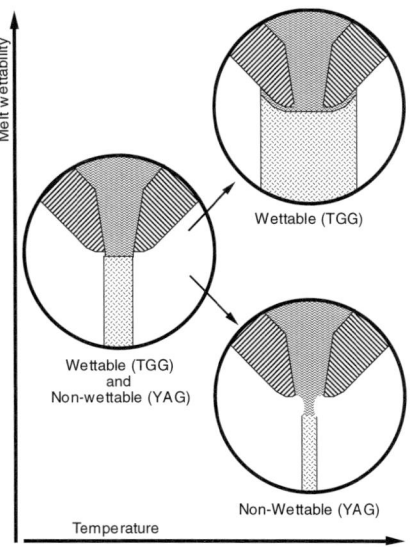

Fig. 5.9. Schematic representation of the dependence of crystal fiber diameter on temperature for wetting (TGG) and nonwetting (YAG) melts.

The opposite tendency was observed also in the dependence of the fiber diameter on the pulling rate. In the case of nonwettable YAG related melts (undoped, Nd-, and Yb-doped), increasing the pulling rate results in an increase of the latent heat of crystallization with corresponding overheating of the melt, increasing of the height of the molten zone, and decreasing the crystal diameter (also similar to the CZ method). The tendency observed in the TGG melt was also opposite; the higher pulling rate resulted in an increase of the crystal diameter.

5.2.4 Seeding and Separation of the Grown Crystal

Where available, oriented rods of about 1–2 mm in cross-section and 30–100 mm in length cut from CZ grown bulk crystals are used as seeds at the first step. The crystallographic orientation of the fibers grown on these seeds normally results from the orientation of the starting seed which is determined using X-ray measurements by observation of unique reflections corresponding to a specific direction. Thereafter the seeds are cut from the fibers grown in previous experiments. Seeding on foreign fiber materials is also possible with consideration of the corresponding phase diagram.

Within the standard ranges of operating parameters, the achievement of macro defect-free growth off a seed crystal was determined in [5] to require the controlled dissolution of part of the original seed prior to initiation of the growth (pulling) process. It is assumed that this necessity had as its basis the tip damage that occurs at the cutting stage. The controlled seed melting is generally made by slow immersion of the seed into the crucible orifice with a pulling-up rate of about 0.1 mm/min. Usually about 1.0–2.0 mm of the seed crystal is melted just before starting the growth (pulling down).

In the case of wettable melts both seeding and separation of the grown crystal from the melt is relatively difficult due to the unusual behavior of the liquid phase discussed in the previous subsection. Partial dissolution of the initial seed fiber cannot be achieved by simply overheating the hot zone. This practice is easily applied exceptionally for fiber growth from nonwettable melts. However, in the case of wettable melts, an increase of temperature, as already mentioned, resulted in an increase of the crystal diameter (see Fig. 5.9). Therefore the application of such a procedure is absolutely impossible in the case of wettable melts. Thus the seeding can be made as follows. First, the seed touches the tip of the crucible at a temperature below the melting point when the starting materials are still in the solid phase. Thereafter the temperature is increased continuously up to the scenery, when the molten zone becomes at least visually observable. Finally pulling is started.

In the case of nonwettable melts, separation of the crystal from the melt was also easily achieved by simple overheating of the melt (see Fig. 5.9). As a result, the diameter of the crystal was decreased up to the condition when the surface energy becomes too large to keep the molten zone in connection with both the crucible and the crystal. The separation was completed automatically.

On the other hand, in the case of wettable melts, separation of the grown fiber in the same way was also impossible for similar reasons. The overheating of the melt resulted in an increase of the crystal diameter. Overcooling of the melt also did not result in a smooth separation. In these circumstances the fiber diameter was first decreased down to the diameter of the orifice. Thereafter solidification of the melt was observed not only in the vicinity of the solid/liquid interface, but also inside the orifice. This way the fiber was bonded to both (i) the solidified melt inside the crucible and (ii) the seed together with the seed holder. Any further pulling resulted in mechanical damage of the crystal. At present, solidification of 100 % of the melt is considered to be the best solution of the above problem. Fibers produced with complete solidification of the melt charged in the crucible are typically of good optical quality.

5.3 $Y_3Al_5O_{12}$ (YAG) Garnet

There has been continuing interest in the development of $Y_3Al_5O_{12}$ (YAG) garnet crystal growth technology because Nd-doped YAG is one of the most important laser hosts for the generation of 1.06 mm infrared radiation. Its good optical, chemical and mechanical characteristics have made it the standard material in industrial applications where reliability is particularly important. The recent demonstration of the growth of garnet fiber crystals by the micro pulling down (μ-PD) technique [5, 9, 13, 14] and the crucibleless laser heating pedestal growth (LHPG) method [1, 2], has led to increased interest in the preparation of uniform fiber garnet crystals for laser applications.

5.3.1 Growth of YAG:Nd Crystals

The results of YAG fiber crystal growth by the μ-PD technique were first reported in Ref. [5]. As a preliminary step YAG and Nd-doped (YAG:Nd) single crystals produced by the conventional Czochralski technique were used as the starting materials. In the YAG:Nd crystals, about 0.7–0.8 at% of Y^{3+} was substituted for Nd^{3+}. Powder raw materials were also prepared from high-purity oxides Y_2O_3, Al_2O_3, and Nd_2O_3 (99.9999%). The constituents were carefully mixed in appropriate ratios together in alumina mortar with ethanol and dried. The mixtures were $Y_{2.9988}Nd_{0.0012}Al_{5.0000}O_{12}$ and $Y_{3.0988}Nd_{0.0012}Al_{4.9000}O_{12}$ in composition.

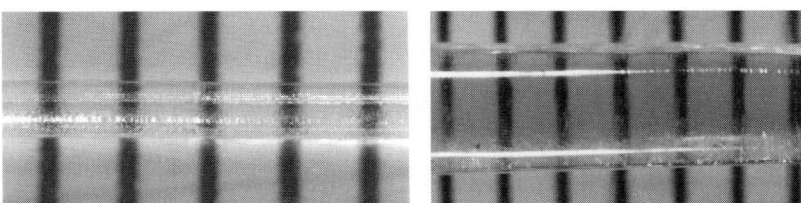

Fig. 5.10. Samples of as-grown YAG (left) and YAG:Nd (right) fiber crystals (scales in mm) [5].

The crucibles were heated from room temperature to about 2000 °C for about one hour. In the case of single-crystal raw materials the seeding procedures were started almost immediately after the melting was completed. However, in the case of powder raw materials the melts were sometimes cooled rapidly to 1600–1800 °C for about 15 min for solidification and preliminary formation of the garnet phase similar to that used in flux growth [15]. After solidification the charge was melted again assuming that the history of the melt prepared in this way was close to the history of single-crystal starting materials.

The crystals were grown at a pulling-down rate of 1–8 mm/min and were 0.3–2.0 mm in diameter. The typical length of the molten zone was about 0.1 mm. Best reproducibility of growth results was found to correspond to a pulling rate of 2–5 mm/min and crystal diameter of 0.5–0.8 mm. The length of the crystals was up to 550 mm, and was limited by the length of the pulling system applied. Sometimes it was possible to produce few crystals of the same length by repeated seeding without cooling of the hot zone and melt solidification. The design of the μ-PD system used in the experiments described in [5, 9] allows seal failure for a short period of time (5–10 min) without oxidation of the crucible material because of the argon gas flowing from above (Fig. 5.5). This way, the just-grown crystal was cut from the seed and removed. However the seed was still kept on the holder for use in the next growth experiment. Thereafter the system was sealed again and the seeding procedure was replicated. Thus, the total yield of the one-growth run carried out with repeated seeding was 1–2 m of fiber crystal material.

Another reason limiting the growth of long crystals was related to vibrations that become especially intensive at the final stage of growth, when the crystal length was more than 400 mm. There is evidence that the amplitude of the elongated fiber crystal oscillations also depended on the crystal diameter. Nevertheless in spite of the instabilities related with the vibrations, the strength of surface forces was observed to be very high. Disconnection of the crystal growing from the molten zone was never observed even in the cases when the amplitude of the oscillations was two–three times greater than the crystal diameter.

All growth processes described in [5] were stooped when the melt was crystallized into the fiber completely, or when the crystal length became greater than 550 mm. After growth the crystal was withdrawn from the melt and cooled to room temperature for about 1 hour. Utilizing the above process, crystals as large as 550 mm in length were obtained. No difference was observed in the behavior of undoped and Nd-doped YAG melts. Figure 5.10 shows typical YAG and YAG:Nd crystals grown using the process outlined here.

5.3.2 Properties of YAG:Nd Crystals

Similar to the results of [13, 14] the as-grown YAG crystals were slightly faceted with an evident hexagonal cross-section that is typical for the <111> garnet growth direction. The best optical quality under microscopic observation and the best fiber cross-section uniformity was observed in crystals grown with the diameter close to the diameter of the crucible opening (0.6–0.8 mm). The crystals were transparent and colourless or slightly coloured due to Nd^{3+}, depending on the composition of the starting materials and/or the growth parameters. The YAG fibers described here were considered to be used as active laser material for elongated laser sources with high surface/volume ratio and low content of Nd^{3+} active ions. Therefore no attempts were made to produce crystals with high concentrations of Nd^{3+} (> 1% with respect to Y^{3+}).

Production of relatively thick (up to 2 mm in diameter) crystals was also found to be possible using the crucible with dimensions shown in Fig. 5.6. An example of such a crystal is shown in Fig. 5.10 (right). Unfortunately in this case structural imperfections and crack formation in the peripheral part of the fibers was occasionally observed. It was assumed that cracking was mainly caused by the high temperature gradient in the vicinity of the molten zone and thermal stresses that occurred because of corresponding thermal expansion of the crystal lattice due to the gradient. The growth of fibers with greater diameter (up to 3 mm) using the crucible described above was also sometimes performed. However the crystals were of very low quality, which was mainly related to the instability of the shape of the molten zone due to the nonwettable properties of the YAG melts, and therefore the difficulty of diameter control.

The fibers produced from starting materials prepared from either YAG bulk single crystals or stoichiometric mixtures were of a single garnet phase that was detected by powder X-ray diffraction analysis. However the crystals produced from a nonstoichiometric mixture slightly enriched with Y_2O_3 contained considerable amounts (3–10%) of the $YAlO_3$ perovskite second phase. The results of YAG fiber growth performed in [5] show that even in rigid conditions such as a high temperature gradient and extremely high growth rate, incorporation of relatively small Y^{3+} cations into the comparatively large octahedral sites of the garnet structure mainly filled with Al^{3+} cations is quite difficult in YAG.

The crystal quality of the YAG fibers produced by the μ-PD technique was estimated by X-ray rocking curve (XRC) measurements made using a multipurpose extra high resolution diffractometer [16]. The XRC profiles were determined with two different optics, such as a 2-bounce Ge(220) channel-cut monochromator + receiving slit optics and a 4-bounce Ge(220) channel-cut monochromator + 2-bounce Ge(220) channel-cut analyzer. For each mode both of a ω-scan and a 2θ-ω scan were measured. The XRC data (ω-scan with high resolution) for the YAG (111) fiber crystal produced at extremely high growth rate (about 8 mm/min) were reported in [5, 9]. The crystal studied was found to be composed of two large blocks with a small deviation angle of about 0.02° from each other. The FWHM values observed for the crystal as a whole and for independent blocks were 0.038 and 0.012°, respectively.

5.3.3 Spatial Distribution of Nd^{3+} in YAG:Nd Crystals

The as-grown YAG:Nd fiber crystals were analyzed for Nd^{3+} dopant concentration. The chemical composition was measured by electron-probe microanalysis (EPMA). Both the CZ grown bulk YAG:Nd single crystal used as a row material and the corresponding fibers produced by the above technique were studied under identical conditions to compare the concentration of Nd^{3+} in the liquid and solid phases, respectively. The results of measurements made along the axial direction were reported in Ref. [5]. It was found that in spite of the exceptionally high growth rate (5 mm/min) the segregation coefficient of Nd^{3+} in this process was about 0.8, and still less than unity, $K(Nd^{3+}) < 1$, as compared with YAG:Nd produced by the CZ method, where K(Nd) is close to 0.2 depending on the growth rate.

EPMA composition analysis made in the radial direction of the YAG:Nd fiber with a step of 10 mm was also reported in [5]. The spatial distribution of Nd^{3+} doping cations in the fiber crystals was measured using an electron probe of 1 μm in diameter. Any detectable decrease of Nd concentration in the periphery region of the crystals as expected according to [1, 17], was not observed. Taking into consideration the extremely small amount of Nd in the crystals and the corresponding accuracy of the measurements it was concluded that there is no tendency toward a nonuniform radial distribution of the Nd dopant. Significant oscillations in Nd concentration observed in [5] followed from the relatively low accuracy of the EPMA measurements especially when the dopant concentration is very low (less than 0.1 at% with respect to all atoms forming the crystal). Similar oscillations were also observed when point measurements were made in the axial direction with a step of 0.5 mm as also illustrated in [5].

To estimate the segregation coefficient of Nd dopant in the above process, the measurements in the radial direction were initially made for the beginning phase of the growth processes at solidification fractions that were very close to $g \approx 0$ [5]. Therefore it was expected that the melt composition at that stage of growth was exceptionally close to or equal to the composition of the YAG:Nd bulk crystal used as a starting material. Average values of Nd concentration in the solid and liquid phases were measured to be 0.097 at% and 0.073 at%, which correspond to about 0.019 and 0.15 atoms per formula unit (AFU) of garnet, respectively. Therefore the effective segregation coefficient of Nd^{3+} for the μ-PD process described in this section was calculated to be $K(Nd^{3+}) \approx 0.8$.

The range of $K(Nd^{3+})$ values for the YAG:Nd bulk crystals produced by the CZ technique at moderate growth rates is well known. Some of these parameters reported elsewhere are listed below:

$K(Nd^{3+}) = 0.15$ at a growth rate of 1.0 mm/h [17],
$K(Nd^{3+}) = 0.172–0.187$ at a growth rate of 0.3–1.6 mm/h [18], and
$K(Nd^{3+}) = 0.20$ at a growth rate of 3.0 mm/h [19].

Moreover, the equilibrium segregation coefficient $k_0(Nd^{3+}) = 0.151$ reported for the CZ growth process in [20] is very close to the $k(Nd^{3+}) = 0.15$ value determined for extremely low growth rates (about 1 μm/min) corresponding to liquid phase epitaxy (LPE) of YAG:Nd thin garnet films reported in [21]. Thus, in spite of quite different growth conditions in the LPE, CZ, and μ-PD systems the main tendency is at least not in conflict with the Barton, Prim and Slichter (BPS) equation [22]. Nd^{3+} dopant incorporation in YAG crystals increases with increasing growth rate [5]. Thus, following this tendency, rapid quenching of the melts of YAG:Nd composition is considered here to correspond to a limiting value of $k(Nd^{3+}) = 1$.

The presence of a unique solid/liquid interface corresponding to the melt/crystal phase boundary seems to be an important advantage of the μ-PD system. In this scheme the growth process can be described as simple solidification of the melt only. The mixed starting materials are first melted and homogenized, resulting in a uniform distribution of constituents (chemical composition) and structural properties. Then the liquid is transformed into a crystalline solid by formation of the corresponding chemical bonds in the vicinity of the solid/liquid interface. This way micro-scale uniform solid material is most probably formed from the micro-scale uniform liquid. This is especially important for melts whose properties depend on the soak time and soak temperature [23]. Regarding μ-PD crystal growth, high uniformity of the melt corresponds well to the CZ technique allowing preparation of high-quality YAG:Nd bulk crystals. On the other hand, the presence of two solid/liquid interfaces corresponding to the feed/melt and melt/crystal phase boundaries seems to be a significant disadvantage of the alternative floating zone (FZ) method [24] that is commonly used for research purposes but not for industrial production of high-quality crystals. Although the μ-PD method is also known to be a powerful laboratory-scale technique, the growth results discussed here show that wide application of the technique for fabrication of miniature laser devices also seems to be possible.

5.3.4 Growth of YAG:Yb Crystals

A μ-PD process employing Ir crucibles and RF heating has also been used to grow single-crystal fibers of Yb-doped $Y_3Al_5O_{12}$ garnet (YAG:Yb) [9]. Although Nd-doped YAG is the best known laser crystal, renewed interest in the development of Yb-doped YAG in recent years is the result of progress in InGaAs diode lasers. Recently, a room-temperature, diode-pumped microchip laser has been demonstrated using YAG:Yb active media. It was proposed that YAG:Yb is a promising laser material which is an alternative to YAG:Nd in high-power systems [21]. In this section the growth and characterization of Yb-doped YAG fiber crystals produced using the same μ-PD system are discussed.

YAG:Yb single crystals grown by the conventional Czochralski technique were used as starting materials in the process. Undoped YAG fibers about 0.7 mm in diameter, produced by μ-PD, were used as seeds. The crystal orientation of the seeds and the fibers grown was determined using X-ray diffraction measurements by observation of the unique (444) reflection corresponding to the <111> direction. The crystals were grown at a pulling down rate of 3–8 mm/min and were 0.5–0.8 mm in diameter. The typical length of the molten zone was about 0.1 mm. The best reproducibility of growth results was found to correspond to a pulling rate of 4–5 mm/min. The length of the crystals was up to 550 mm, and was limited by the length of the pulling system only. The repeated seeding technique [9] was utilized to produce a few crystals of similar length without cooling of the hot zone to room temperature. The melts were cooled to about 100–200 °C below the growth temperature between two growths to decrease the risk of crucible oxidization. All growth processes were stopped when all the melt was crystallized into the fiber, or when the crystal length became greater than 550 mm. After growth, the crystals were withdrawn from the melt and cooled to room temperature for about 1 h.

The crystals were slightly faceted with an evident hexagonal cross-section which is typical for the <111> garnet growth direction. The best optical quality under microscopic observation and the best diameter uniformity was observed in the crystals grown with the diameter close to the diameter of the crucible opening (0.6 ± 0.1 mm). The crystals were transparent and colourless or slightly green depending on the Yb^{3+} concentration. The fibers were single garnet phase, as was detected by powder X-ray diffraction analysis. It is pointed out that in the general case, 100% of the melt was solidified into the fibers. No shifting of melt composition resulting in a second phase formation was detected during growth. Therefore, the end parts of the crystals were single garnet phase materials of reasonable optical quality, as shown in Fig. 5.11.

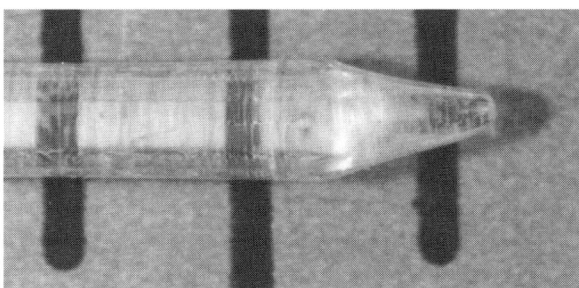

Fig. 5.11. The "tail" part of the YAG:Yb fiber crystal produced at solidification fraction g = 99.9 ± 01 % (scale in mm). The growth rate was 3.8 mm/min.

Some variations in the fiber diameter were observed because no automatic diameter control system was applied. The diameter was controlled manually by smoothly increasing or decreasing the RF power, followed by decreasing or increasing of the diameter, respectively (Fig. 5.9). Two crystals grown by the repeated seeding technique [9] were selected for composition and absorption characterization.

The crystals grown from the 3 at% Yb-doped melt (approximate composition $\{Y_{2.91}Yb_{0.09}\}Al_5O_{12}$) were analyzed for the distribution of Yb^{3+} concentration along the growth axis. The fiber produced at a pulling rate of 4.4 mm/min was studied by an EPMA technique similar to that discussed for YAG:Nd fibers. The distribution of Yb^{3+} doping cations in the fiber crystals was measured using an electron probe 10 mm in diameter. The EPMA composition analysis was made with a step of 1 mm for the fiber samples produced for different solidification fractions. Generally, except for the starting point of the growth, any decrease or increase of Yb concentration along the growth axis was not detected. Taking into consideration the very small amount of Yb in the crystal and the corresponding accuracy of the measurements, it was concluded that there was no tendency towards a nonuniform axial distribution of the Yb dopant. Thus, the segregation coefficient of Yb^{3+} was reported to be not far from unity, $k(Yb^{3+}) \approx 1$. Considering the average concentration of Yb^{3+} in the initial part of the growth, the reported value $k(Yb^{3+}) = 0.95$ [9] well represents the segregation in the above process.

The transmittance measurements were performed on the samples cut from the fibers produced at different stages of the growth process to understand the dependence of the optical quality on the solidification fraction. The samples of about 100 mm in length were irradiated with a He:Ne laser beam 250 μm in diameter with an input power of 535 μW along the crystal axis. The intensity of the output light was measured by a power detector.

Usually, the optical and crystallographic quality of crystals produced from the melt is degraded with increasing solidification fraction. Shifting of the melt composition from the starting mixture during growth due to segregation, vaporization, and other phenomena leading to instability in the melt is considered as the main cause of the degradation. This degradation is often accompanied by nonuniformity of crystal growth and second phase formation. This is the general case. However, the results obtained for the YAG:Yb fiber crystals [9] were quite the opposite. Improvement of the optical quality with an increase of the solidification fraction was observed for all crystals produced using the μ-PD system.

Two reasons seem to be suitable to explain this tendency. One of them is related to growth instabilities at the initial stage of each growth. From this point of view, the adjustment of suitable and stable growth parameters (the pulling rate and RF power supply) just after seeding with corresponding variations of the parameters have an effect on the crystal quality. At this stage, the heat transfer through the crystal is not constant and therefore some fluctuation of composition and crystal inhomogeneity can occur. On the other hand, this observation may be related to soaking time. It is known that, in general, some time is necessary for the complete melting of raw materials and melt homogenization. The longer the time, the better the uniformity of the melt. From this point of view, the YAG:Yb melt can be considered as one whose properties depend on soak time and soak temperature, similar to the case described in Ref. [23]. Thus, it was concluded that the optical quality of the YAG:Yb crystals produced by the μ-PD process increased considerably with increase of the soaking time.

After each growth the melt was cooled to about 100–200 °C according to the procedure described above for the repeated seeding technique. At this stage the melt had probably solidified. This was impossible to control the state of the melt at this step because the liquid phase was not observed after the separation of the crystal from the melt. It was related to the low wetting of the YAG melt with respect to the crucible material (Ir). Therefore, the melt was moved up into the crucible orifice just after separation. Thus, the second and following crystals were probably grown using starting conditions similar to those for the starting crystal. However, decrease of the Yb concentration in the starting stage similar to that detected for the first crystal was not observed in following fibers.

Unfortunately, it was difficult to apply a long soaking time to check this assumption in the process described because the equipment used normally did not allow a very long experimental time. Heating of the system for more than 5 hours resulted in partial recrystallization and damage to the quartz tube used in the standard apparatus to seal the crucible from contact with the oxidizing air atmosphere (Figs. 5.5 and 5.7). Even in the case of partial recrystallization, the transparency of the tube was considerably reduces. Thereafter, visual control of the growth process became difficult or even impossible.

5.4 Tb₃Ga₅O₁₂ (TGG) Garnets

Terbium-gallium garnet, $Tb_3Ga_5O_{12}$ (TGG), has been shown to be paramagnetic down to 0.8 °K [26], and has a high Verdet constant in the visible and near-infrared spectral region [27] and low absorption coefficient [28]. Therefore TGG is a promising material for fabrication of Faraday isolators for optical communication systems. TGG bulk crystals were reported [28, 29] which could be grown by the Czochralski (CZ) technique. However the fiber crystals are considered here to be more attractive materials for mass production of miniature Faraday isolators. This way, the Faraday units can be prepared by simple cutting of the one-dimensional crystal fiber of required diameter perpendicular to the growth axis, as shown schematically in Fig. 5.12.

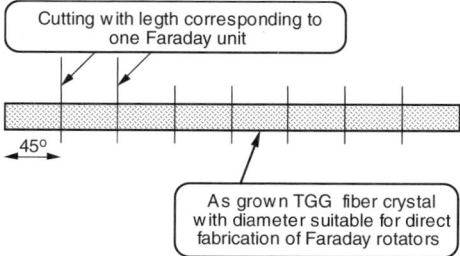

Fig. 5.12. Schematic of the application of fiber-shaped crystals in mass production.

The TGG melt performance was found [10] and is different from that of YAG [5, 9] and YAG related materials (Al₂O₃/YAG eutectic composites) [4]. In contrast to YAG based melts, Ir crucibles are wetted by the TGG melt. Therefore unexpected problems were noticed when the first attempts were made to produce TGG fiber crystals by the μ-PD technique. At first sight this performance was considered to be an important disadvantage of the TGG melt. However, this property made it particularly attractive for the fabrication of bulk single crystals with diameters up to 10 mm using standard μ-PD apparatus and crucibles with orifice less than 1 mm in diameter.

5.4.1 Tb₃Ga₅O₁₂ Fiber Crystal Growth

As reported in Ref. [10] the chemicals used for TGG fiber crystal growth were Tb_2O_3 and Ga_2O_3 (both 99.99 % pure). The starting mixtures were prepared according to the $Tb_3Ga_5O_{12}$ chemical formula assuming that the terbium oxide used was Tb_2O_3. However, Tb_4O_7 is known [30] to be the most stable terbium oxide in air. Therefore simple recalculation of the mixture assuming that Tb_4O_7 was a starting oxide was necessary. The resulted $Tb_{2.94}Ga_5O_{12-x}$ chemical formula was considered to correspond well to the true starting melt composition used in these experiments [10]. It should be noted that relatively high evaporation of Ga_2O_3 was observed during the growth runs. Therefore the negligible (about 2 %) excess of

Ga_2O_3 following from the $Tb_{2.94}Ga_5O_{12-x}$ formula was anticipated to be lost through volatilization. Thus the starting melts used were probably slightly enriched with Ga_2O_3 at the beginning of each growth process. Following the growth run the Ga_2O_3 content was decreased. Then the melt reached a composition close to the congruently melting one [29, 31].

Three mixtures of the same composition were prepared for different growth runs to check the reproducibility of the growth procedure, as given in Table 5.1, and no differences were observed in the growth results. The TGG crystals were grown using conventional μ-PD apparatus with RF heating described previously. A crucible of about 35 mm in height by 16 mm in diameter (Fig. 5.8) was used to produce the crystals. The diameter of the orifice made in the bottom of the crucible was 600 mm. The crucible was charged for about 100 vol% of the unreacted mixture of the starting oxides and pressed. The undoped YAG and TGG fibers of about 0.7 mm in diameter produced by the same technique were used as seeds. The crystals were grown at a pulling down rate of 1.5–3.0 mm/min and were 1.0–4.0 mm in diameter. The typical length of the molten zone (the distance between the tip of the crucible and solid/liquid interface) was about 0.1 mm. The best reproducibility of the growth results was found to correspond to a pulling rate of about 2 mm/min. Certainly, decreasing the growth rate was assumed to result in better crystal quality. However high volatilization of Ga_2O_3 was considered to be an additional limiting parameter, which does not allow unlimited decrease of the pulling rate with the corresponding prolongation of the growth run for more than 3–5 hours. A view of the μ-PD fibers can be seen in Fig. 5.13.

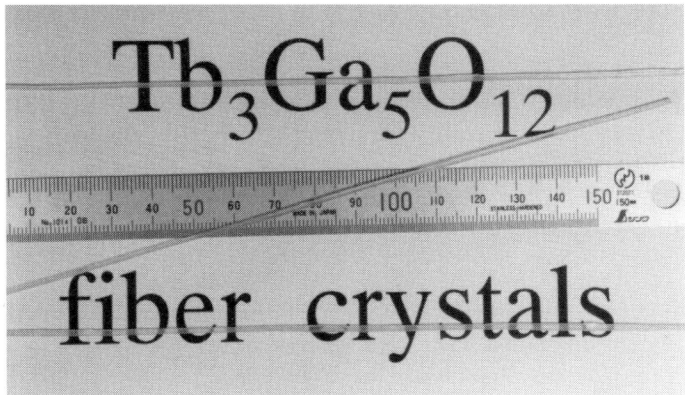

Fig. 5.13. TGG fiber crystals No. 1-3, 1-4, 2-1 (from top to bottom). The crystals are numbered according to Table 5.1. Scale in mm

Fibers about 400 mm long, which are listed in Table 5.1, were produced with complete solidification of the melt charged in the crucible. After growth of the fibers of about the desired length the temperature of the melt was increased, which resulted in an increase of the crystal diameter. This way the remaining melt was taken away from the crucible with a relatively high rate up to complete removal. Thus, the crystals were separated from the crucible by complete solidification of the melts and cooled to room temperature for about 1 hour. In spite of good wetting, the melt was removed completely without any melt drops inside the crucible, as detected under microscopic observation. Sometimes a negligible amount of small melt drops remained in the crucible. In this case it was cleaned with an additional charge of Al_2O_3/YAG eutectic mixture, pulling the corresponding fiber through the orifice. A Al_2O_3/YAG mixture was selected for this purpose (i) due to its nonwetting properties and (ii) due to the fact that the segregation coefficients of all the cations in the vicinity of the eutectic composition are close to unity.

Table 5.1. Growth conditions and dimensions of $Tb_3Ga_5O_{12}$ crystals.

Crystal No.	Pulling rate, mm/min	Length, mm	Diameter, mm
1-1*	0.4	8	0.4
1-2	0.6	75	0.7–0.8
1-3	2.5	195	0.7–1.1
1-4	2.0	435	0.9–1.2
2-1	2.2	418	0.5–1.2
2-2	2.5	415	0.5–1.2
2-3	3.0	140	1.0–9.0
2-4	1.5	400	0.5–3.0
3-1	1.0	29	1.0–10.0

* first digits represent the starting mixture

5.4.2 $Tb_3Ga_5O_{12}$ Fiber Characterization

Similar to other garnets [3, 5, 6, 9, 10], TGG crystals were slightly faceted and had an hexagonal cross-section that is typical for the <111> garnet growth direction. The crystals were transparent and colourless or slightly yellow. A summary of the crystal growth results by the above technique is given in Table 5.1. The fibers were single garnet phase, which was detected by powder X-ray diffraction analysis. The XRD data for the TGG compound were not found in the JCPDS data file [32] and therefore were indexed according to Card No. 42-0136 ($Gd_3Ga_5O_{12}$ = GGG).

It is pointed again that in all cases 100 % of the melt was solidified into the fibers. Nevertheless, no shifting of the melt composition, which could result in a second phase formation, was detected during the growth runs because no difference was observed between the X-ray data made using the starting (solidification fraction 0–1 wt%) and final (99-100 wt.%) fraction of the crystals. The end (tail) parts of the crystals were also single garnet phase materials of reasonable optical quality. Some variations (± 10 %) of the fiber diameter were observed because no automatic diameter control system was applied in the procedure reported.

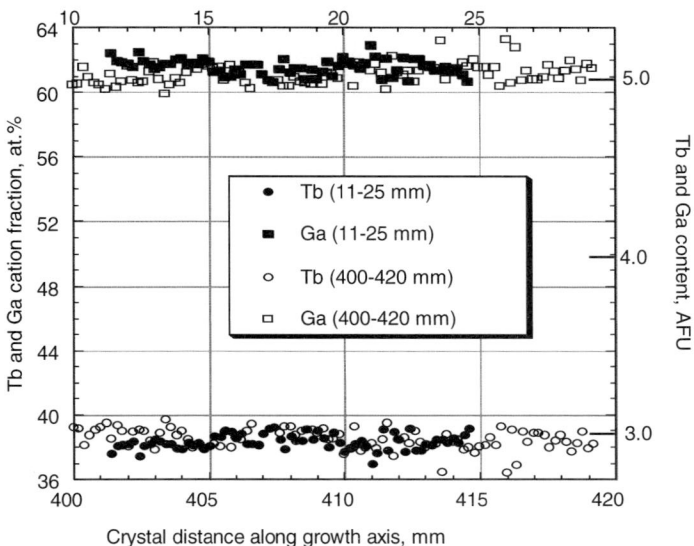

Fig. 5.14. Distribution of Tb and Ga concentration (point data) along the growth direction for No. 1–4 fiber crystal (Table 5.1) in cation fraction (at.%) and in atoms per garnet formula unit (AFU). The distance from the seed for each point measured is given in the legend and both X-axes [10].

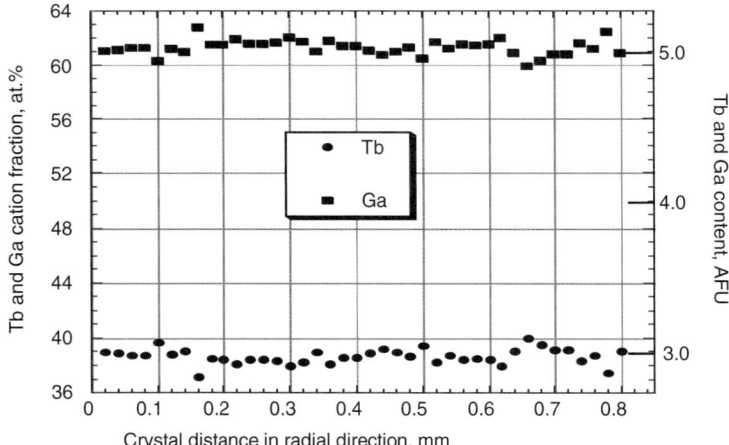

Fig. 5.15. Distribution of Tb and Ga concentration (point data) in the radial direction for No. 1–4 fiber crystal (Table 5.1) in cation fraction (at%) and in atoms per garnet formula unit (AFU) [10].

The crystal composition was measured by electron-probe microanalysis (EPMA). The results of the measurements made along the axial direction are given in Fig. 5.14. The distribution of Tb and Ga in the fiber crystals was measured using an electron probe of 10 mm diameter in steps of 200 mm. Any detectable variations of the concentration along the growth axis were not reported in [10]. Results of similar measurements in the radial direction are given in Fig. 5.15, which also demonstrates high uniformity of the crystals. The measured fiber No. 1–4 was grown at a pulling rate of 2 mm/min with a cross-section that was about twice as large as the cross-section of the orifice. Thus, the resulting velocity of the melt flow through the crucible orifice was estimated to be about 4 mm/min, which is relatively high for any possible segregation of the crystal-forming cations especially for a compound which does not contain any foreign dopants.

5.4.3 $Tb_3Ga_5O_{12}$ Bulk Crystal Growth

An attempt was also made to check the ability of the pulling down system to produce $Tb_3Ga_5O_{12}$ crystals with large diameter (> 2.0 mm) [10], which we call here bulk crystals. Surprisingly the attempt was successful. Increasing the crucible temperature resulted in increasing both the thickness and cross-section of the melt film between the crucible and the crystallized material due to good wetting of the melt. This way increasing of the total volume of the molten zone allows better mass transport from the melt located inside the crucible to the crystal/liquid interface through the orifice. A schematic of these growths is illustrated in Fig. 5.9. As a result the maximal diameter of the crystal was increased up to 10 mm. It is noted that further increase of the crystal diameter up to the maximal outer diameter of the crucible (16 mm) was generally possible. The bulk TGG crystals are shown in Fig. 5.16. In spite of the very high growth rate (1 mm/min), the crystals were of reasonable optical quality as observed under an optical microscope. Sometimes cracking of the crystals was detected. It is assumed that cracking can be prevented by appropriate optimization of temperature gradients inside the crystals and of the cooling rate [33].

Thus the pulling down technique is also promising for the growth of bulk crystals. It is evident that chemical uniformity of such crystals in the axial direction should be very high. Simple calculations show that in order to achieve a crystal diameter of about 10 mm at a pulling rate of 1 mm/min it requires a velocity of the melt flow through the orifice (0.6 mm in diameter) of about 300 mm/min. Therefore any diffusion and convection flow inside the orifice is practically impossible.

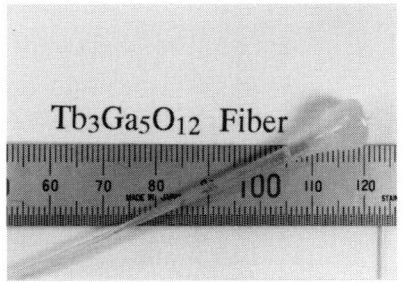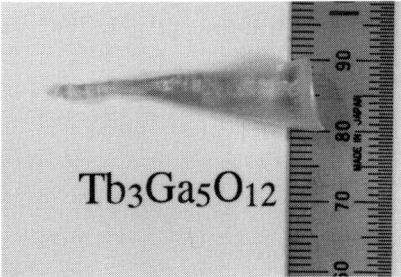

Fig. 5.16. Bulk TGG crystals No. 2-3 (left) and 3-1 (right) (scale in mm).

5.5 $Tb_3Al_5O_{12}$ (TAG) Based Mixed Garnets

It was shown long ago that terbium-aluminum garnet, $Tb_3Al_5O_{12}$ (TAG) is another suitable material for use as a Faraday effect optical isolator [34]. Almost 50 years have passed since that time. However, up to now no reports about crystal growth of perfect TAG single crystals suitable for device fabrication have been found. Recently the phase diagram of the Tb_2O_3-Al_2O_3 binary system was reported [35], and it was shown that the TAG compound melts incongruently with a prime crystallization of $TbAlO_3$ perovskite (TAP) phase from the melt corresponding to the stoichiometry of TAG (Fig. 5.17). It was observed that single garnet phase formation in the above system is only possible in the presence of some excess of Al_2O_3. According to the data reported [35] a minimal amount of Al_2O_3 as a flux corresponds to the approximate melt composition of $Tb_3Al_5O_{12} + 0.3\ Al_2O_3$. It is evident that application of such a melt for TAG mass production is almost impossible because the difference between the melt and the crystal composition is very high and the segregation coefficients of both Tb^{3+} and Al^{3+} in such a process are very far from unity.

An attempt was made in [36] to produce TAG fibers from a nonstoichiometric Al_2O_3-rich melt using the same apparatus with RF heating. However the pulling rate required to produce fibers of high optical quality was found to be very low (0.15 mm/min). The TAG fibers produced from the stoichiometric melt according to the procedure described in [37] were of low quality (Fig. 5.18) with precipitation of a second phase. It was confirmed that the growth rate resulting in defect-free TAG single-crystal fibers should be very low. The experimental procedure described below was proposed to present in some detail how the modification of the melt by mixing of corresponding isostructural end member compounds can be established to modify the phase diagram of the Tb_2O_3-Al_2O_3 system [37, 38].

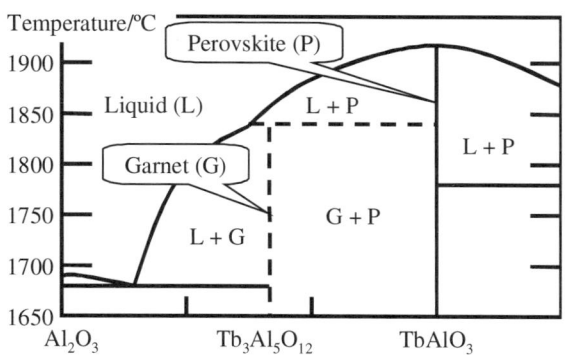

Fig. 5.17. Fragment of the phase diagram for the Al_2O_3-Tb_2O_3 system according to [35].

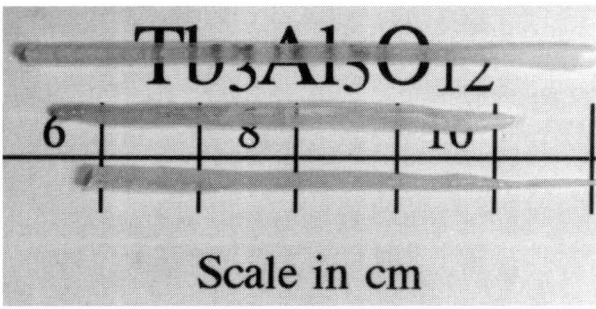

Fig. 5.18. TAG fibers produced from a stoichiometric $Tb_3Al_5O_{12}$ melt at a pulling rate of about 1 mm/min.

5.5.1 Crystal Chemistry of Mixed Garnets

It is common knowledge, that the growth of single crystals "by mixing simple end members is not the right way to obtain homogeneous mixed garnet crystals suitable for technical applications" [39]. This statement is evident, because segregation between the end member compounds and their constituents results in a non-uniform distribution of ingredients along the growth axis, the variation of lattice parameters and significant mechanical stresses. As a result the mixed (or solid solution) crystals produced from the melt are generally of low structural and optical quality with precipitation of second phases. This is the usual understanding of the behavior of mixed systems following from the phase diagram type given in Fig. 5.19a. The other diagrams (5.19b and 5.19c), which also describe systems with unlimited solubility of two isostructural end members in liquid and solid states, are usually considered as an exotic kind of behavior in pseudo-binary systems and can in be general found in fundamental research on thermodynamics [40]. It is well known that the growth of high-quality uniform crystals from the melt of the same composition is possible only from mixtures which melt congruently with a corresponding invariant point on the phase diagram and segregation coefficients of all constituents equal to unity.

On the other hand, it is known that the segregation coefficients of RE cations in garnet crystals are occasionally equal [41, 42]. The best illustration of the behavior of assorted RE cations with respect to the garnet structure is the dependence of segregation coefficients of various RE in gallium and iron garnets reported in [42, 43]. Briefly, $(\Sigma RE)_3Fe_5O_{12}$ and $(\Sigma RE)_3Ga_5O_{12}$ garnet films were grown by liquid phase epitaxy from melts containing equal amounts of each RE oxide, and the film composition was measured with subsequent calculation of the segregation coefficients. The resulting data are given in Fig. 5.20.

Corresponding data for aluminum garnets, $(\Sigma RE)_3Al_5O_{12}$ are not available. However, based on the simple assumption that the size of the most stable cation with maximal segregation coefficient (Ho^{3+} for Ga-garnets and Y^{3+} for Fe-garnets, respectively) is proportional to the lattice parameter of the host garnet structure, it was estimated that the Er^{3+} cation is the best one for the subfamily of Al-garnets. In this way it was possible [37, 38] to select appropriate couples of RE cations whose segregation coefficients are similar or equal in Al-garnets as follows: Tb^{3+}/Lu^{3+}, Tb^{3+}/Yb^{3+}, Tb^{3+}/Tm^{3+}, and possibly Tb^{3+}/Er^{3+}.

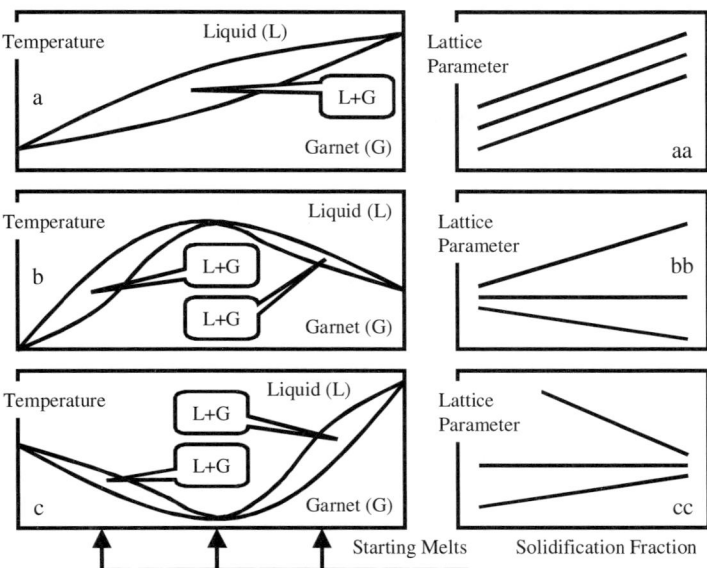

Fig. 5.19. Three possible types (**a**)–(**c**) of phase diagram for systems with unlimited solubility in solid and liquid phases for two isostructural compounds and corresponding dependencies of the lattice parameters of the crystals produced from the different starting melts on the solidification fraction (**aa**)–(**cc**).

Each garnet structure is characterized by values of equilibrium cation sizes [41–43], which can be found for any crystallographic sublattice. Similar tendencies were also observed for other non-garnet structures [44–49]. The segregation coefficients of the cations whose size is close to equilibrium are maximal for a given sublattice (Fig. 5.20). Y^{3+} and Ho^{3+} cations have the highest segregation coefficients in dodecahedral sites of $RE_3Fe_5O_{12}$ and $RE_3Ga_5O_{12}$ garnets, respectively, within the range of all other RE from Lu^{3+} to La^{3+} as shown in Figs. 5.20 and 5.21. Estimation of the equilibrium RE cation for the $Ca_3Nb_{1.6875}Ga_{0.1875}Ga_3O_{12}$ (CNGG) garnet with lattice parameter $a = 12.509$ Å was made in Ref. [43]. It was found that this value corresponds to hypothetical RE cation with radius $R_{CN=8}(RE) = 102.4$ pm. The data for these three materials are plotted in Fig. 5.21 similar to that reported in [43]. Here and later, Shannon's cation radii [50] for coordination number CN = 8 are used for comparison of the behavior of various RE in the garnet structure.

The dependence of the size of the equilibrium cation on the lattice parameter of the host garnet crystal was linearly extrapolated in Fig. 5.21 from $Yb_3Al_5O_{12}$ [JSPDS Card-73-1369] [32] with $a = 11.931$ Å to $Bi_3Fe_5O_{12}$ with $a = 12.623$ Å [51]. According to the model discussed in [43] the size of each site and the size of the equilibrium cation corresponding to the garnet with a different volume of unit cell are changed linearly on the lattice parameter due to expansion of the garnet lattice. Two TAG lattice parameters are plotted in Fig. 5.21. One of them was reported in [JSPDS Card-17-0735]; another one was measured in [37] using a polycrystalline fiber, produced by the μ-PD technique from a stoichiometric TAG melt

(Fig. 5.18). Thus, the size of the equilibrium cation corresponding to the TAG garnet was estimated using the average value of these two lattice parameters and the linear dependence, as given in Fig. 5.21.

According to [42, 43], the greater the radii misfit between a RE cation and an equilibrium one, the smaller the segregation coefficient. Therefore, the paradoxical conclusion follows from Figs. 5.20 and 5.21, that the segregation coefficient of the host Tb^{3+} cation in the TAG crystal should be small compared with that of some guest equilibrium cation because Tb^{3+} is too large for TAG. The result of this paradox is low structural stability of TAG compared with that of the corresponding perovskite structure TAP and incongruent melting of TAG.

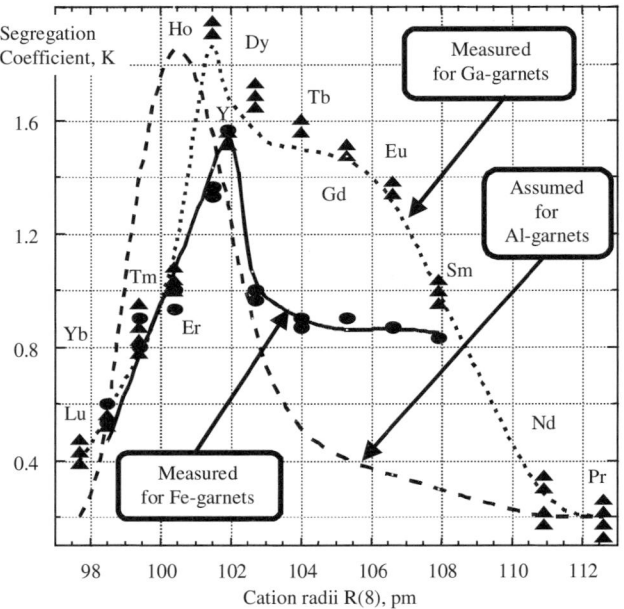

Fig. 5.20. Dependence of segregation coefficients of rare-earth elements RE^{3+} in $\{\Sigma RE\}_3 Ga_5 O_{12}$ and $\{\Sigma RE\}_3 Fe_5 O_{12}$ garnet films on cation radii [42, 43]. The films were grown [42] from melts containing equal amounts (mol.) of 13 or 10 rare-earth cations. The dependence for $\{\Sigma RE\}_3 Al_5 O_{12}$ is proposed assuming that the maximum points of the dependencies are proportional to the garnet lattice parameter.

Therefore it was reasonable to select another coupled RE cation, which is small compared with that of the equilibrium one in the same degree as Tb^{3+} is large. It was expected that such a cation has the same segregation coefficient as that of Tb^{3+} according to Fig. 5.20. Moreover decreasing the garnet lattice parameter following from the substitution with the small guest RE will probably result in automatic optimization between the three variables: the lattice parameter and the segregation coefficients of both Tb^{3+} and doping small RE. In this case the growth of $(Tb,RE)_3 Al_5 O_{12}$ from the stoichiometric melt will become easy because the segregation coefficients of all three cations, including Al^{3+}, present in the system will become equal. As can be seen from the plot in Fig. 5.20, few RE with size smaller than 100.3 pm can be considered as suitable dopants. In particular, Lu^{3+} and Yb^{3+} cations were selected for this purpose in [37, 38].

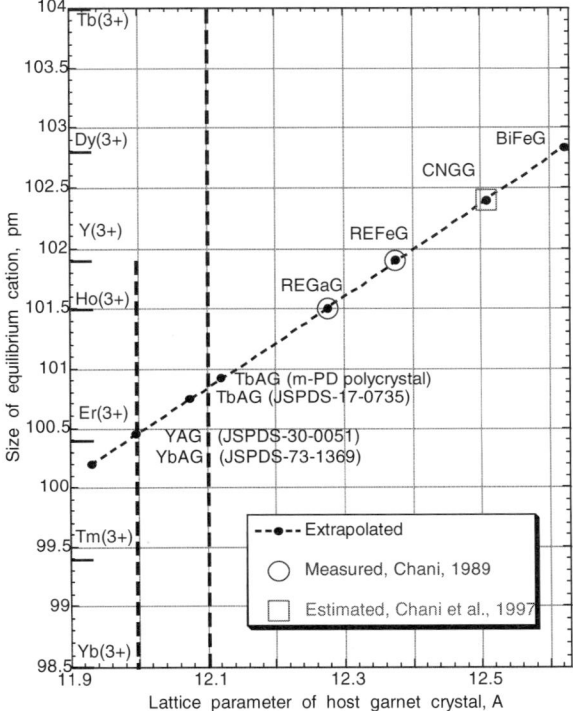

Fig. 5.21. Dependence of the size of the equilibrium cation in dodecahedral sites of the garnet structure on the lattice parameter of the host garnet according to [42, 43, 50]. The two vertical dotted lines represent the lattice parameters of $Y_3Al_5O_{12}$ and $Tb_3Al_5O_{12}$, respectively.

5.5.2 $(Tb,Lu)_3Al_5O_{12}$ (TLAG) Fiber Crystals

The host/guest substitution in $(Tb,Lu)_3Al_5O_{12}$ mixed garnet crystals have been studied by solidification of the corresponding stoichiometric melts using the μ-PD apparatus [38]. It was found that with substitution of Tb^{3+} with Lu^{3+}, the tendency of $TbAlO_3$ perovskite phase formation in undoped $Tb_3Al_5O_{12}$ decreases. Although growth of mixed crystals in general does not allow a uniform distribution of constituents in the resulting material, an invariant point was found to exist in the $Tb_3Al_5O_{12}$-$Lu_3Al_5O_{12}$ system in the vicinity of the $Tb_{2.2}Lu_{0.8}Al_5O_{12}$ composition.

The chemicals used were Tb_4O_7 (99.99), Lu_2O_3 (99.9) and Al_2O_3 (99.999). The melt compositions are given in Table 5.2. The raw materials and the crystals (Fig. 5.22) were prepared according to the procedure described earlier. The crystals were grown at a pulling down rate of 1.2 mm/min and were 1.3 mm in diameter. The diameter was measured in situ using a scale attached to the monitor with resolution better than 0.1 mm. Adjustment of desired fiber diameter was made by control of the RF power supply. The typical thickness of the molten zone was about 0.1-0.2 mm.

In spite of the very high growth rate (1–5 mm/min), segregation in the μ-PD process is still detectable [5]. Therefore it was interesting to compare the chemical uniformity of the crystals fabricated from different $(Tb,Lu)_3Al_5O_{12}$ melts. The accuracy of the electron-probe microanalysis (EPMA) technique widely used to study the spatial distribution of the constituents is relatively low (about 1 at.%). On the other, the hand accuracy of the lattice parameters measurements by X-ray diffraction is very high. For that reason the uniformity of the $(Tb,Lu)_3Al_5O_{12}$ crystals reported in [38] was studied by measurement of the lattice parameters. The XRD peaks were indexed according to JCPDS card No.76-0111 for undoped $Tb_3Al_5O_{12}$ [32]. The ten strongest peaks in the range $2\theta = 60–80°$ were selected for the calculation of the lattice parameters. The accuracy of the measurements was estimated to be 0.001–0.002 Å.

All the Lu-doped TAG fiber crystals discussed here were observed to be single garnet phase (Table 5.2). The XRD data collection for the end part of the best $(Tb,Lu)_3Al_5O_{12}$ crystal was reported in [38]. The fiber produced from undoped $Tb_3Al_5O_{12}$ contained a $TbAlO_3$ perovskite phase which is in agreement with [35]. Thus the segregation of both Tb^{3+} and Lu^{3+} in the mixed crystals produced from the melts given in Table 5.2 was studied through the lattice parameters measurements. The results given in Table 5.2 and Figs. 5.22 and 5.23 show that the best crystal quality and the best lattice parameter uniformity was achieved for the fibers produced from the melts close to $Tb_{2.20}Lu_{0.80}Al_5O_{12}$ composition.

Table 5.2. $(Tb,Lu)_3Al_5O_{12}$ crystal growth. Melt composition (mol.%) and growth results.

Melt No.	Melt composition	Length, mm	Second phase	Colour	Fiber diameter	Fiber surface
TAG-01	$Tb_3Al_5O_{12}$	210	TAP	White	Very unstable	Rough
TAG-13	$Tb_{2.44}Lu_{0.56}Al_5O_{12}$	474	No	Yellow	Uniform	Rough
TAG-11	$Tb_{2.20}Lu_{0.80}Al_5O_{12}$	390	No	White	Uniform	Smooth
TAG-17	$Tb_{1.96}Lu_{1.04}Al_5O_{12}$	443	No	Yellow	Partly unstable	Smooth

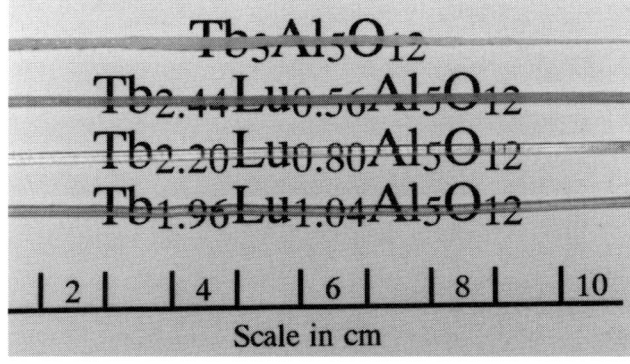

Fig. 5.22. $(Tb,Lu)_3Al_5O_{12}$ fibers produced from different melts at a pulling rate of 1.2 mm/min according to Table 5.2.

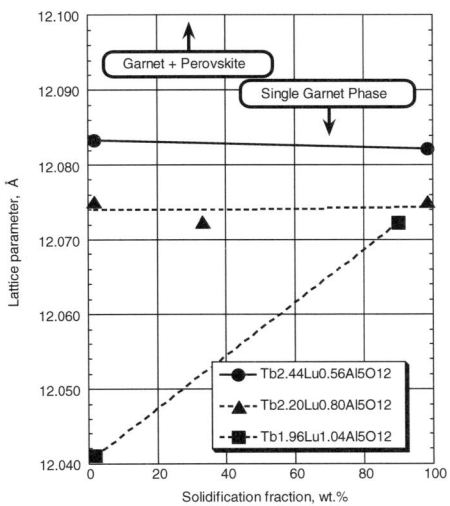

Fig. 5.23. Dependencies of the lattice parameter on solidification fraction for the $(Tb,Lu)_3Al_5O_{12}$ fibers produced at a pulling rate of 1.2 mm/min according to Table 5.2.

The melting points of the $(Tb,Lu)_3Al_5O_{12}$ mixtures were not reported in [38]. However a detectable decrease of the RF power was mentioned for the growth of the $Tb_{2.20}Lu_{0.80}Al_5O_{12}$ fiber compared with the others listed in Table 5.2. These results allowed the authors of [33] to propose the phase diagram of the $Tb_3Al_5O_{12}$-$Lu_3Al_5O_{12}$ pseudo-binary system because none of the data were in conflict with the diagram given in Fig. 5.19c. Noting that the TAG compound melts incongruently, the diagram shown in Fig. 5.19c was modified as illustrated in Fig. 5.24. The phase transition temperatures for the end members of $Lu_3Al_5O_{12}$ and $Tb_3Al_5O_{12}$ used in Fig. 5.24 were borrowed from the data reported in [35] and [52], respectively. The existence and position of the invariant point follow from the results of Tab. 5.2 and Figs. 5.22 and 5.23.

Of course, additional optimization of the Tb_2O_3/Yb_2O_3 ratio is possible to increase the reproducibility of the growth results and the crystal quality. The composition of the end parts of the $(Tb,Lu)_3Al_5O_{12}$ crystals grown from different melts should be closest to the material produced from the melt corresponding to the invariant point on the diagram (Fig, 5.24). The content of Al_2O_3 can almost be fixed not far from stoichiometry of the garnet, because it is well known that incorporation of relatively large RE cations into octahedral sites of the garnet structure instead of Al^{3+} is almost impossible [5]. At the same time localization of small Al^{3+} in the relatively large dodecahedral sites with coordination number CN = 8 was also never observed [50].

Fig. 5.25 gives a simple explanation of the above results using the images of the garnet structure introduced in [43]. In the case of the $Tb_3Al_5O_{12}$ garnet the Lu^{3+} cation is considered as a dopant, whose size does not fit well into the structure of the host (TAG) crystal. Therefore the bonding of Lu^{3+} with the whole structure is weaker than that of Tb^{3+} host cation. As a result, substitution of the Tb^{3+} with Lu^{3+} is accompanied by decreasing of the melting point. The same conclusion can be formulated for another end member, $Lu_3Al_5O_{12}$. On the other hand, in the case of the mixed $(Tb,Lu)_3Al_5O_{12}$ crystal both RE cations do not fit the dodecahedral sites of the garnet in equal degree.

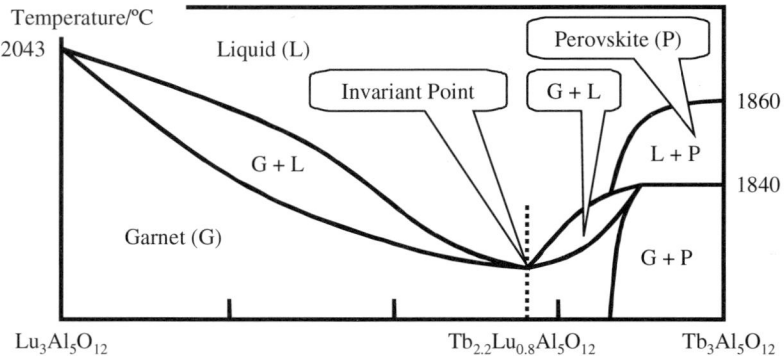

Fig. 5.24. Proposed phase diagram for the $Lu_3Al_5O_{12}$-$Tb_3Al_5O_{12}$ pseudo-binary system using the data of [35, 38, 52].

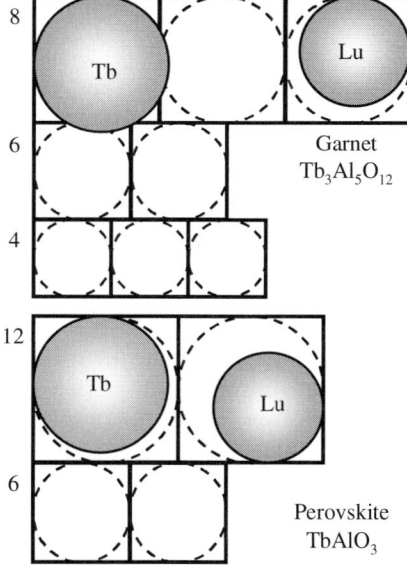

Fig. 5.25. Motives of $(Tb,Lu)_3Al_5O_{12}$ garnet and $(Tb,Lu)AlO_3$ perovskite structures the coordination numbers for each sublattice are given on the left side of the image.

Finally the segregation coefficients of both Tb^{3+} and Lu^{3+} become equal for the composition corresponding to the invariant point in Fig. 5.24. Therefore minimization of the melting point in the diagram is a simple result of the weak bonding of both Tb^{3+} and Lu^{3+} cations occupying the same sublattice with coordination number $CN = 8$. In this way substitution of Tb^{3+} with Lu^{3+} was found to stabilize the garnet phase in the Tb_2O_3-Al_2O_3 binary system. The results show that an invariant point, when the liquid and solid have same composition [40], exists in the $Tb_3Al_5O_{12}$-$Lu_3Al_5O_{12}$ pseudo-binary system. Five independent phenomena allow us to make similar conclusion: (1) the segregation coefficients of Tb^{3+} and Lu^{3+} in aluminum garnets should be close to or equal depending on the lattice parameter in in accordance with [42,43] and Fig. 5.20; (2) uniformity of the fiber crystals produced from the melts close to $Tb_{2.2}Lu_{0.8}Al_5O_{12}$ is the best one, within the set of experiments according to Fig. 5.23; (3) convergence of the crystal lattice parameters with increasing solidification fraction for the fibers produced from different melts given in Table 5.3 (Figs. 5.19 and 5.23), (4) detectable decrease of the RF power supply necessary for the growth of fibers of the same diameter (1.3 mm) with the same pulling rate (1.2 mm/min) for the crystals produced from the melt of the $Tb_{2.2}Lu_{0.8}Al_5O_{12}$ composition; (5) best optical quality and stability of growth conditions was observed for fibers produced from melts close to $Tb_{2.2}Lu_{0.8}Al_5O_{12}$ (Table 5.2 and Fig. 5.22).

Based on the above results, the existence of other mixed garnet crystals which melt congruently was proposed [38] from the data given in Figs. 5.19 and 5.20. About 50 new compounds with the garnet structure was estimated, taking into consideration all possible combinations of "small" and "large" RE cations for the Al, Ga , and, probably, Fe garnets.

5.5.3 (Tb,Yb)$_3$Al$_5$O$_{12}$ (TYAG) Fiber Crystals

The growth of $(Tb,Yb)_3Al_5O_{12}$ (TYAG) fiber single crystals was considered in [37] as another attempt to stabilize garnet phase formation by partial substitution of Tb^{3+} with small guest RE cations. According to the phase diagram of the Yb_2O_3-Al_2O_3 binary system [53], the formation of $YbAlO_3$ perovskite was also not observed. Therefore, doping of TAG with Yb^{3+} was also used to decrease the probability of perovskite phase formation in the starting $Tb_3Al_5O_{12}$ melt because $YbAlO_3$ perovskite is an unstable phase. Similar to Lu^{3+}, Yb^{3+} is unsuitable for any site of the perovskite structure. It is too small for sites with coordination number $CN = 12$ and too large for the octahedral positions ($CN = 6$) in the perovskite. Thus, Yb^{3+} was selected as an additional dopant allowing modification of the Tb_2O_3-Al_2O_3 binary system.

The phase diagram of the Tb_2O_3-Yb_2O_3-Al_2O_3 ternary is unknown. Nevertheless, a preliminary estimation of this diagram was made in [37] based on two known diagrams of Tb_2O_3-Al_2O_3 [35] and Yb_2O_3-Al_2O_3 [53]. In this way, the composition of about $Tb_2O_3 : Yb_2O_3 : Al_2O_3 = 2.4 : 0.6 : 5.0$ was considered to be a starting composition allowing solidification of perovskite-less fiber crystals from the melt corresponding to the stoichiometry of the garnet. The chemicals used were Tb_4O_7, Yb_2O_3 (both 99.99 % pure), and Al_2O_3 (99.999 % pure). The starting oxides were carefully mixed together in the above ratios and dried. The $(Tb,Yb)_3Al_5O_{12}$ crystals were grown using the conventional μ-PD apparatus described previously.

The diameter of the orifice made in the bottom of the crucible was 0.5–0.6 mm. The shape of the crucible was slightly modified as shown in Fig. 5.26. The tip of the crucible was polished. Therefore the growth system used for the $(Tb,Yb)_3Al_5O_{12}$ fiber growth was considered as a combination of μ-PD and EFG [54, 55] growth techniques. The die diameter (cross-section of the horizontal surface in the vicinity of the crucible tip) was about 1.8 mm, as can bee seen from Fig. 5.26. It is noted that the weight of the empty crucible was about 45 g with dimensions of about 44 mm in height by 14 mm in diameter. Growth results show that the fibers produced with the modified crucible were much more uniform in diameter than those grown using the conventional crucibles shown in Figs. 5.6 and 5.9. The crystal diameter was not very sensitive to fluctuations of the RF power when the crystal diameter was adjusted to be equal or close to the diameter of the die as illustrated in Fig. 5.27. The diameter was measured in situ using a scale attached to the monitor with an accuracy of about 0.1 mm.

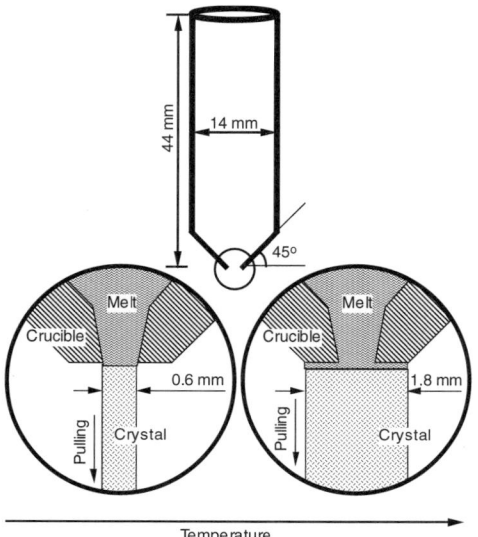

Fig. 5.26. Schematic of μ-PD crucible and the growth procedure for the modified crucible [37].

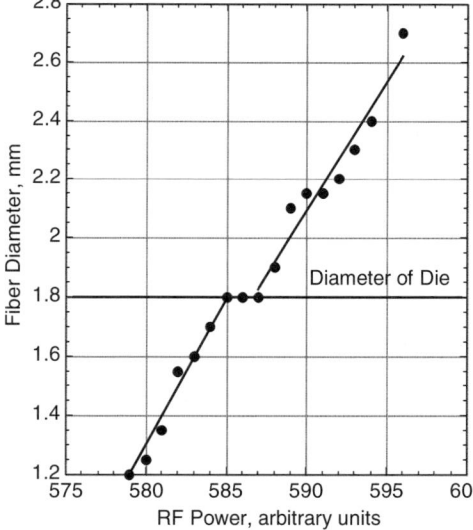

Fig. 5.27. Dependence of the crystal diameter on the RF power applied to the modified crucible during growth [37].

Table 5.3. $(Tb,Yb)_3Al_5O_{12}$ melts compositions (mol.%).

Melt No.	$0.5(Tb_4O_7)$	Yb_2O_3	Al_2O_3	Tb/(Tb+Yb), at.%
1	33.96	0	66.04	100
2	36.00	0	64.00	100
3	37.00	0	63.00	100
4	34.00	3.50	62.50	90.67
5	30.00	7.00	63.00	81.08
6	27.00	10.00	63.00	72.97
6a	27.20	9.80	63.00	73.51
7	28.50	8.50	63.00	77.03
8	26.00	11.00	63.00	70.27
9	27.75	9.25	63.00	74.32

At the beginning of these experiments the polycrystalline fibers produced from the $Tb_3Al_5O_{12}$ undoped melt (see Table 5.3, No.1 and Fig. 5.18) were used as seeds to avoid contamination of the system with foreign substances. At the later stage, the (111)-oriented YAG single-crystal fibers grown by the same technique [5, 9] were also used. Firstly, the seed touched the tip of the crucible at the temperature certainly below the melting point. Thereafter the temperature was increased continuously until the molten zone became observable through the CCD camera. Finally the pulling was started, and the growth parameters (RF power and the pulling rate) were adjusted to fit stable growth conditions.

In contrast to $Y_3Al_5O_3/Al_2O_3$ eutectic [4] and $Y_3Al_5O_{12}$ [5, 9] melts which are nonwettable with respect to Ir, the wettability of TAG melts in a temperature range close to the melting point was not evident. In the case of an undoped TAG melt (No. 3, Table 5.3) the dependence of the fiber diameter on the pulling rate was observed to be similar to nonwettable melts. Increasing the pulling rate resulted in an increase of the latent heat of crystallization with corresponding overheating of the melt, increasing of the thickness of the molten zone, and decreasing of the crystal diameter, as shown in Fig. 5.28.

Evolution of the Yb-doped TAG melt behavior in the μ-PD system with the modified crucible was similar to that described in [10, 12] for the wettable melts. An increase of the crucible temperature resulted in an increase of the fiber diameter at least when the fiber diameter was less than the diameter of the horizontal plane surface made at the vicinity of the tip (Fig. 5.26, left). It is still unclear what mechanism is responsible for this behavior. Chemically both the $Y_3Al_5O_{12}$ and $(Tb,Yb)_3Al_5O_{12}$ melts are very similar. Therefore the interaction of both liquids with iridium should not be very different in the same range of temperatures. However, according to the phase diagram given in Fig. 5.24 and a similar diagram reported in [37] for the $Y_3Al_5O_{12}$-$(Tb,Yb)_3Al_5O_{12}$ system, partial substitution of Tb^{3+} with Yb^{3+} results in e decrease of the melting temperature with possible change of the wetting properties.

On the other hand, the crucible used in the $(Tb,Yb)_3Al_5O_{12}$ fiber growth experiments [37] discussed here was also modified according to Fig. 5.26. Therefore it is difficult to compare the behavior of these melts which were used in different experimental conditions. Nevertheless, it is evident that the driving force resulting in raising the melt level through the outer surface of the crucible with heating was suppressed by gravity in the case of the conventional crucible (Fig. 5.9.). In the case of the modified crucible (Fig. 5.26a) raising the melt level is not generally necessary because the crucible/liquid interface is horizontal at least when the diameter of the molten zone is less than the diameter of the die. Thus, control of the shape of the solid/liquid interface for the modified crucible was somewhat different from that described for the conventional shaped crucibles (Figs. 5.6 and 5.9).

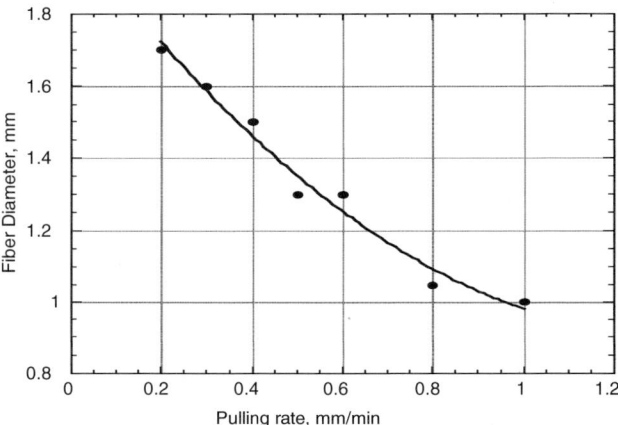

Fig. 5.28. Dependence of $Tb_3Al_5O_{12}$ fiber crystal diameter on the pulling down rate at constant RF power (case of nonwettable melts) [37].

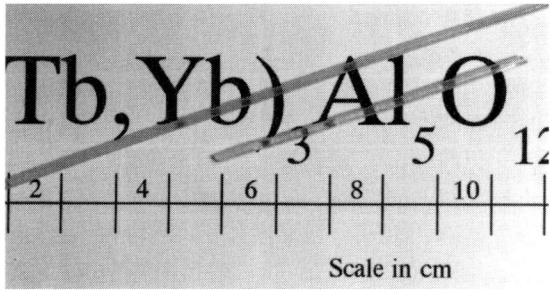

Fig. 5.29. $(Tb,Yb)_3Al_5O_{12}$ fiber crystals No. 6-1, produced from melt No. 6, according to Table 5.3.

The maximal cross-section of the molten zone and the diameter of the crystals grown from the modified crucible were not completely limited by the size of the plane surface on the bottom of the crucible, as follows from Fig. 5.26. The growth of fibers with a greater diameter was also possible (Fig, 5.27). The crystals grown with maximal diameter up to 3.0 mm had a tendency to crack. To prevent mechanical damage fibers of improved optical quality were grown with minimal possible overheating of the melts. This strategy allows us to minimize the thickness of the molten zone and the fiber diameter (Fig. 5.27). Views of the TYAG fibers produced at a pulling rate of about 0.5–1.2 mm/min are shown in Fig. 5.29.

The separation of the just-grown crystals from the melt was not easily achieved by simple overheating of the melt as reported for the nonwettable melts and conventional μ-PD crucibles [5, 9]. In the case of a conventional crucible the diameter of the crystal was decreased with overheating up so the surface energy became too great to maintain the molten zone in contact with both the crucible and the crystal (Fig. 5.9). For the $(Tb,Yb)_3Al_5O_{12}$ melts and the crucible described at this point (Fig. 5.26) the overheating resulted in an adjustment of the crystal cross-section to the cross-section of the flat surface of the die. At superior overheating a greater increasing of the fiber diameter was observed, as shown in Fig. 5.27. Therefore in the main the crystals reported here were grown with complete solidification of the melt. In the case of overcharging the crucible before these experiments, the growth processes were completed with overheating of the melt up to a crystal diameter of more than 3.0 mm. In thin case consumption of the melt increased considerably and discharge of the crucible was completed much faster.

The crystals were grown at a pulling down rate of 0.2–2.0 mm/min with diameter 0.5–1.7 mm. The typical length of the molten zone was in the range 0.1–0.4 mm. The best reproducibility of growth results was found to correspond to a pulling rate of about 1 mm/min. Certainly, decreasing the growth rate will result in better crystal quality. However the duration of the growth process also results in higher segregation of RE cations between the liquid and solid phases with subsequent chemical nonuniformity of the fibers. It was assumed that the segregation coefficients of both Tb^{3+} and Yb^{3+} in the crystals produced from melts No. 6, 6a, 7, and 8 (Table 5.3) were close, but not equal. Future detailed scanning of the phase diagram in the vicinity of these compositions is necessary to determine the exact position of the invariant point, when the liquid and the solid have the same composition [40].

The length of the crystals was up to 400 mm which was mainly determined by the length of the pulling system and the amount of starting material. The fibers were produced using a crucible with size given in Fig. 5.26. The crystals were transparent and colourless or slightly green due to the presence of Yb^{3+}. The Yb-substituted TAG fibers were single garnet phase which was detected by powder X-ray diffraction analysis. 100% of the melt was solidified into the fibers. Nevertheless, no significant shifting of the melt composition resulting in a second phase formation was detected during growth even at the final stage of growth. XRD data for the starting (solidification fraction 0–1 wt.%) and final (99–100 wt.%) parts of the crystals corresponded to the single garnet phase. The end (tail) parts of the crystals were of reasonable optical quality.

Detectable variations (± 5 %) of the fiber diameter were observed in crystals grown with a thickness of about 1 mm because no automatic diameter control system was applied. However, the uniformity of the fiber diameter was even better when the crystal diameter was adjusted to be close to the diameter of the die, as shown in Figs. 5.26 and 5.27. The results also show that the best crystal quality was observed for fibers produced from melts No. 6 and 8 (see Table 5.3) and their vicinity.

Generally it was important to prove the phase diagram type given in Fig. 5.24 for both $Tb_3Al_5O_{12}$-$Yb_3Al_5O_{12}$ and $Tb_3Al_5O_{12}$-$Lu_3Al_5O_{12}$ systems. However, the facilities used in [37] did not allow direct measurements of the melting point of the corresponding mixtures. Nevertheless detectable decreasing of the RF power was found to be necessary for the growth of the fibers with a relatively high content of Yb_2O_3. The results plotted in Fig. 5.30 illustrate the tendency. The magnitude of the RF power which is necessary to produce fibers of the same diameter (about 1.3 mm) with the same pulling rate (1.0–1.2 mm/min) was considered to be a suitable parameter, which corresponds well to the melting point of the particular starting mixture.

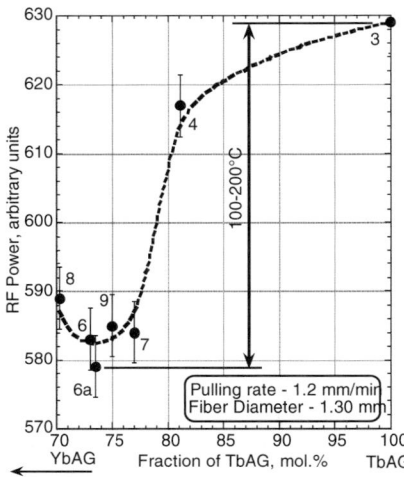

Fig. 5.30. Dependence of the RF power supply resulting in the growth of fibers of 1.3 mm in diameter at a pulling rate of 1.0–1.2 mm/min at stable conditions on the Tb/Yb melt ratio.

The measurements of the power in Fig. 5.30 were made at established and stable conditions. Therefore the heat and mass transfer in these experiments were assumed to be very similar for all the melts listed. As can be seen from the plot, starting from an undoped $Tb_3Al_5O_{12}$ melt, significant decreasing of the melting temperature was observed with increasing Yb:(Tb+Yb) ratio. The dependence passed through a minimum point in the range of compositions corresponding to Yb:(Tb+Yb) = 0.22–0.30. Notably, the difference between the maximal and minimal RF powers given in Fig. 5.30 corresponds to a melting temperature of 100–200 °C.

The results discussed in the previous paragraph and the experimental data reported in Refs. [35–37, 53] allowed the authors of [37] to assume that the phase diagram of the $Tb_3Al_5O_{12}$-$Yb_3Al_5O_{12}$ pseudo-binary systems is similar to that proposed for the $Tb_3Al_5O_{12}$-$Lu_3Al_5O_{12}$ system (Fig. 5.24). Nevertheless further optimization of the Tb_2O_3/Yb_2O_3 ratio is necessary to determine the precise position of the invariant point and to increase reproducibility of the growth results and the crystal quality. The content of Al_2O_3 can almost be fixed not far from stoichimetry of the garnet. Therefore, the original Tb_2O_3-Yb_2O_3 - Al_2O_3 ternary system can be simplified to a binary one. Thus, the total number of independent variables can be decreased from three to two. Therefore, the number of experiments necessary for such optimization can be decreased drastically. Thus, $(Tb,Yb)_3Al_5O_{12}$ (TYAG) garnet fibers of good optical quality were also grown by the μ-PD technique from the $Tb_{3-x}Yb_xAl_5O_{12}$ melt (x = 0.8–1.0). The best crystal the $Tb_{3-x}Yb_xAl_5O_{12}$ melt ($x = 0.8–1.0$). The best crystal quality and the best stability of the melt behavior were observed for crystals produced from the melts No. 6 and No. 8 (Table 5.3). Therefore the invariant point should be not far from the $Tb_{3-x}Yb_xAl_5O_{12}$ ($x=0.8–0.9$) composition range.

The approach discussed in this and the previous subsections can be used to modify normally incongruently melting compounds for crystal growth from a stoichiometric melt. Substitution of Tb^{3+} in $Tb_3Al_5O_{12}$ crystals with other relatively small guest RE cations (Tm^{3+} or Er^{3+}) was also noted [37] and is suitable for the fabrication of stabilized TAG crystals. This way the data plotted in Figs. 5.20 and 5.21 can be used for the reference. In particular, substitution of Tb^{3+} with Tm^{3+} was also found to stabilize the garnet phase and can be described by a phase diagram similar to that shown in Fig. 5.24. However the amount of Tm^{3+} necessary for the preparation of uniform fibers was estimated to be greater than that of Lu^{3+} or Yb^{3+}. Therefore use of $(Tb,Tm)_3Al_5O_{12}$ in Faraday rotator applications is not desired, because of the relatively low content of Tb^{3+} in the crystals, and therefore the low value of rotation of the polarization plane in the material.

Future experiments will probably show the applicability of the approach described in this section for synthesis of other mixed crystals using other combinations of large and small RE. Generally, a decrease of the content of small RE in $(Tb,RE)_3Al_5O_{12}$ crystals is very important because the Tb^{3+} cation is responsible for the high magnitude of the Faraday rotation in Tb garnets.

5.6 KNbO$_3$ (KN) Based Perovskite Crystals

Potassium niobate ($KNbO_3$ or KN) is a well-known ferroelectric material for electro-optic, nonlinear optic, and photorefractive applications [4, 11, 56, 57]. It is an efficient material for doubling the frequency of near-infrared (Ga,Al)As diode lasers used for recording and reading data from optical compact disks. The information density of optical systems arranged with a frequency doubling crystal is expected to be four times greater than those of nonarranged ones. However, it is difficult to grow these crystals because $KNbO_3$ melts incongruently at a temperature above 1000 °C [58–61]. Therefore the crystals have to be grown from K_2O-rich, non-stoichiometric melts.

Moreover $KNbO_3$ is known to exist in three phases. The high-temperature phase crystallizes in the cubic perovskite structure. Within the temperature range 225–435 °C $KNbO_3$ has a tetragonal structure. At room temperature it is isostructural with the distorted perovskite form of $BaTiO_3$ and has an orthorhombic structure with two formula units per unit cell. Therefore it is also difficult to obtain high-quality crystals because of structural reordering that occurs during crystal cooling.

The flux growth technique is widely used to grow $KNbO_3$ crystals from the melts containing K_2O excess [56, 57]. However reproducibility of growth results is difficult to control because of nonstoichiometry of the starting mixtures and easy volatilization of K_2O from the melt and crystal surface [57, 58]. It was also reported [58] that single-crystal fibers of $K(Ta,Nb)O_3$ solid solutions were grown by the laser heated pedestal growth method (LHPG). The source rods were enriched with K_2O to prevent formation of the $K_4Nb_6O_{17}$ phase, which is stable in the $KNbO_3$ stoichiometric melt without excess of K_2O [59–61]. Similarly to the µ-PD system the temperature gradient in this method was high enough to prevent constitutional supercooling and therefore to avoid spontaneous nucleation on the liquid/solid interface. In the case of $K(Ta,Nb)O_3$ mixed crystals [58] the addition of a gas blower to the apparatus was necessary to increase and to control the gradient in the vicinity of the molten zone.

The µ-PD technique with resistive heating (Figs. 5.3 and 5.4) was also reported to be a versatile method of preparation of high quality fiber crystals of other niobate materials [62]. Therefore application of this technique for the growth of $KNbO_3$ fiber crystals from the melts of nonstoichiometric composition was studied in [8, 11]. The experimental procedure and the growth parameters which allow fabrication of relatively large $KNbO_3$ crystals are presented in the following sections.

5.6.1 $KNbO_3$ Fiber Crystals

Colourless, transparent, and crack-free crystals were produced from melts containing an excess of K_2O as a flux [8, 11]. The growth of relatively large-size (up to 2 mm in diameter and up to 30 mm in length) single crystals was found to be possible using crucibles with a corresponding nozzle size (up to 2.0 mm in outer diameter). The crystals were of reasonable optical quality, and second harmonic generation (SHG) was observed in crystals irradiated by fundamental beam with wavelength about 860 nm along the growth axis (*a*-axis).

The starting materials were made using K_2CO_3 and Nb_2O_5 powders both of 99.99 % purity. The constituents were carefully weighed and mixed by grinding in ethanol and dried at 100 °C for 5–10 hr. Special attention was paid to prepare water-free K_2CO_3. Therefore preliminary annealing of the starting K_2CO_3 at a temperature of 350–400 °C for 5–10 hr was absolutely necessary to control the composition of the starting mixture. The compositions of the melts used in the growth experiments are given in Table 5.4.

The schematic diagram of the µ-PD system and the details of the experimental technique are similar to those described above and in Ref. [62]. The crystals were grown under air atmosphere. The melt was contained in a crucible, which was made of Pt plate of 0.1 mm thickness and Pt pipe of 1.0–2.0 mm in outer diameter and a wall thickness of 0.05–0.01 mm, as shown in Figs. 5.3 and 5.4. Below, the modified arrangement (large nozzle diameter of about 2 mm) is called a pulling down (PD) system to distinguish it from the conventional (µ-PD) technique.

In the main, two variations of the seeding technique were used in the experiments described here. Solidification of the melt was started (i) on Pt wire of 0.5 mm in diameter or (ii) on Pt pipe of 0.4 mm in outer diameter and a wall thickens of 0.05 mm. Use of a $KNbO_3$ single-crystal seed was also possible, but it was difficult to keep the fiber quality at the seeding stage because of cracking which occurred during heating or seeding. In the case of seeding on the Pt wire one crystallographic direction of high growth rate was developed by applying of relatively high pulling-down rate (0.50–1.00 mm/min) and necking procedure (Fig. 5.2). Further precise orienting of the crystal was made during the growth process by manipulations of the micro X-Y horizontal stage. This was possible due to faceting of the crystals observed in situ using an optical microscope. The construction of the apparatus used in [8, 11] allows smooth change and precise adjustment of the seed orientation during the growth run without stopping the pulling process. Best results were achieved in the runs where the Pt tube was used as a seed similar to that described elsewhere [63] for crystal growth by the Czochralski method. As a first step the tube was inserted into the crucible nozzle and kept there for about 1 min. During that time overheating of the crucible was necessary to prevent solidification of the melt inside the nozzle because of the high thermoconductivity of platinum. Thereafter pulling down was started with a rate close to that used for crystal growth as given in Table 5.4.

Table 5.4. $KNbO_3$ crystal growth conditions: melt composition (mol.%), outer and inner diameters of the crucible nozzle (mm), seed material, pulling down rate (mm/min), and crystal length (mm) [11].

No.	K_2O	Nb_2O_5	Nozzle	Seed	Pulling rate	Length
12-2	40	60	1.2·1.0	Pt wire	0.35	35
1-2	50	50	1.2·1.1	Pt wire	0.11	32
4-2	50	50	1.2·1.0	Pt wire	0.10	8
3-3	52	48	1.2·1.0	Pt wire	0.08	10
7-1	54	46	1.2·1.0	Pt tube	0.30	50
7-2	54	46	1.4·1.3	No. 7-1	0.15	34
6-1	55	45	1.2·1.0	Pt wire	0.13	25
8-1	56	44	1.2·1.0	Pt wire	0.09	22
8-2	56	44	1.2·1.0	Pt tube	0.20	45
8-3	56	44	1.4·1.3	Pt tube	0.12	30
8-4	56	44	1.4·1.3	KTN (flux)	0.13	37
15-1	57	43	1.7·1.6	Pt tube	0.10	30
15-3	57	43	2.0·1.9	Pt tube	0.10	19
11-1	58	42	1.2·1.0	Pt wire	0.11	13
11-2	58	42	1.2·1.0	Pt tube	0.16	27
11-4	58	42	1.2·1.0	Pt tube	0.30	70
11-5	58	42	2.0x1.9	Pt tube	0.10	20
13-1	62	38	1.2x1.0	Pt tube	0.15	18

Fig. 5.31. KNbO₃ fiber crystal No. 11-2 of Tables 5.4 and 5.5 (scale in mm).

Crack formation was sometimes observed in the fibers listed in Table 5.5. The extremely high temperature gradient under the crucible nozzle and the phase transitions mentioned above were considered to be the main origin of the cracking. K(Nb,Ta)O$_3$ crystals grown by conventional flux technique were also used as seeds to prevent these disadvantages. In particular the crystal grown on K(Nb,Ta)O$_3$ seed (sample No. 8-4 of Tables 5.4 and 5.5) was crack-free. All growth processes were stopped after observation of any kind of crystal imperfection. After that the crystals were separated from the molten zone and pulled down with a velocity corresponding to the cooling rate of about 30 °C/min. Thereafter the crystals were removed from the seed holder. Figure 5.31 shows the as-grown KNbO$_3$ crystal. The fibers grown had a habit corresponding to published data [56]. The crystals showed simple crystallographic {100} faces because of the presence of flux. In the main, the rod-like crystals had four-fold symmetry corresponding to the [100] orientation of the pseudo-cubic high-temperature phase [64]. Similarly to the crystals grown by top-seeded solution growth, the µ-PD crystals had very flat cubic faces because progressive nucleation on cubic planes is known to be difficult [65]. As a second stage of each experiment the remaining melt was removed from the crucible using the same seeding material with a pulling rate of about 0.50 mm/min. The deposition of small drops of flux (K$_2$O) was sometimes observed on the surface of the crystals. The drops were removed using warm water.

According to [8, 11] the crystals were blue and colourless depending on the melt composition and pulling rate. In general optimization of crystal growth conditions was necessary for all of the melts, because at least a light blue coloration following from the presence of oxygen vacancies [56, 58] was observed in all crystals grown at a relatively high pulling rate. Almost all crystals were transparent, as shown in Figs. 5.2 and 5.31. The typical size of crystals was about 1–2 mm in cross-section, depending on the diameter of the nozzle, and a few centimeters in length. In the main about 70–80 vol.% of the melt was crystallized into KNbO$_3$ crystals. The maximum yield (crystal/melt volume ratio) achieved was about 90 vol.% for melts containing an insignificant excess of K$_2$O. It was also possible to grow the KNbO$_3$ single crystals with an extremely high pulling rate of about 1–2 mm/min. In this case the crystals were also single KNbO$_3$ phase and had typical four-fold symmetry. However these crystals were dark blue in colour (Fig. 5.32).

Table 5.5. KNbO$_3$ fiber crystal growth results: weight (W), diameter (D), colour, and phase of the crystals grown and remaining melts [11]

No.	W (mg)	D (mm)	Colour	Crystal	Remain Melt	Cracks
12-2	150	1.0	colourless	K$_4$Nb$_6$O$_{17}$	K$_4$Nb$_6$O$_{17}$	Yes
1-2	–	1.0	colourless	K$_4$Nb$_6$O$_{17}$	K$_4$Nb$_6$O$_{17}$	-
4-2	–	1.1	blue	KNbO$_3$	–	-
3-3	39	1.1	colourless	KNbO$_3$	K$_4$Nb$_6$O$_{17}$	Yes
7-1	206	1.1	colourless	KNbO$_3$	–	Yes
7-2	171	1.3	dark blue	KNbO$_3$	–	Yes
6-1	94	1.1	colourless	KNbO$_3$	K$_4$Nb$_6$O$_{17}$	No
8-1	77	1.1	light blue	KNbO$_3$	K$_4$Nb$_6$O$_{17}$	No
8-2	207	1.1	colourless	KNbO$_3$	–	No
8-3	177	1.3	colourless	KNbO$_3$	–	No
8-4	236	1.4	light blue	KNbO$_3$	–	No
15-1	246	1.8	colourless	KNbO$_3$	K$_4$Nb$_6$O$_{17}$	No
15-3	239	2.1	light blue	KNbO$_3$	K$_4$Nb$_6$O$_{17}$	No
11-1	52	1.2	light blue	KNbO$_3$	K$_4$Nb$_6$O$_{17}$	Yes
11-2	93	1.1	colourless	KNbO$_3$	KNbO$_3$	No
11-4	255	1.1	light blue	KNbO$_3$	–	No
11-5	231	2.1	colourless	KNbO$_3$	–	No
13-1	57	1.0	blue	KNbO$_3$	KNbO$_3$	Yes

Phase homogeneity of the crystals grown and the melts remaining after growths were studied by X-ray powder diffraction analysis. In the main the crystals grown were KNbO$_3$ single phase (JSPDS data card No. 32-822) [32] as given in Table 5.5. However, the remaining melts were found crystallized as either KNbO$_3$ or K$_4$Nb$_6$O$_{17}$ depending on the composition of the starting mixture. In the melts corresponding to the vicinity of the stoichiometric composition of KNbO$_3$ crystallization of the second phase was often observed. It was assumed this the phase is K$_4$Nb$_6$O$_{17}$.

Fig. 5.32. KNbO$_3$ crystal with high concentration of oxygen vacancies (scale in mm).

5.6.2 Flux Growth of KNbO$_3$ Fiber Crystals

A comparison of the most common features of PD and related crystal growth techniques is given in Table 5.6. The two most important advantages of the μ-PD technique modified by increasing of the diameter of the capillary channel (PD) are discussed below briefly. One of the most significant results of the study reported in Ref. [11] seems to be related with the ability of the PD technique to produce single crystals in the presence of considerable amounts of flux in the starting melts. For example, recalculation of the melts used for the growth of crystals No. 11-5 and No. 13-1 (Table 5.4) gives KNbO$_3$: K$_2$O = 84 : 16 and KNbO$_3$: K$_2$O = 76 : 24 molar ratios, respectively. Thus, the systems correspondingly contained 16 mol.% and 24 mol.% of K$_2$O flux.

In the case of conventional μ-PD growth reported elsewhere [62] the crucibles were fabricated with a nozzle diameter less than 1 mm. In this case segregation was not observed because of the low mass transport inside the narrow capillary channel. Therefore the intensity of the cations exchanged between the melt positioned in the vicinity of the solid/liquid interface and the melt as a whole was also very low. The same phenomenon is usually observed in LHPG crystal growth [58]. In both these methods the segregation coefficients reported were close to unity: $k \approx 1$. However, in the modified PD arrangement examined in [8, 11] the diameter of the capillary channel has been increased considerably up to 2 mm which is close to the size of the crucible (about 10 · 5 · 2 mm, as shown in Figs. 5.3 and 5.4). In this way, the rate of natural convection has also been increased, and segregation on the liquid/solid interface was observed to become possible ($k \neq 1$). Thus the modification discussed here results in the unusual possibility of flux growth with a comparatively high pulling rate (0.1–1.0 mm/min).

Another important result is related with the examination of macro-limitations of the μ-PD system. In all previous reports concerning oxide crystal growth by the μ-PD technique with resistive heating the diameter (or cross-section) of the fibers reported did not exceed 1 mm. The maximal size of the KNbO$_3$ crystals grown in this study was greater than 2 mm. Therefore fields of application of the PD technique and the crystals discussed here are assumed to increase considerably. The main problems of KNbO$_3$ fiber growth were related with the control of the reproducibility of growth results because of (i) nonstoichiometry of the starting mixtures, (ii) easy volatilization of K$_2$O [57, 58], and (iii) phase transitions observed in KNbO$_3$. Some of these disadvantages can be avoided by appropriate modification of the crystal and melt compositions discussed below.

Table 5.6. Comparison of PD and related techniques.

Methods	μ-PD and LHPG*	PD	Flux Growth
Crystal Diameter	~ 0.5 mm	≥ 2.0 mm	≥ 10 mm
Segregation Coefficients	$k \approx 1$	$k \neq 1$	$k \neq 1$
Flux Growth	Difficult because K ≈ 1	Possible	Possible
Growth Rate (mm/min)	Very high (0.1–1.0)	Very high (0.1–1.0)	Very low (< 0.01)
Growth control	Easy	Easy	Difficult

*LHPG – laser heating pedestal growth

5.6.3 Growth of Ta- and Li-Doped KNbO₃ Fiber Crystals

In spite of the improved structural properties, K(Ta,Nb)O$_3$ (KTN) mixed crystals grown by the LHPG method contained inclusions and cracks, and were dark blue or black [58] in color. Therefore an attempt was made to produce these crystals by the μ-PD technique also [8]. The starting materials made using K$_2$CO$_3$, Li$_2$CO$_3$, Ta$_2$O$_5$, and Nb$_2$O$_5$ of 99.99 % purity were mixed together (10–15 g) in ethanol and dried at 100 °C similar to those used for the growth of undoped KNbO$_3$ fibers. The melt compositions are summarized in Table 5.7. The crystals were grown under air atmosphere with a pulling down rate ranging from 0.1 to 1.0 mm/min. The crystal diameter was adjusted manually by selection of suitable temperatures of the crucible, afterheater and the pulling rate. The nozzle/crystal diameter ratio was in the range 0.2–1.0. The crystal growth parameters are given in Table 5.8.

No significant difference between KN and KTN fiber growth conditions and results was observed at least at relatively low substitution of 5–10 % Ta^{5+} instead of Nb^{5+}. Melts No. 18 and 19 from Table 5.7 with the corresponding growth parameters in Table 5.8 give examples of the crystals grown. The KTN crystal No. 19-1 was not of uniform color ranging from colourless to dark blue. In spite of the coloration, this crystal was successfully used as a seed material for the growth of colourless Li-containing crystals described later. In the case of growth of Li-doped fibers using KTN seed the coloration of the resulting fiber was successfully suppressed almost immediately after seeding at about 1 mm of the crystal length.

Table 5.7. Ta- and Li-doped KNbO$_3$ growth conditions and results: melts compositions (mol.%), phase of the crystals grown and remaining melts (two-phase), and color of the crystals [8].

No	K₂O	Ta₂O₅	Nb₂O₅	Li₂O	Crystal	2-phase	Colour
16	54	–	43	3	KN	KN	Colourless
17	52	–	46	2	KN	KN	Colourless
18	58	2	40	–	KN	KN	Blue, colourless
19	56	4	40	–	KN	KN	Blue, colourless
20	51	–	47	2	KN	KN	Blue, colourless
21	50	–	46	4	KN	KN	Blue, colourless
22	52	–	44	4	KN	KN	Colourless
23	52	–	42	6	KN	KN	Colourless
24	50	–	44	6	KN	KN	Colourless

Table 5.8. Ta- and Li-doped KNbO$_3$ growth parameters and results: outer and inner diameters of the crucible nozzle (mm), seeds, maximal pulling rate (mm/min), resulting macroperfect crystal growth, average crystal diameter (mm), and crystal length (mm) [8].

Crystal*	Nozzle Diameter	Seed	Pulling Rate	Crystal Diameter	Crystal Length
Li:KN-16-3	1.5 × 1.4	KN:10%Ta	0.15	1.0	35
Li:KN-17-5	1.6 × 1.5	KN:10%Ta	0.30	0.9	100
KTN-18-3	1.2 × 1.0	KN:5%Ta	0.30	0.8	45
KTN-19-1	1.7 × 1.6	KN:5%Ta	0.30	0.9	45
Li:KN-20-1	1.5 × 1.4	KN:5%Ta	0.14	1.4	25
Li:KN-21-1	1.7 × 1.6	KN:10%Ta	0.15	0.9	30
Li:KN-22-1	1.5 × 1.4	KN:10%Ta	0.40	0.7	55
Li:KN-23-1	1.5 × 1.4	KN:10%Ta	0.40	0.8	170
Li:KN-24-1	1.2 × 1.0	KN:10%Ta	0.40	0.9	100

* The first number represents the melt composition given in Table 5.7.

The blue coloration of KN crystals caused by the formation of oxygen vacancies [8, 56, 58, 66–68] is accompanied by the formation of Nb^{4+} [8, 58, 66]. Therefore, it is evident that suppression of the coloration can be achieved by substitution of Nb^{5+} cations located in octahedral sites of the perovskite structure with those whose charge is less than 5+. Fortunately there are many cations in nature which satisfy both the following conditions: low charge (C < 5+) and size suitable for coordination number CN = 8 [50]. The addition of a small amount of Li$_2$O into the starting mixture was reported [8] to be a useful approach preventing both the blue coloration [68] and K$_4$Nb$_6$O$_{17}$ second phase formation generally observed in KNbO$_3$ crystals grown from undoped K$_2$O-Nb$_2$O$_5$ binary melts [8, 69]. A view of the Li-doped KN fibers is given in Fig. 5.33.

The phase diagram of the K$_2$O-Nb$_2$O$_5$-Li$_2$O ternary system together with the range of melt compositions allowing the growth of colourless Li-doped KNbO$_3$ crystals at pulling down rates of up to 0.50 mm/min was reported in [8]. It was found that substitution of 1–3 mol.% of Nb$_2$O$_5$ with Li$_2$O is generally necessary to avoid blue coloration of the crystals. Some details of the crystal growth procedures for Li-doped KN fibers can be found in Tables 5.7 and 5.8.

Thus, the growth of KN crystals from melts with compositions different from those of crystals (flux growth) is considered to be one of the most important conclusions following from fiber growth experience described in Refs. [8, 11]. Some of the melts (especially Nos. 11, 13, and 23) are located relatively far from the stoichiometric point of KNbO$_3$ (Tables 5.4 and 5.7), which corresponds to a significant difference between the starting liquid and resulting solid phases, and therefore to the flux nature of systems applied for crystal growth.

Thus, similar to YAG:Nd and YAG:Yb fiber crystal growth discussed above [5, 9], the segregation was confirmed to be possible in the μ-PD method with resistive heating also. Another important result is related with the examination of macro-limitations of the μ-PD method following from the same increase of the nozzle diameter. It was also suggested in Ref. [8] that it is possible to produce KN crystals by the Edge-defined, Film-fed Growth (EFG) method [54] because the μ-PD technique can be considered as a download oriented EFG arrangement with a narrow capillary channel.

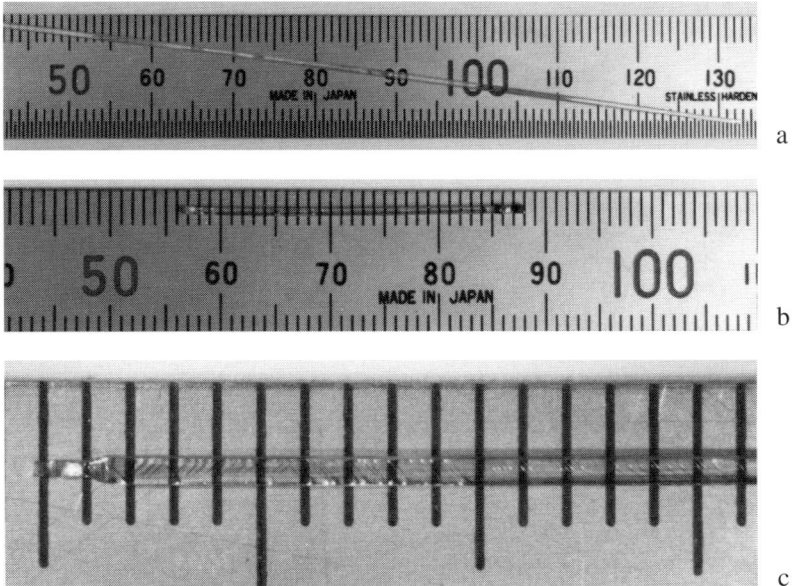

Fig. 5.33. Crystals of Li:KN-17-5 (**a**), Li:KN-21-1 (**b**), and Li:KN-22-1 (**c**) from top to bottom (scales in mm) grown according to the procedure described in [8]. The melt compositions and growth conditions are given in Tables 5.7 and 5.8.

5.7 β-Ga$_2$O$_3$ Fiber Crystals

Gallium oxide, β-Ga$_2$O$_3$ (β-gallia), is another optical material that was shown can be produced in fiber-like form by the μ-PD technique with RF heating. β-Ga$_2$O$_3$ is a wide band gap compound with a forbidden energy gap of about 4.8–4.9 eV, which exhibits semiconductor and luminescence properties. It is generally an n-type semiconductor due to the presence of oxygen vacancies. The structure of β-Ga$_2$O$_3$ is monoclinic with $a = 12.23$, $b = 3.04$, $c = 5.80$Å, $b = 103.7°$ [70, 71]. Ga^{3+} cations were reported [72] to be located in two crystallographically nonequivalent cation sites with octahedral and tetrahedral geometry in equal quantities. Four Ga$_2$O$_3$ molecules are contained in each unit cell. β-Ga$_2$O$_3$ is a transparent conducting oxide, which has potential applications in optoelectronic devices. It is also a promising sensor material for detecting oxidizing and reducing gases [73]. This crystal has high thermal and chemical stability.

There are very few reports on melt growth of β-Ga$_2$O$_3$ crystals. They include Verneuil [74], floating zone (FZ) [70, 75, 76], Czochralski (CZ) [77], and flux growth [72, 78] techniques. Attempts to produce β-Ga$_2$O$_3$ bulk crystals by edge-defined film fed growth (EFG) [79] were unsuccessful due to the unusual behavior of the melt. It is also known [74] that the crystals can be easily cleaved along the (100) plane into plate-like samples of few mm thick. An additional problem of β-Ga$_2$O$_3$ crystal growth is related to intensive evaporation of the Ga$_2$O$_3$ melt at a melting temperature of about 1900 °C.

The micro pulling down (μ-PD) technique with an RF heating system is considered here to be an alternative technique, which allows fabrication of high-quality β-Ga$_2$O$_3$ fiber crystals [12]. The β-Ga$_2$O$_3$ crystals described here were grown by a procedure similar to that introduced earlier for YAG and TGG single-crystal fiber growth [5, 9, 10]. Ga$_2$O$_3$ powder of 99.999 % purity was used as the starting material. Growth was performed at pulling rates in the range of 1–2 mm/min using Ir crucibles with an orifice of 0.6 mm in diameter in an argon atmosphere. A schematic of the procedure was given earlier in Figs. 5.5–5.7.

As a first step the YAG fibers, produced by the same technique were used as a seed. Thereafter the crystals were grown on the β-Ga$_2$O$_3$ fibers produced in previous experiments. It is pointed out that the growth of the first β-Ga$_2$O$_3$ fiber was extremely difficult because of the lack of a suitable seed material. Immersion of the YAG fiber seed into the Ga$_2$O$_3$ melt resulted in fast dissolution of the seed. The melting point of the resulting mixture (YAG + Ga$_2$O$_3$) was visually observed to be much lower than the melting point of both YAG and Ga$_2$O$_3$ origin materials probably due to the eutectic nature of the YAG-Ga$_2$O$_3$ system. Therefore great overheating of the resulting melt was observed with subsequent separation of the seed from the melt.

Typical fiber crystals are shown in Fig. 5.34. The fibers were grown up to 400 mm in length and 1–2 mm in diameter. In the main, the crystals were slightly blue in color because of the presence of oxygen vacancies, which are easily formed when the material is prepared under a reducing atmosphere [71, 75]. At the tail part of the fibers the coloration of the crystals was negligible. The fibers were studied using scanning electron microscopy (SEM), and were found to be uniform in composition along the radial direction. Any inclusions of the second phase following from reduction of Ga$_2$O$_3$ because of atmospheric conditions or phase transformations were not detected under SEM observation (Fig. 5.35). The fibers were slightly bent as can be seen in Fig. 5.34. The reasons for the bending are not evident. Except the tail part of about 50–100 mm, the as-grown fibers were coated with β-Ga$_2$O$_3$ powder, which was deposited through volatilization of the material from the melt surface, followed by transport of the vapor in the argon gas flow to the surface of the just-grown fibers. The powder was easily removed from the fiber surface with ethanol and water.

Main problem of β-Ga$_2$O$_3$ crystal growth was associated with the relatively high wetting of the Ga$_2$O$_3$ melt on the Ir crucible. This property leads to significant growth instability, which affects crystal diameter control. Controlled minimization of the melt overheating, however, allows decreasing of the thickness of the molten zonel with corresponding decreasing of the fiber diameter. Therefore, the diameter control was similar to that reported for the growth of TGG garnet fiber crystals from the wettable melt [6].

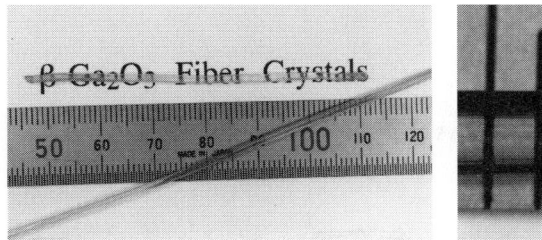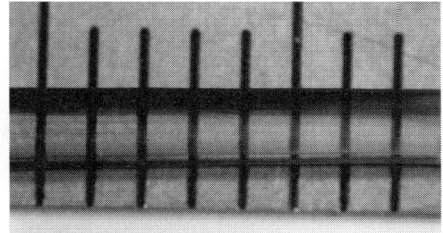

Fig. 5.34. β- Ga$_2$O$_3$ fiber crystals (scales in mm)

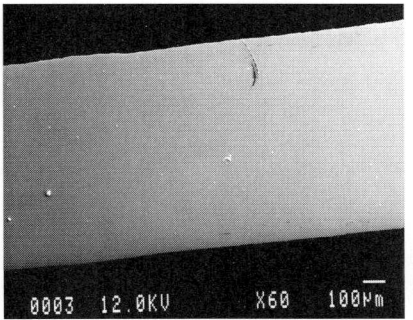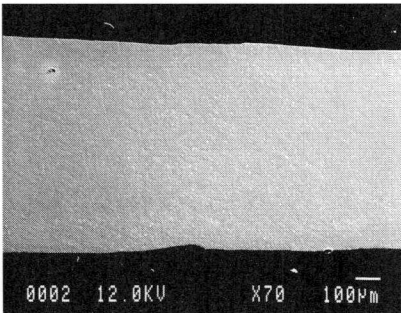

Fig. 5.35. Images of β-Ga$_2$O$_3$ fiber crystals made using a scanning electron microscope (SEM).

5.8 Other Materials

The fiber crystal growth discussed above was mainly considered as an alternative technique that allows the production of materials with new shape and sometimes unusual physical and technical parameters. It was assumed that this technique could be used in the production of novel optical materials. However, the results related with the study of segregation, phase diagrams, and development of novel oxide crystals discussed above show that the μ-PD system can be considered as a very useful materials science tool. The conduct of preliminary growth experiments allowing pilot examination of the melt properties before the growth of bulk crystals by conventional techniques was found to be a very useful strategy for most materials introduced above. At present there are a lot of examples of such application of the μ-PD technique. Many new crystals were first obtained in the shape of μ-PD fibers with further growth of large bulk single crystals using conventional bulk crystal growth methods.

Langasite (La$_3$Ga$_5$SiO$_{14}$; LGS) type crystals are good examples of the application of this scheme. Langasite and its isostructural aliovalent analogs, La$_3$Nb$_{0.5}$Ga$_{5.5}$O$_{14}$ (LNG) and La$_3$Ta$_{0.5}$Ga$_{5.5}$O$_{14}$ (LTG) crystals have received much attention as new candidates for piezoelectric device applications [3, 43, 80–85]. The LGS related materials have a Ca$_3$Ga$_2$Ge$_4$O$_{14}$-type structure with space group P321.

There are four kinds of cation sites in this structure. Therefore, this structure can be described by the chemical formula $A_3BC_3D_2O_{14}$. In this formula, A and B represent the decahedral (twisted Thomson cube) site coordinated by eight oxygen anions, and an octahedral site coordinated by six oxygen anions, respectively. C and D represent tetrahedral sites coordinated by four oxygens; the size of the C site is slightly larger than that of the D site.

Various isovalent and aliovalent substitutions in a given structure are quite interesting, and could, perhaps, also result in the appearance of useful physical properties (optical, piezoelectric, etc). It is evident that materials of this structural family have a wide range of possible substitutions due to the wide variety of possible cation localization in the structure. Therefore there is a good possibility of finding new compounds that probably will result in improved properties necessary for advanced industrial applications.

At the same time, langasite-type crystals can be grown from Pt or IR crucibles using the conventional CZ technique with RF heating. High-quality LGS, LNG and LTG single crystals of approximately 2 and 3 inches in diameter were grown by the CZ technique. These crystals were obtained at a relatively high pulling rate of 1 mm/h free of cracks and inclusions.

However, both the development of new crystals and the optimization of growth procedure for known materials require a lot of time and resources in the case when exceptionally expensive techniques such as CZ are used. Examples of optimization of melt composition using the CZ method are given in [80, 82], where the crystals were produced using a full charge of crucibles (in the range of about 1 kg of starting materials) and standard growth runs (5–7 days). On the other hand, preliminary results can be easily obtained using growth of micro-crystals by the μ-PD method using a negligible amount of high-quality starting materials (1–5 g only) and a short period of time (3–8 hours). Examples of such application of the μ-PD method for the development of new langasite-type crystals and a preliminary test of the crystal growth conditions are given in [83, 84].

The result of growth of Lu- and Yb-substituted TAG fiber crystals given in Figs. 5.22, 5.23, 5.30, and Table 5.2 can also be considered as preliminary optimization of the melt composition and growth parameters for further application of the results in conventional bulk crystal growth. Below some other examples of application of the μ-PD as a materials science instrument are described.

5.8.1 $Ca_3(Li,Nb,Ga)_5O_{12}$ (CLNGG) Garnet Crystals

The development of $Ca_3(Li,Nb,Ga)_5O_{12}$ garnet bulk crystal growth using the preliminary fiber crystal growth results reported in [13] is another example of application of the μ-PD technique as a research tool. The purpose of that study was to develop a new substrate material for liquid phase epitaxy of Bi-containing iron garnet films with the general formula $RE_{(3-x)}Bi_x(Fe,Ga)_5O_{12}$, where RE = rare-earth elements. These films are important materials for application in various magneto-optical devices due to the high magnitude of Faraday rotation [42, 86], and are normally produced by liquid phase epitaxy using multicomponent flux systems [87–89]. The incorporation of a large Bi^{3+} cation into the garnet structure demands preparation of nonmagnetic garnet substrates with high transparency and large lattice parameters.

To develop new bulk garnet crystals the starting materials were first examined using μ-PD growth processes. The desired quantities of $CaCO_3$, $SrCO_3$, Nb_2O_5, and Li_2CO_3 of at least 99.99 % purity were weighed and mixed. The mixtures were then pressed to pellets and heated for about three hours at 1350 °C or for 10 min at 1460 °C. The phase purity and lattice parameters of all ceramic materials obtained through this thermal treatment were studied using X-ray powder diffraction. Concentration of the second phase was estimated using a relative intensity ratio $I_{(2nd)}/I_{(420)}$, where $I_{(2nd)}$ is the intensity of the largest peak of the nongarnet phase, and $I_{(420)}$ is the intensity of the greatest peak of the garnet phase (420). It was assumed that this ratio is proportional to the amount of nongarnet phase in the resulting material, and therefore was used to estimate the applicability of the corresponding mixture and melt for the preparation of single garnet phase fiber. The composition, lattice parameter and $I_{(2nd)}/I_{(420)}$ ratio for selected starting mixtures are given in Table 5.9.

Table 5.9. Garnet phase formation in the $Ca_3(Li,Nb,Ga)_5O_{12}$ system: chemical composition, lattice parameter and relative intensity ratio $I_{(2nd)}/I_{(420)}$ of second phase in the garnets synthesized by solid state reaction [13].

Composition	Lattice Parameter (Å)	$I_{(2nd)}/I_{(420)}$
$Ca_3Li_{0.00}Nb_{1.50}Ga_{3.5}O_{12}$	12.51	0.01
$Ca_3Li_{0.20}Nb_{1.75}Ga_{3.0}O_{12}$	12.52	0.01
$Ca_3Li_{0.30}Nb_{1.80}Ga_{2.9}O_{12}$	12.54	0.00
$Ca_3Li_{0.50}Nb_{2.0}Ga_{2.5}O_{12}$	12.55	0.02
$Ca_3Li_{0.75}Nb_{2.25}Ga_{2.0}O_{12}$	12.55	0.19
$Ca_3Li_{1.75}Nb_{3.25}Ga_{0.0}O_{12}$	–	Non-garnet

It is evident that the $Ca_3Li_{0.30}Nb_{1.80}Ga_{2.9}O_{12}$ composition obtained by partial substitutions illustrated in Table 5.9 is the best one in accordance with the requirements for possible bulk crystal growth. The high phase purity and low melting point ($T_m \sim 1450$ °C) detected at the stage of solid state synthesis allowed the authors of [13] to continue study of the formation of this material from the liquid phase by μ-PD and CZ methods using Pt or Pt/Rh crucibles. Details of garnet phase formation and fiber and bulk crystal growth from the melts with composition close to that of $Ca_3Li_{0.3}Nb_{1.8}Ga_{2.9}O_{12}$ are given below.

First the fiber single crystals of $Ca_3Li_xNb_{(1.5+x)}Ga_{(3.5-2x)}O_{12}$ (CLNGG), where x = 0.0–0.5, were grown by the μ-PD method using a Pt/Rh crucible in an air atmosphere. The shape of the μ-PD crucible was modified similarly to that discussed for the KN fiber crystals to allow intensive mass transport between different parts of the melt. The diameter of the capillary channel was increased up to 0.8 mm, and the length of the nozzle was decreased considerably to be about 1 mm. A moderate pulling down rate of 3–18 mm/hr was applied to produce the fibers listed in Table 5.10. The pre-calcinated or pre-melted polycrystals prepared by solid-state reaction were used as starting materials. The parameters of the fiber crystals grown are given in Table 5.10. Examples of the CLNGG fiber crystals grown by the μ-PD method from the melts of different composition are given in Fig. 5.36.

Generally use of the raw materials containing significant amounts of second phase ($I_{(2nd)}/I_{(420)} > 0.03$ according to Table 5.9) did not result in high-quality fiber crystals. In particular, a crystal of 15 mm in length and 0.8 mm in diameter produced from the $Ca_3Li_{0.30}Nb_{1.80}Ga_{2.9}O_{12}$ melt was a transparent and crack-free single garnet phase fiber. However, second phase formation was detected in the solidified melt remaining after the growth.

In the main formation of the non-garnet phases was not observed in the fibers listed in Table 5.10. These phases were detected only in the starting raw materials and in the solidified melt remaining in the crucible. Single garnet phase formation was observed in the mixtures with concentration of Li^+ cations between 0.27 and 0.30. In the solidified remaining melts, the amount of second phase was minimal for the growth performed using the $Ca_3Li_{0.275}Nb_{1.775}Ga_{2.95}O_{12}$ mixture. Thus, the $Ca_3Li_{0.275}Nb_{1.775}Ga_{2.95}O_{12}$ melt was found to have minimum content of non-garnet phases in (i) the starting raw material, (ii) grown fiber crystal, and (iii) solidified remaining melt.

Table 5.10. $Ca_3(Li,Nb,Ga)_5O_{12}$ μ-PD fiber crystal growth: melt composition, maximal growth (pulling) rate f (mm/h), density of cracks D, total crystallization fraction C (wt. %), average lattice parameter a (Å), and relative intensity ratio $I_{(2nd)}/I_{(420)}$ of the second phase for the raw materials (RM) and solidified remaining melts (SM) after growth [13].

Melt Composition	f	D	C	a	RM	SM
$Ca_3Li_{0.00}Nb_{1.50}Ga_{3.5}O_{12}$	3	Very high	21	–	0.011	0.193
$Ca_3Li_{0.25}Nb_{1.75}Ga_3O_{12}$	12	High	49	12.541	0.014	0.012
$Ca_3Li_{0.2625}Nb_{1.7625}Ga_{2.975}O_{12}$	12	Low	37	12.542	0.005	–
$Ca_3Li_{0.27}Nb_{1.77}Ga_{2.96}O_{12}$	12	Low	51	12.541	0.000	0.005
$Ca_3Li_{0.275}Nb_{1.775}Ga_{2.95}O_{12}$	15	Very low	63	12.540	0.000	0.005
$Ca_3Li_{0.2875}Nb_{1.7875}Ga_{2.925}O_{12}$	12	High	27	12.541	0.000	0.132
$Ca_3Li_{0.30}Nb_{1.80}Ga_{2.9}O_{12}$	12	Low	16	12.540	0.000	–
$Ca_3Li_{0.50}Nb_{2.0}Ga_{2.5}O_{12}$	6	Low	8	12.542	0.022	0.037

Fig. 5.36. Fig. 5.36. View of the μ-PD fiber crystals grown from $Ca_3Li_{0.25}Nb_{1.75}Ga_3O_{12}$ (left) and $Ca_3Li_{0.275}Nb_{1.775}Ga_{2.95}O_{12}$ (right) melts (scales in mm).

It is evident that as a result of segregation, the amount of non-garnet phase in the solidified melt was much higher than that of raw materials. The impurities that form non-garnet phases were rejected in the μ-PD system from the solid-liquid interface, and traveled toward the melt in the crucible passing through the capillary tube. Thus, the conditions allowing transport of the particles and segregation were considered in [13] to be similar to those observed in conventional CZ growth. Moreover, the melt fraction solidified into a single crystal was also maximal (more than 60 %) for the particular $Ca_3Li_{0.275}Nb_{1.775}Ga_{2.95}O_{12}$ composition.

In this way, three generally independent parameters were observed to the be best ones for the $Ca_3Li_{0.275}Nb_{1.775}Ga_{2.95}O_{12}$ melt as follows: (1-2) maximal growth rate and maximal solidification fraction resulting in high-quality transparent single-crystal growth, and (3) best garnet phase purity of the raw material and remaining melt. For these reasons it was concluded [13] that the $Ca_3Li_{0.275}Nb_{1.775}Ga_{2.95}O_{12}$ compound is closest to congruently melting composition within the mixtures listed in Tables 5.9 and 5.10. Therefore this starting mixture was selected as the basic one for CZ bulk crystal growth. Two types of CLNGG crystals were produced by the CZ technique from slightly different $Ca_3Li_{0.25}Nb_{1.75}Ga_3O_{12}$ and $Ca_3Li_{0.275}Nb_{1.775}Ga_{2.95}O_{12}$ initial melts to compare the growth results. The crystals were transparent (Fig. 5.37) and slightly yellow. It was reported [13] that the crystals are suitable for substrate applications because no optical absorption was observed in the 400–1200 nm wavelength range.

Both crystals had some cracks depending on the melt composition. However, the density of cracks in the crystal grown using the $Ca_3Li_{0.275}Nb_{1.775}Ga_{2.95}O_{12}$ starting mixture was lower than that using $Ca_3Li_{0.25}Nb_{1.75}Ga_3O_{12}$. The phase composition of solidified melts remained in the crucibles was reported [13] to be in the range 0.05 < $I_{(2nd)}/I_{(420)}$ < 0.06 for the $Ca_3Li_{0.25}Nb_{1.75}Ga_3O_{12}$ melt and $I_{(2nd)}/I_{(420)} = 0$ for the $Ca_3Li_{0.275}Nb_{1.775}Ga_{2.95}O_{12}$ starting melt. Bubbles and inclusions were not detected in both crystals shown in Fig. 5.36. Thus it was proved that the results of both μ-PD and CZ techniques are well matched each to other. In this way it was again shown that preliminary results of μ-PD fiber growth can be used to predict and to optimize the growth conditions for further bulk crystal growth.

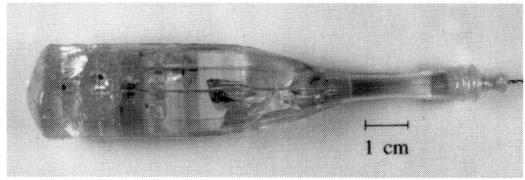

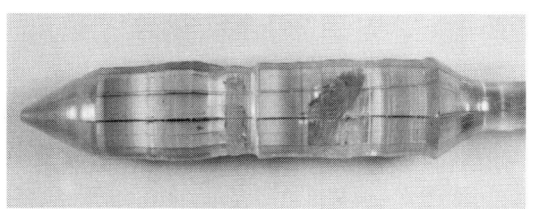

Fig. 5.37. View of the CZ bulk crystals grown from $Ca_3Li_{0.25}Nb_{1.75}Ga_3O_{12}$ (above) and $Ca_3Li_{0.275}Nb_{1.775}Ga_{2.95}O_{12}$ (below) melts.

1 cm

5.8.2 Vanadium Garnet Crystals

Some members of the family of complex vanadates $\{A_3\}[B_2](V_3)O_{12}$ with the garnet structure [14, 43, 90–95] are other examples of solid state materials produced in the form of fiber single crystals using the μ-PD method. In the above general formula the cations that occupy the c, a, and d positions with coordination numbers 8, 6, and 4 are enclosed in braces, brackets, and parentheses, respectively. Unfortunately there are very few reports concerning the crystal growth of vanadium garnets.

In particular the $NaCa_2Mg_2V_3O_{12}$ crystal reported in Ref. [90] has a garnet structure and melts congruently at a temperature of about 1165 °C. Details of the crystal growth were not found in the literature. Therefore an attempt was made to produce vanadium garnet fiber crystals with the defective structure highly concentrated by cation vacancies using the μ-PD system [14]. The precalcinated polycrystals prepared from corresponding oxides and carbonates with a purity of 99.99 % were used as starting materials. Growth of the fiber crystals was performed with a pulling down rate of 1.2–6.0 mm/hr, using a Pt/Rh crucible with 0.8 mm nozzle diameter in an air atmosphere. The melt compositions and crystal growth parameters are given in Table 5.11. In particular, the crystals of stoichiometric $NaCa_2Mg_2V_3O_{12}$ garnet (No. 4 of Table 5.11) grown in air had a gray colour. It was assumed that the coloration was observed due to the presence of V^{4+} cations in the garnet structure [90].

As an attempt to reduce V^{4+} formation, fiber crystals were produced from different melts, listed in Table 5.11. Na-rich crystals had a gray colour and low transparency. On the contrary, the $Na_xCa_{2.5-x/2}Mg_2V_3O_{12}$ crystals grown from the Ca-rich melts (x = 0.4, 0.2, and 0.0) were transparent and colourless. The pulling-down rate was 6 mm/h, and the formation of non-garnet phases was not observed in the crystals.

Variations of the lattice parameter on solidification fraction were measured using X-ray diffraction. The lattice parameters of the fibers grown from $Na_xCa_{2.5-x/2}Mg_2V_3O_{12}$ melts (x = 0.0–0.2) were practically constant along the growth axis [14]. However, those of the crystals grown from the Na-rich melts (x = 0.4, 0.9, and 1.0) continuously decreased with increasing solidification fraction. Evaporation of the sodium from the melt was assumed to be the main reason of the decreasing of the parameters of the crystals grown from the melts with $0.4 < x < 1.0$. In this way, both the crystal quality and lattice parameter uniformity increased considerably with increasing [Ca]/[Na] concentration ratio in the melt. Large lattice parameters at low solidification fraction observed in the fibers produced from the Na-rich melt followed from incorporation of large Na^+ cations into dodecahedral sites of the garnet structure in accordance with the stoichiometric formula of $NaCa_2Mg_2V_3O_{12}$. However, at higher solidification fraction the higher amount of Na was evaporated. In the limit, the melt composition became $Ca_2Mg_2V_3O_{11.5}$ which corresponds to a $\{Ca_{2.25}Mg_{0.25}\}Mg_2V_3O_{12}$ melt containing negligible excess of V_2O_5 as a flux. Flux growth in the μ-PD system as discussed above is possible. Therefore, the lattice parameters of the fibers produced at the final stage of growth were less than those measured for stoichiometric $NaCa_2Mg_2V_3O_{12}$ [94] garnet compounds due to the presence of small Mg^{2+} cations in dodecahedral sites. In another limiting case of $Ca_{2.5}Mg_2V_3O_{12}$ (No. 10 of Table 5.11) the melt composition was stable throughout the procedure and no changes of lattice parameters were detected.

Table 5.11. $Na_xCa_{2.5-x/2}Mg_2V_3O_{12}$ μ-PD fiber crystal growth: melt composition, average diameter (D), length (L), and average lattice constant (a) of fiber crystals grown at a pulling rate of 6 mm/h [14].

No.	Melt Composition	D, mm	L, mm	a, Å	Colour
1	$Na_{1.1}Ca_{1.9}Mg_2V_3O_{11.95}$	1.1	24	12.468	Grey
2	$Na_{1.05}Ca_{1.95}Mg_2V_3O_{11.975}$	1.0	23	12.467	Grey
3	$Na_{1.05}Ca_2Mg_2V_{2.95}O_{11.9}$	1.0	25	12.473	Grey
4	$NaCa_2Mg_2V_3O_{12}$	0.8	77	12.469	Grey
5	$Na_{0.9}Ca_{2.05}Mg_2V_3O_{12}$	0.8	22	12.471	Grey
6	$Na_{0.8}Ca_{2.1}Mg_2V_3O_{12}$	1.2	20	12.472	Grey
7	$Na_{0.6}Ca_{2.2}Mg_2V_3O_{12}$	1.0	19	12.474	Grey
8	$Na_{0.4}Ca_{2.3}Mg_2V_3O_{12}$	0.8	34	12.468	Colourless
9	$Na_{0.2}Ca_{2.4}Mg_2V_3O_{12}$	0.8	37	12.470	Colourless
10	$Ca_{2.5}Mg_2V_3O_{12}$	0.8	45	12.469	Colourless

Thus, it was found that the transparency and uniformity of the lattice parameters of the crystals increases with decreasing Na^+ concentration in the melt. Therefore crack-free, transparent, and colourless fiber crystals with a size of 0.8~1.2 mm in diameter and 20~80 mm in length were grown from the melts of $Na_xCa_{2.5-x/2}Mg_2V_3O_{12}$ where $x = 0.0-1.0$. It was found that the uniformity of the lattice parameters on the solidification fraction increase with decreasing Na content in the starting melts, and best the uniformity was observed in $Ca_{2.5}Mg_2V_3O_{12}$ crystals with deviations less than ± 0.001 Å.

5.8.3 Rare-Earth Vanadates

The rare-earth orthovanadates $REVO_4$ (RE = Lu, Er, Dy, and/or Gd) are also well-known functional materials because of their laser and magnetic properties. These crystals are utilized in efficient diode-laser pumped solid-state lasers and may be useful in magnetic refrigeration applications. In the main the crystals of this family crystallize with tetragonal space group $I4_1/amd$ and are isomorphous with zircon $ZrSiO_4$. It is also known that all these compounds melt congruently. Therefore the Czochralski technique is widely used for the fabrication of these crystals. In spite of the fact that all these materials can be produced from stoichiometric melts as single crystals, the wide application of orthovanadates is still limited because of problems related with the unusual behavior of the solid/liquid interface during crystal pulling in the CZ system and complexity of control of the crystal shape. These difficulties prevent the production of large and high-quality crystals suitable for industrial applications.

The EFG technique is known [55] to result in the growth of vanadate crystals with good reproducibility of the crystal shape and quality. The similarity of the μ-PD and EFG systems was discussed above for the $(Tb,Yb)_3Al_5O_{12}$ crystals (Figs. 5.26 and 5.27). Therefore the μ-PD method was considered in [12] as an alternative crystal growth technique that will possibly allow control of the crystal shape similar to that produced using the EFG arrangement. According to preliminary results of fiber crystal growth reported in Ref. [12] undoped $GdVO_4$ crystals were grown by the μ-PD method using RF heating apparatus from an iridium crucible with a diameter of 16 mm and a height of 44 mm.

The crystals were produced using high purity powders (99.99 %) of the corresponding oxides Gd_2O_3 and V_2O_5 as starting materials. The mixture of stoichiometric composition was charged into the crucible and heated up to complete melting at a temperature of about 1800 °C. The crystals were first grown using YAG fiber crystal seeds under an Ar atmosphere. The crystals grown were slightly colored. The length of the crystals was in the range of a few mm. The diameter was less than 1 mm. It was difficult to control the crystal shape of the fibers and step faceting was also observed similar to the crystals produced by the conventional Czochralski method. The wettability of the $GdVO_4$ melt was observed to be high [12], and control of the solid/liquid interface was quite difficult. Nevertheless it was concluded that growth of vanadate fiber crystals is generally possible. A view of a $GdVO_4$ fiber crystal produced using the μ-PD apparatus is given in Fig. 5.38. As can be seen from the photograph, the quality of the crystal is very far from that of garnet fibers discussed above. Nevertheless, it was declared in [12] that the quality of the fibers can be improved in future.

Fig. 5.38. $GdVO_4$ crystal produced by the μ-PD technique [12].

5.9 Summary

A wide range of complex oxide materials was successfully produced by the micropulling-down (μ-PD) technique. Generally the optical quality of these fibers is not yet high enough for industrial application compared with the quality of materials produced using the Czochralski and other conventional melt growth systems. However some of the parameters are much better than that of commercially available crystals (uniformity, dopant distribution, etc.). In particular a high concentration of normally unsuitable guest cations (dopants) can be easily achieved due to the nature of the method (high temperature gradient and extraordinarily high growth rate).

The μ-PD growths are generally conducted in conditions that are very far from thermodynamical equilibrium. Therefore novel materials with advanced properties can probably be grown using μ-PD apparatus. The method is still very young and immature compared with traditional methods used in mass production. Therefore, it is evident that much effort is necessary to understand all the details of the process when a relatively thin fiber ingot is produced at a very high pulling rate in conditions that are not well understood.

On the other hand the technique has already proved to be a very powerful laboratory-scale technique. A very small amount of starting materials and very short growth runs are considered to be very important features of this method. At the same time the tendencies observed for the growths described here show that many physical and chemical phenomena (segregation, phase formation, flux growth, etc.) could be studied using the apparatus established at a very miniature scale. In spite of the fact that many important crystals cannot be produced by the μ-PD technique, it is already evident that most crystal growth laboratories should be equipped with this apparatus for preliminary study of melt behavior and solidification, and especially for educational purposes in universities.

It was not the aim of this review to mention crystal growth of all optical materials produced in fiber-like form up to now. In particular, many details of fiber growth of $LiNbO_3$ [3, 62, 96], $Ba_2NaNb_5O_{15}$ [97] and other important optical crystals were not discussed here [6, 8, 98–101]. The reader can find details of those studies in the corresponding original papers and reports.

Acknowledgements

The author is indebted to his colleagues from Japan (Prof. T. Fukuda, Prof. T. Shishido, Dr. A. Yoshikawa, Dr. K. Shimamura, Dr. H. Takeda, Dr. Y. Kuwano, Dr. H. Machida, Dr. K. Inaba, K.Hasegawa), Germany (Prof. P. Rudolph, Dr. B. M. Epelbaum, Dr. Y. Tomm), Korea (Dr. Y. M. Yu, Dr. D. H. Yoon, Dr. J. M. Ko, Dr. J. H. Lee), France (Prof. G. Boulon, Dr. K. Lebbou), Russia (Prof. B. Mill, Dr. V. V. Kochurikhin), Spain (Dr. E. G. Villora) and other countries for their ideas, suggestions, and practical assistance in both the development of the fiber growth techniques and solving of material science problems related with prediction, preparation, and characterization of novel optical crystals. Their enthusiasm and collaboration in past decade is greatly appreciated.

References

1   R.S. Feigelson, J. Cryst. Growth 79 (1986) 669.
2   R.S. Feigelson, Mater. Sci. Eng. B1 (1988) 67.
3   T. Fukuda, V.I. Chani, K. Shimamura, in: Recent Development in Bulk Crystal Growth edited by Isshiki, Research Signpost, India, 1998, p. 191.

4   B.M. Epelbaum, A. Yoshikawa, K. Shimamura, T. Fukuda, K. Suzuki, Y. Waku, J. Cryst. Growth, 198/199 (1999) 471.

5   V.I. Chani, A. Yoshikawa, Y. Kuwano, K. Hasegawa, T. Fukuda, J. Cryst. Growth, 204 (1999) 155.

6   V.I. Chani, K. Nagata, T. Kawaguchi, M. Imaeda, T. Fukuda, J. Cryst. Growth, 194 (1998) 374.

7   V.I. Chani, K. Nagata, M. Imaeda, T. Fukuda, Ferroelectrics, 218 (1998) 187.

8   V.I. Chani, K. Nagata, T. Fukuda, Ferroelectrics, 218 (1998) 9.

9   V.I. Chani, A. Yoshikawa, Y. Kuwano, K. Inaba, K. Omote, T. Fukuda, Mater. Res. Bull. 35 (10) (2000) 1615.

10   V.I. Chani, A. Yoshikawa, H. Machida, T. Satoh, T. Fukuda, J. Cryst.Growth, 210(4) (2000) 663.

11   V.I. Chani, K. Shimamura, T. Fukuda, Cryst. Res. Technol. 34 (1999) 519.

12   V.I. Chani, A. Yoshikawa, H. Machida, T. Fukuda, in: Extended Abstracts (The 60th Autumn Meeting, 1999) The Jap. Soc. of Appl. Phys., September 1–4, Kobe, Japan, 3a-S-2.

13   Y.M. Yu, V.I. Chani, K. Shimamura, T. Fukuda, J. Cryst. Growth, 171 (1997) 463.

14   Y.M. Yu, V.I. Chani, K. Shimamura, K. Inaba, T. Fukuda, J. Cryst. Growth, 177 (1997) 74.

15   V.I. Chani, K. Shimamura, S. Endo, T. Fukuda, J. Cryst. Growth, 171 (1997) 742.

16   K. Inaba, K. Omote, private communication.

17   R.F. Belt, R.C. Puttbach, D.A. Lepore, J. Cryst. Growth, 13/14 (1972) 268.

18   R.R. Monchamp, J. Cryst. Growth, 11 (1971) 310.

19   C. Belouet, J. Cryst. Growth, 15 (1972) 188.

20   T. Nishimura, T. Omi, Jpn. J. Appl. Phys. 14 (1975) 1011.

21   B. Ferrand, D. Pelenc, I. Chartier, Ch. Wyon, J. Cryst. Growth, 128 (1993) 966.

22   J.A. Burton, R.C. Prim, W.P. Slichter, J. Chem. Phys, 21 (1953) 1987.

23   S.A. Markgraf, S. Kimura, T. Sawada, M. Göbbels, J. Cryst. Growth, 135 (1994) 253.

24   W. Class, J. Cryst. Growth, 3/4 (1968) 241.

25   J.J. Zayhowski, J. Harrison, in: Handbook of Photonics edited by M.C. Gupta, CRC Press, Boca Raton, FL 1996, p. 326.

26   U.V. Valiev, G.S. Krinchik, S.B. Kruglyashov, R.Z. Levitin, K.M. Mukimov, V.N. Orlov, B. Yu. Sokolov, Sov. Phys. Solid State 24 (1982) 1596.

27   M.Y.A. Raja, D. Allien, W. Sisk, Appl. Phys. Lett. 67 (1995) 2123.

28   B. Sugg, H. Nürge, B. Faust, E.Ruza, R. Niehüser, H.-J. Reyher, R.A. Rupp, L. Ackerman, Optical Materials 4 (1995) 343.

29   R.D. Shannon, M.A. Subramanian, T.H. Allik, H. Kimura, M.R. Kokta, M.H. Randles, G.R. Rossman, J. Appl. Phys. 67 (1990) 3798.

30   R Horyn, Z. Bukowski, M. Wolcyrz, J. Solid State Chem, 122 (1996) 321.

31   C.D. Brandle, R.L. Barns, J. Cryst. Growth, 26 (1974) 169.

32   See powder diffraction file, JCPDS cards.

33   J.C. Brice, J. Cryst. Growth, 42 (1977) 427.

34   C.B. Rubinstein, L.G. Van Utert, W.H. Grodkiewicz, J. Appl. Phys. 35 (1964) 069.

35   S. Ganschow, D. Klimm, P. Reiche, R. Uecker, Cryst. Res. Technol. 34 (1999) 615.

36   S. Ganschow, B.M.Epelbaum, D. Klimm, A.Yoshikawa, T. Fukuda, Proc. Int. Soc. Opt. Eng. 3724 (1999) 52.

37   V.I. Chani, A. Yoshikawa, H. Machida, T. Fukuda, Mater. Sci. and Engineering B 75(1) (2000) 53.

38 V.I. Chani, A. Yoshikawa, H. Machida, T. Fukuda, J. Cryst. Growth, 212 (3–4) (2000) 469.

39 D. Mateika, E. Völkel, J. Hisma, J. Cryst. Growth, 102 (1990) 994.

40 F. Rosenberger, Fundamentals of Crystal Growth, Springer-Verlag, Berlin, 1979.

41 V.I. Chani, Sov. Phys.Tech.Phys. 31(1) (1986) 114.

42 V.I. Chani, Proc.SPIE, 1125 (1989) 107.

43 V.I. Chani, K. Shimamura, Y.M. Yu, T. Fukuda, Mater. Sci. and Eng., R20 (1997) 281.

44 V.I. Chani, K. Shimamura, K. Inoue, T. Fukuda, K. Sugiyama, J. Cryst. Growth 132 (1993) 173.

45 V.I. Chani, M.I. Timoshechkin, K. Inoue, K. Shimamura, T. Fukuda, Inorg. Mater.30(12) (1994) 1570.

46 V.I. Chani, K. Shimamura, S. Endo, T. Fukuda, J. Cryst. Growth, 173 (1997) 117.

47 V.I. Chani, K. Shimamura, T. Fukuda, J. Mater. Res., 12 (1997) 2470.

48 V.I. Chani, K. Shimamura, S. Endo, T. Fukuda, J. Mater. Res., 14(6) (1999) 2458.

49 V.I. Chani, T. Fukuda, J. Cryst. Growth, 206 (1999) 245–248

50 R.D. Shannon, Acta Crystallogr. A32 (1976) 751.

51 T. Okuda, T. Katayama, K. Satoh, T. Oikawa, H. Yamamoto, N. Koshizuka, in: Recent Advances in Magnetism and Magn. Mater., Proc. of the Fifth Symp. on Magn. and Magn.Mater., Taipei, Taiwan, 19–20 Apr, 1989, World Scientific, p. 61.

52 P. Wu, A.D. Pelton, J. Alloys and Comp., 179 (1992) 259.

53 M. Mizuno, T. Noguchi, Yogyo Kyokaishi, 88 (1980) 322.

54 H.E. LaBelle, Jr, J. Cryst. Growth, 50 (1980) 8.

55 B.M. Epelbaum, K. Shimamura, K. Inaba, S. Uda, V.V. Kochurikhin, H. Machida, Y. Terada, T. Fukuda, Crys. Res. Tech, 34 (1999) 301.

56 T. Fukuda, Y. Uematsu, Jpn. J. Appl. Phys. 11 (1972) 163.

57 U. Flückiger, H. Arend, J. Cryst. Growth 43 (1978) 406.

58 T. Imai, S. Yagi, Y. Sugiyama, I. Hatakeyama, J. Cryst. Growth 147 (1995) 350.

59 E. Irle, R. Blachnik, B. Gather, Thermochim. Acta 179 (1991) 157.

60 A. Reisman, G.F. Holtzberg, J. Am. Chem. Soc., 77 (1955) 2117.

61 R.S. Roth, Progr. Solid State Chem, 13 (1980) 159.

62 D.H. Yoon, I. Yonenaga, T. Fukuda, N. Ohnishi, J. Cryst. Growth 142 (1994) 339.

63 H. Kimura, T. Numazawa, M. Sato, J. Cryst. Growth 165 (1996) 408.

64 H.C. Zeng, J. Cryst. Growth 173 (1997) 446.

65 J. Hulliger, R. Gutmann, H. W est, J. Cryst. Growth 128 (1993) 897.

66 P. W. Whipps, J. Cryst. Growth 12 (1972) 120.

67 T. Fukuda, Y. Uematsu, T. Ito, J. Cryst. Growth 24/25 (1974) 450.

68 R. Hofmeister, A. Yariv, A. Agrant, J. Cryst. Growth 131 (1993) 486.

69 V.I. Chani, K. Shimamura, T. Fukuda, Proc. of Intern. Symp. on Laser and Nonlinear Optical Materials, edited by T. Sasaki, November 3–5, 1997, Singapore, p.301.

70 N. Ueda, H. Hosono, R. Waseda, H. Kawazoe, Appl. Phys. Lett. 70 (1997) 3561.

71 L. Binet, D. Gourier, J. Phys. Chem. Solids, 59 (1998) 1241.

72 J. Åhman, G. Svensson, J. Albertsson, Acta Cryst. C52 (1996) 1336.

73 M. Fleischer, L. Höllbauer, E. Born, H. Meixner, J. Am. Ceram. Soc., 80 (1997) 2121.

74 T. Harwig, G.J. Wubs, G.J. Dirksen, Solid State Comm., 18 (1976) 1223.

75 L. Binet, D. Gourier, G. Minot, J. Solid. State Chem., 113 (1994) 420.

76 Y. Tomm, J.M. Ko, A. Yoshikawa, T. Fukuda, Solar Energy Mater. and Solar Cells 66 (2001) 369.

77   Y. Tomm, P. Reiche, D. Klimm, T. Fukuda, J. Cryst. Growth 220 (2000) 510.
78   V.I. Chani, K. Inoue, K. Shimamura, T. Sugiyama, T. Fukuda, J. Cryst. Growth 132 (1993) 335.
79   Y. Tomm, H. Machida, T. Fukuda, unpublished, 1999.
80   H. Takeda, K. Shimamura, V.I. Chani, T. Fukuda, J. Cryst. Growth, 197 (1999) 204.
81   V.I. Chani, H. Takeda, T. Fukuda, Mater. Sci. and Egineering B, B60 (1999) 212.
82   H. Takeda, K. Shimamura, V.I. Chani, T. Kato, T. Fukuda, Crys. Res. Tech, 34 (1999) 1141.
83   H. Takeda, T. Kato, V.I. Chani, H. Morikoshi, K. Shimamura, T. Fukuda, J. Alloys and Comp, 290 (1999) 79.
84   H. Takeda, T. Kato, V.I. Chani, K. Shimamura, T. Fukuda, J. Alloys and Comp. 290 (1999) 244.
85   T. Fukuda, K. Shimamura, V.V. Kochurikhin, V.I. Chani, B.M. Epelbaum, S.L. Baldochi, H. Takeda, A. Yoshikawa, J. Mater.Sci.: Mater. Electron. 10 (1999) 571.
86   P. Hansen, J.P. Krumme, Thin Solid Films 114 (1984) 69.
87   V.I. Chani, A.M. Balbashov, J. Cryst. Growth 66 (1984) 616.
88   A.M. Balbashov, V.I. Chani, Inorg. Mater.20 (12) (1984) 1762.
89   V.I. Chani, Inorg. Mater.23 (8) (1987) 1215.
90   V. Halv'cek, P. Novák, M. Vichr, Phys.stat.sol.(b) 44(1971)K21.
91   G. Ronniger, B.V. Mill, Sov.Phys.Crystallogr., 16 (1972) 902.
92   G. Ronniger, B.V. Mill, Sov.Phys.Crystallogr., 18 (1973) 76.
93   G. Ronniger, B.V. Mill, Sov.Phys.Crystallogr., 18 (1973) 339.
94   G. Ronniger, B.V. Mill, Sov.Phys.Crystallogr., 18 (1973) 187.
95   G. Ronniger, B.V. Mill, V.I.Sokolov, Sov.Phys.Crystallogr., 19 (1974) 219.
96   D.H. Yoon, T. Fukuda, J. Cryst. Growth, 144 (1994) 201.
97   K. Lebbou, V.I. Chani, A. Yoshikawa, T. Fukuda, M.Th. Cohen-Adad, G. Boulon, M. Ferriol, in: CLEO/Pacific Rim'99, The Pacific Rim Conf. on Lasers and Electro-Optics, Aug. 30–Sept. 3, 1999, Seoul, Korea, p. 100.
98   T. Fukuda, V.I. Chani, K. Shimamura, in: Encyclopedia of Materials: Science and Technology, Elsevier Science, 2001, p. 4603.
99   V.I. Chani, in: International Forum on Science and Technology of Crystal Growth, March 4–5, 2002, Tohoku Univ., Sendai, Japan, P-M-05.
100 H. Sato, V.I. Chani, Y. Kagamitani, A. Yoshikawa, H. Machida, T. Fukuda, in: International Forum on Science and Technology of Crystal Growth, March 4–5, 2002, Tohoku Univ., Sendai, Japan, P-M-06.
101 P. Rudolph, T. Fukuda, Crys. Res. Tech. 34 (1999) 3.

6 Oxide Eutectic Crystals for High-Temperature Structural Application

Akira Yoshikawa

The micro pulling down method has been adapted to grow oxide eutectic crystals. Details of known and novel oxide eutectic crystal growth using the micro pulling down method are described. Optimization of crystal growth parameters for eutectic crystals are outlined.

6.1 Introduction

Directionally solidified oxide alloys (natural oxide composites) have attracted strong research interest over the past four decades. Because of their unique combination of high mechanical strength at elevated temperatures and excellent thermal stability, they should by now have found wide use as advanced structural materials in various aerospace and industrial applications. However, this has not yet happened, mainly because of the limited control available over final material properties in the high-temperature solidification process. So far, oxide eutectics have been produced by gradient solidification, Bridgman, and floating molten zone techniques, i.e. various methods of bulk growth [1].

Tiller et al. (1957) [2] were the first to propose a formal theory for the solidification of eutectics. Diffusion equations were formulated and solved by resorting to dimensional arguments and physical reasoning. In analyzing eutectic growth, Hunt and Jackson (1966) [3] solved the diffusion conditions for a flat, uniformly advancing liquid/solid interface.

Viechnicki and Schmid (1969) [4] studied the eutectics of sapphire/YAG and sapphire/ZrO_2. Using a Bridgman furnace with molybdenum crucibles and a thermal gradient of approximately 200 °C/cm only nonuniform types microstructures were obtained. Borodin et al. (1987) [5] investigated sapphire/ZrO_2 eutectic by the Edge-defined Film-fed Growth (EFG) method at rates between 20 and 100 mm/h, and also a nonuniform microstructures. They suspected that the specific features (nonuniformities) of the microstructure of eutectics are caused mainly by loss of stability of the (originally) plane crystallization front because of constitutional superseding. Borodin et al. (1990) [6] studied sapphire/ZrO_2 eutectic by the Verneuil growth method, as well. But, a homogeneous microstructure was never obtained. In their case, they concluded that the irregularity of the powder (starting materials) supply to the crystallization zone may result in a situation. Their research interest was mainly centered on strength properties and they succeeded in getting samples which show tensile strength values from 320 to 425 MPa at room temperature and from 130 to 281 MPa at 1300 °C in air. However, in such nonuniform structures, the strength-controlling microstructure parameter appears closely related to the coarse size of the colony or grain boundary areas. This makes the real strength much lower than expected from the characteristic microstructure or lamellar size [7]. Mechanical strength requires materials with highly uniform morphologies.

Though there were a lot of studies about oxide eutectic systems, no parameters for the control of the microstructure have yet been clarified. The reason seems to be directly related to the limited availability of phase equilibrium data and the extraordinary high temperature of many of the eutectics.

Although bulk oxide eutectic samples often exhibit certain areas with fine, well-aligned structure, actual overall structures generally include many complex arrangements at different scales, namely grains, colonies of various shapes, and, at the finest scale, the inherent eutectic microstructure (which may also be of various types, depending on growth conditions). In such nonuniform structures, the strength-controlling microstructural parameter appears to be the coarse size of the colony or grain boundary areas. This makes the real strength much lower than expected from the characteristic microstructural or lamellar size [7]. In addition, the structural applications mentioned require highly reliable materials with uniform properties. Consequently, the investigation of eutectic morphologies in oxide systems in the context of growth conditions remains of primary importance.

One more interesting object in studying eutectic systems is the discovery of new shapes of the microstructure. Already in 1988 Feigelson [8] speculated on fiber diameters comparable with the size of eutectic phases leading to self-cladding structures if the surface energy of one phase shows a significantly larger value than that of the other one and, hence, will form up at the crystal periphery. But no experiments are known to have been done so far. To investigate this self-cladding structure, growth attempts for thin fibers 150 μm in diameter were carried out in 1999 [9]. A thin Ir wire 150 μm in diameter as an additional shaping element was inserted into the crucible orifice. The thin Ir wire was suspended from a wire holder to keep the meniscus at the bottom of it. The diameter of the grown fiber in this method was 150 μm.

The sum properties can be considerably improved in very thin, long and crystallographically controlled eutectic specimens. From this point of view, the orientation relationship between eutectic components should be of special interest. However, from a study of the literature, it can be concluded that even the combination of structural perfection and fiber shape has not yet been explored completely and it may be only at the beginning stage.

For fundamental research, the composite sum or product qualities of eutectics are not fulfilled due to the phase separation after growth. Therefore, the question arises: Which composite properties will appear in directional solidified eutectic fibers? What will occur if the diameter of the growing rod approaches the eutectic needle dimension? Until now these questions have not yet been clarified.

In the following section, a YAG/sapphire eutectic system (Fig. 6.1) was adopted as a research object to investigate the key parameter for controlling the microstructure in the oxide eutectic system. The growth stability, microstructure morphology and the orientation relationship between the two phases have been mainly traced. The self-cladding structures, which Feigelson speculated in 1988 [8], were discovered by the combined investigation of the fiber growth stability mechanism and the microstructure control parameter.

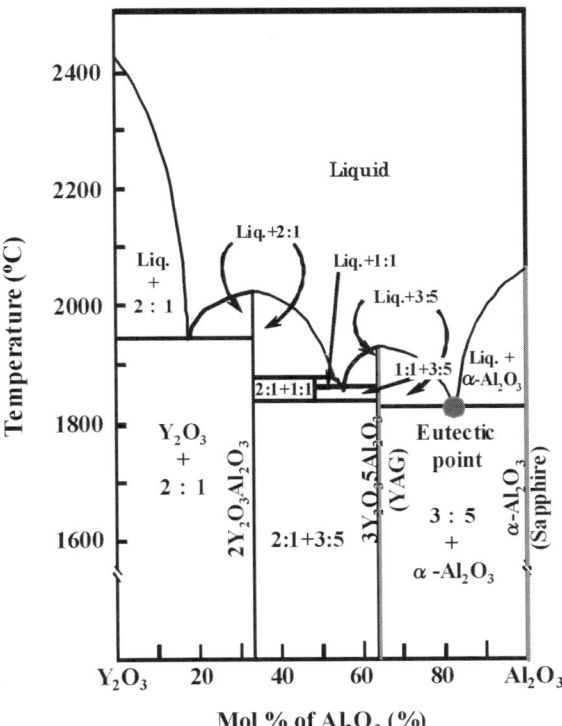

Fig. 6.1. Phase diagram of psuedo-binary system between Y_2O_3 and Al_2O_3.

The technique chosen in this book for the production of oxide eutectics was the micro pulling down (μ-PD) method [10–14]. This method allows the growth of fiber-like samples with diameters in the range 150–2000 μm and of virtually unlimited length, at widely variable pulling rates and temperature gradients at the solidification interface, while other processing parameters (fiber diameter, melt composition, very small radial temperature gradient) are kept constant. Significant features of the μ-PD process as applied to oxide eutectic studies are a high axial temperature gradient (of the order of 3–5×10^3 °C/cm) and the absence of macrosegregation phenomena. Small-diameter fibers are essentially free of grains, resulting in simplification of the morphology analysis.

In this section, several aspects of the μ-PD growth method as applied to Al_2O_3/$Y_3Al_5O_{12}$(YAG) [15, 16] and Al_2O_3/GdAlO$_3$(GdAP) [17] eutectic growth will be discussed. It also describes the microstructural characteristics of the fibers grown, with a focus on the problems of material homogeneity and fiber diameter uniformity.

It is well known that especially fiber crystals show an ultra-high yield strength (> 1 GPa) in pure monocrystalline sapphire fibers at room temperature (R.T.). This is due to their crystalline perfection and small dimensions, which minimize the occurrence of the defects that are responsible for the low strength of materials in bulk form. Moreover, it has been demonstrated in numerous other fibers that the tensile strength increases with decreasing fiber diameter [17, 18]. Therefore, it is quite an interesting task to investigate the fiber growth of eutectic composites and compare these with single-phase filaments and their bulk properties.

6.2 Growth Apparatus and Growth Procedures

6.2.1 Starting Materials and Compositions

The starting materials and compositions used in this work are listed in Table 6.1. These powders were dried, measured in proper proportions and mixed directly in the crucible. The eutectic composition of $Al_2O_3/Y_3Al_5O_{12}$ was taken from reference [15], and those of the other materials from reference [19].

6.2.2 Fiber Growth Assembly and Procedure

Micro Pulling Down Method

Hot Zone Setup. The eutectic fibers were grown by the micro pulling down (μ-PD) method from an iridium crucible with a 275 μm∅ hole, as shown schematically in Fig. 6.1. Induction heating at a frequency of 5 kHz, specially designed for the fabrication of crystals and composites at high temperature (above 1800 °C), was employed. In order to observe the hanging meniscus and growing fiber during the experiments, small windows for a CCD camera were inserted in the afterheater and heat shield. The growth process was controlled manually by changing the pulling rate and heating power. The fibers were grown in an Ar atmosphere (gas flow 2 l/min) to avoid oxidation of the crucible. This apparatus is described in greater detail in references [9, 13, 14].

Seeding and Growth Procedure. Sapphire [0001] oriented single crystal fiber grown by the edge-defined film-fed growth (EFG) method was used as a seed crystal. The seed was attached to the holder provided with an X-Y manipulator for final alignment to the crucible bottom. In order to observe the hanging meniscus and growing fiber during the experiments, two small in-tandem arranged windows for a CCD camera are inserted in the afterheater and heat shield. Figure 6.2a shows a photo just before the seeding and Fig. 6.2b shows a view directly after the seeding and during the steady-state pulling down process with constant diameter is shown in Fig. 6.2c. All photos were taken from the camera coupled TV screen.

The growth process was controlled manually by changing the pulling rate and using as a clue the ratio of the meniscus height to fiber diameter. Experimentally a reasonable ratio was found to be 1:5–10. Nonwetting was observed between the Ir and sapphire/YAG eutectic melt. Therefore, stable anchoring of the melt meniscus at the hole edge took place through all the experiments. Even in the case of wetting, the melt came out through the orifice because of the capillary phenomenon. The length of the grown crystal was only limited by the length of the pulling system applied. The diameter of each individual fiber was constant within 5–10 % accuracy. It is possible to produce several crystal fibers of the same length by repeated seeding without cooling of the hot zone and melt solidification. Since in our μ-PD machine, Ar gas was flown from the above direction, it allowed seal failure for a short time (5–10min) without oxidation of the crucible material (Ir). The grown fiber was cut from the seed and removed. Thereafter the system was sealed again and the seeding procedure was replicated. Thus, the total yield of the one growth run carried out with replicated seeding was more than 1 m of fiber crystal material.

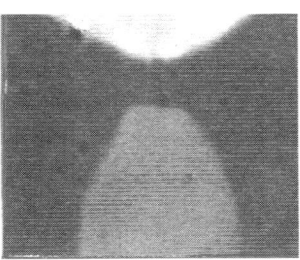

Fig. 6.2a. Photo just before the seeding;

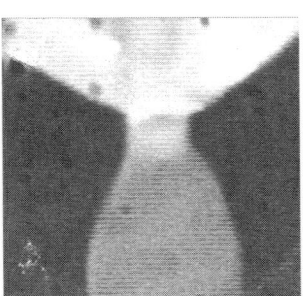

Fig. 6.2b. just after the seed touch;

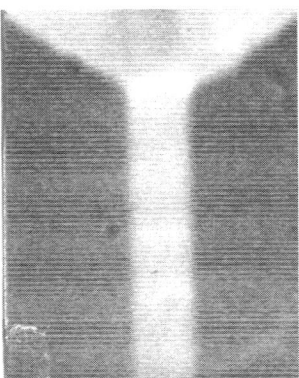

Fig. 6.2c. Steady-state pulling down process with constant diameter.

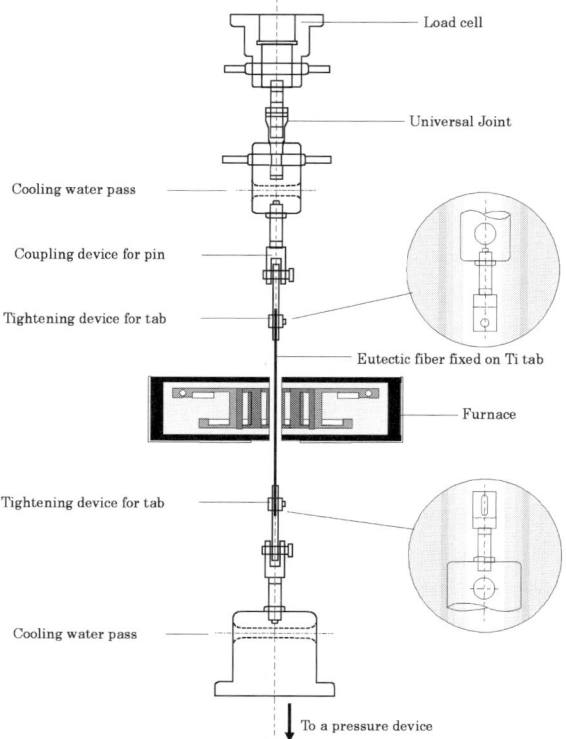

Fig. 6.3. Schematic assembly for tensile strength measurement.

6.2.3 Evaluation Techniques

A scanning electron microscope (SEM) was used for the observation of all samples and a back-scattered electron image (BEI) was used to distinguish the second phase. Since the brightness of the BEI strongly depends on the atomic nucleus, it is very easy to distinguish YAG and sapphire. Electron probe micro analysis (EPMA) was used for quantitative composition analysis for all samples, especially, for the investigation of Nd^{3+}, Yb^{3+} segregation phenomena in YAG, JEOL; JMX-862MX.

A transmission electron microscope (TEM), JEM-2010 by JEOL, was used to get higher magnification image. Electron diffraction patterns were also obtained to analyze the orientation relationship of the eutectic phase.

Powder X-ray diffraction analysis was carried out to identify the obtained phase using RAD by RIGAKU under the conditions of 35kV-25mA with $CoK\alpha$ and $CuK\alpha$ radiations.

Oscillation photographs were taken using a four-circle diffractometer (AFC5R) with imaging plate (IP). The X-ray source was $MoK\alpha$ with a power of 40kV, 30mA and monochromatized by graphite. The camera focal length was 60 mm and the oscillation angle was ±15°. The samples (fibers) were oscillated both along the longitudinal axis and vertical axis.

Tensile strength was measured at room temperature and also at high temperature using the uniaxial tension–compression facility at the Japan Ultra-high Temperature Materials Research Center (JUTEM) MTS808 system using samples of length 200 mm; the applied tension was parallel to the fiber axis (pulling direction). The strain rate was 10^{-4} N s^{-1} and tests were performed in vacuum to reduce heat transfer to the Ti sample holder. The eutectic fiber was fixed on this tab using Cu foil as a spacer. A schematic of the assembly for the tensile strength test is shown in Fig. 6.3. At the top part, there is a load cell to measure the amount of loading. The load data was divided by the fiber cross-section and regarded as tensile strength data.

6.3 Oxide Eutectic Crystal Growth, Morphological and Structural Properties

In this section, several combinations of Al_2O_3 based oxide eutectic crystals will be introduced from the viewpoint of morphological and structural properties.

Pattern formation in crystal growth has been studied since 1960s. The cellular and dendritic microstructures in directional solidification were first observed in directional solidified alloys.

6.3.1 YAG/Al$_2$O$_3$ Eutectic

Microstructure of YAG/Al$_2$O$_3$ Eutectic Fibers

Figure 6.4a shows the YAG/Al$_2$O$_3$ eutectic fibers, grown by the μ-PD method. The fiber diameter was controlling from approximately 200 μmφ to 2 mmφ, with a length of about 500 mm. It can be seen in this photo that the diameter is nearly constant, i.e. the diameter change is less than 5%. Stable growth was observed from the rate of 0.1 mm/min to 15 mm/min.

Figure 6.4b shows a back-scattered electron image (BEI) of the fiber microstructure grown at the rate of 15 mm/min. From electron probe micro analysis (EPMA), the bright region was identified as YAG and the dark region as sapphire. The YAG and sapphire phases form a three-dimensional network. Although we used an <0001>-oriented sapphire seed, in this eutectic system, they exhibit no anisotropy in the growth direction.

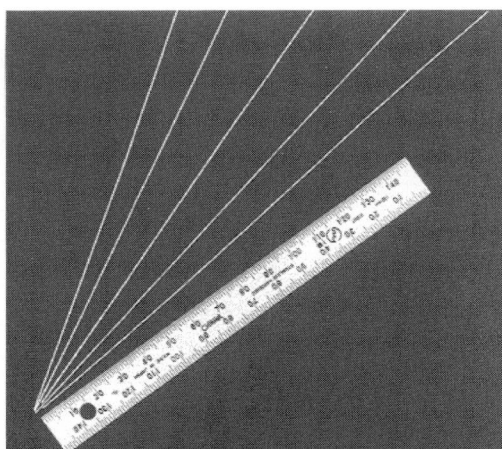

Fig. 6.4a. YAG/Al$_2$O$_3$ eutectic fibers grown by the μ-PD method.

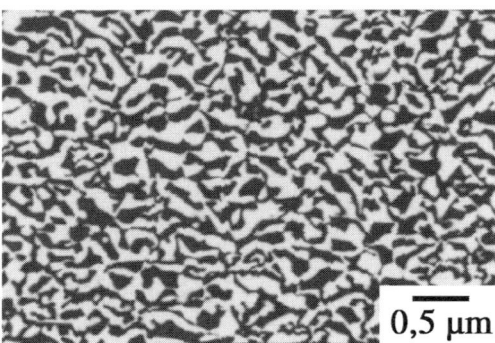

0,5 μm

Fig. 6.4b. BEI of YAG/Al$_2$O$_3$ eutectic fiber cross-section growth rate was 15 mm/min.

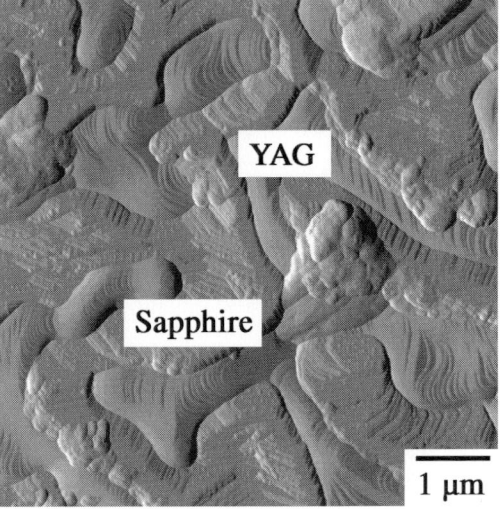

YAG

Sapphire

1 μm

Fig. 6.5. AFM images of the surface of a YAG/sapphire eutectic fiber grown at a pulling rates of 2 mm/s. This sample was lightly etched at 1300 °C for 30 min (this surface is longitudinal to the growth axis from left to right).

Topographic Study by Atomic Force Microscopy (AFM)

I attributed the two morphologies to the different behavior under etching of the garnet and sapphire phases. As seen in Fig. 6.5, one component typically had a relatively smooth and rounded surface, while the other displayed a partially faceted or stepped appearance, with planes parallel to certain directions. The latter morphology is similar to that of the individual alumina grains in the polycrystalline samples studied by Baretzky et al. (1996) [20]. I also compared those of the eutectics with images of pure YAG and sapphire fibers that had received the same etching treatment. In general, YAG fibers displayed rounded surfaces while sapphire fibers displayed facetted ones. However, at certain orientations or in certain regions of the fiber, the YAG could also show facets. Because of these variations, some caution in the phase identification was necessary, though processing of the images into maps showing the distribution of garnet and sapphire was reasonably straightforward.

Though AFM is, strictly speaking, a purely surface-sensitive technique, it is actually capable of analyzing the bulk of a sample through examination of internal surfaces exposed by conventional polishing or by fracture. These can be studied in the same way as the outer surfaces, and provide complementary information. The domain structure within the sample, characterized by various parameters as described in the following section, could thus be compared with that at the surface. In addition, fractographic analysis can be applied to study fracture mechanisms and material response to extreme stress.

Analysis of AFM Images

The first step in the analysis of the AFM images was segmentation, i.e. precise identification of the distinct regions corresponding to the garnet and sapphire phases. Segmentation and subsequent image analysis was performed using the public domain program NIH Image (developed at the US National Institute of Health and available at http://rsb.info.nih.gov/nih-image/), with additional operations programmed by myself for this specific purpose. The measurements of the relative phase area in the two dimensional AFM images were carried out by the modified NIH program.

The next step is to characterize the average domain size. This is not a trivial matter, as there is no simple concept of domain size for the complicated "Chinese script" pattern. Furthermore, especially for images of the surface or longitudinal sections, domains may be elongated, and a significant proportion of the domains may extend beyond the borders of a particular image. I chose as characteristic parameter the distance to the nearest phase boundary, averaged over all points of a given domain. This mean distance-to-boundary d_m can be calculated for any type of structure, and separately for each phase. It carries the important physical meaning of the average diffusion distance of a species in the melt required to realize the given phase segregation. In a perfect lamellar structure, the mean distance equals one fourth the usual lamella thickness. Indeed, to facilitate comparison with previously published data, I determined that the mean distance-to-boundary was 0.2–

0.25 times the more subjective estimate of lamellar spacing performed on the same pictures in the usual manner by EM images.

To characterize the domain shape, several parameters are available. Of particular physical significance is the total length L of the boundaries between domains, since processes important for both crystal growth and mechanical properties occur along these boundaries. To obtain a scale-invariant parameter, i.e. one that depends only on domain shapes and not sizes, I defined the quantity $\alpha = L\, d_m\, /A$, where A is the area of the image under analysis. Larger values of α correspond to more compact domains, smaller values to more extended ones.

The Microstructure Information Available from AFM

It is worth noting, first of all, that this observation was the first to show that the characteristic microstructure of YAG/sapphire eutectic fibers (Fig. 6.6 a–d), and probably of the class of such fibers, extends all the way to the outer surface. This finding is important because of the impact of surface characteristics on overall fiber properties.

More surprising is that the garnet area fraction (assumed equal to the volume fraction, as discussed previously) of the fiber surfaces changed in a regular way as a function of the pulling rate, as shown in Fig. 6.7. The solid straight line is a least-squares fit to the surface data, and the dashed line represents the theoretical fraction for the eutectic composition, which fits the bulk data well. Furthermore, its magnitude was much smaller than that of 0.57 found in the bulk of the fibers, independent of pulling rate. This bulk value is close to the theoretical garnet volume fraction of 0.56 for eutectic composition ($Y_2O_3:Al_2O_3 = 18.7:87.3$ mole %), as well as to the experimental value of 0.55 obtained from analysis of electron micrographs of fiber sections. This agreement, incidentally, confirms that I have correctly identified the two phases. It is especially remarkable that the predominance of sapphire at the surface becomes stronger as the pulling rate is decreased. This accords with recent EM findings that the surface may even consist entirely of the sapphire phase at extremely low pulling rate [9].

Orientation Relationship of this Eutectic System

From the microstructure study using BEI and AFM, it can be seen that the interface between YAG and sapphire forms not only a rounded boundary but also a boundary defined by straight lines. And these lines for different domains tend to be parallel. It is not surprising that there exists a preferred orientation relationship between the YAG and sapphire during eutectic growth since both phases are considered as single crystals, and a single crystal has anisotropy. In this section, this relationship between YAG and sapphire will be discussed. The oscillation photographs were taken both along the longitudinal axis (Fig. 6.8a) and the perpendicular axis (Fig. 6.8b). In the photographs, diffraction spots were observed, which indicates that this fiber was not polycrystalline but consisted of a single crystal.

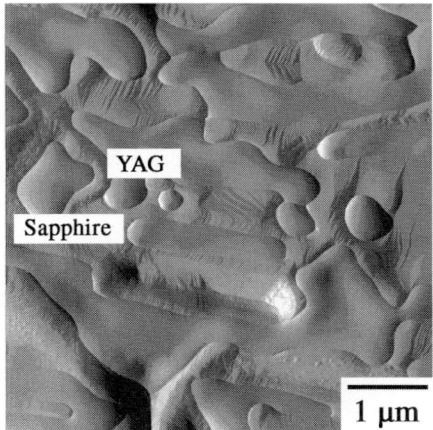

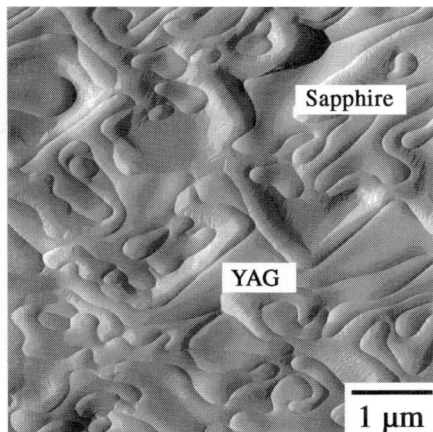

Fig. 6.6 (a) AFM images of the surfaces of YAG/sapphire fibers grown at pulling rates of 1 mm/min. The samples were thermally etched at 1500 °C for 30 min. (this surface is longitudinal to the growth axis from left to right).

(b) AFM images of the surfaces of YAG/sapphire fibers grown at pulling rates of 2 mm/min. The samples were thermally etched at 1500 °C for 30 min. (this surface is longitudinal to the growth axis from left to right).

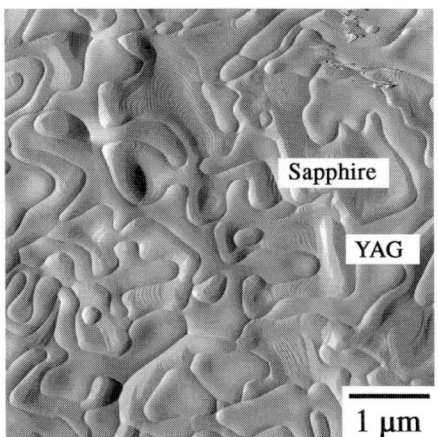

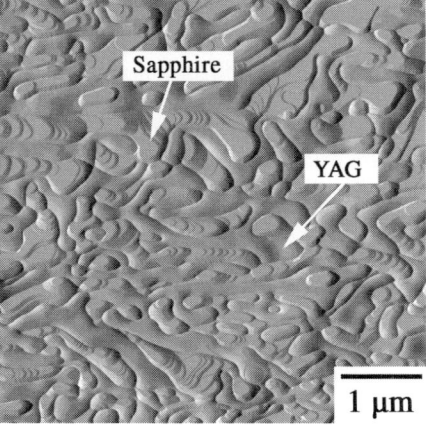

(c) AFM images of the surfaces of YAG/sapphire fibers grown at pulling rates of 4 mm/min. The samples were thermally etched at 1500 °C for 30 min. (this surface is longitudinal to the growth axis from left to right).

(d) AFM images of the surfaces of YAG/sapphire fibers grown at pulling rates of 9 mm/min. The samples were thermally etched at 1500 °C for 30 min. (this surface is longitudinal to the growth axis from left to right).

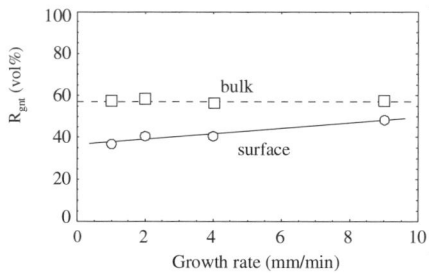

Fig. 6.7. Garnet area fraction (assumed equal to the volume fraction) of the fiber surfaces changed in a regular way as a function of the pulling rate.

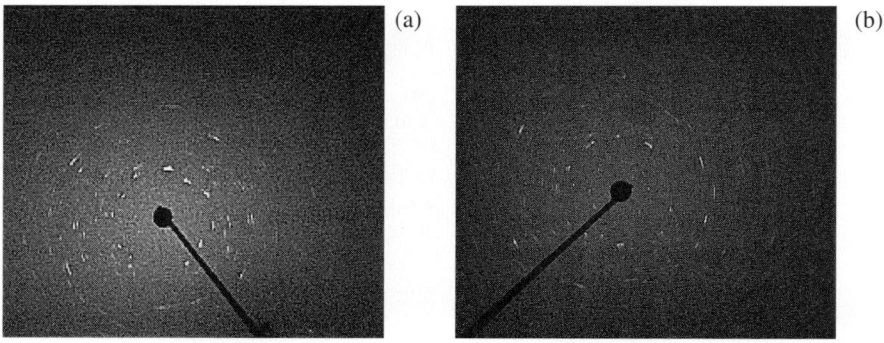

(a)   (b)

Fig. 6.8. (a) X-ray diffraction oscillation photograph of YAG/sapphire eutectic. The fiber was oscillated with the axis which is longitudinal to the growth direction. The oscillation angle was ±15°, and monochromatic MoKα radiation was used. **Fig. 6.8 (b)** X-ray diffraction oscillation photograph of YAG/sapphire eutectic. The fiber was oscillated with the axis which is perpendicular to the growth direction. The oscillation angle was ±15°, and monochromatic MoKα radiation was used.

The TEM image and its Fourier transformation images are shown in Fig. 6.9(a). It was confirmed that the left region corresponds to YAG, and the right region to sapphire, according to the results of energy dispersive X-ray spectroscopy.

In the Fourier transformation image, O, A_Y, B_Y, C_Y, A_s, B_s and C_s were denoted as Fig. 6.9b. The lengths of OA_Y, OB_Y, OC_Y, OA_s, OB_s, OC_s were OA_Y = 9.0mm, OB_Y = 9.0mm, OC_Y = 14.5mm, OA_s = 19.0mm, OB_s = 10.5mm, OC_s = 21.5mm. The camera constant is L = 45, so the experimental d values were calculated to be d_{OAY} = 5.0Å, d_{OBY} = 5.0Å, d_{OCY} = 3.10Å, d_{OAs} = 2.37Å, d_{OBs} = 4.29Å, d_{OCs} = 2.09Å. The d_{hkl} values of both YAG and sapphire were calculated.

From comparison of the observed and calculated d values, the most likely assignments are:

for YAG, OA_Y: d_{112} , OB_Y : d_{211}, OC_Y: d_{321} and d_{400}
for sapphire, OA_s: d_{015} , OB_Y : d_{101}, OC_Y: d_{006}

From consideration of the three-dimensional relationship among their reciprocal lattices, the Fourier transformation patterns were indexed as follows; A_Y is $11\overline{2}_{YAG}$, B_Y is 211_{YAG}, C_Y is $32\overline{1}_{YAG}$, A_s is $0\overline{1}5_{sapphire}$, B_s is $011_{sapphire}$ and C_s is $006_{sapphire}$. It was found that the zone axis is $[\overline{3}51]YAG//[100]sapphire$.

From this TEM image and the indexed pattern, it was found that the interface between the YAG and sapphire phases corresponds to (211) of YAG and also to (011) of sapphire. These planes form the hetero-epitaxy interface.

Tensile Strength

Although the fiber is brittle at room temperature, about 10 % plastic deformation was observed at 1500 °C, as shown in Table 6.1. The tensile strength is about three times as high as that of the bulk material (~200 MPa [17]).

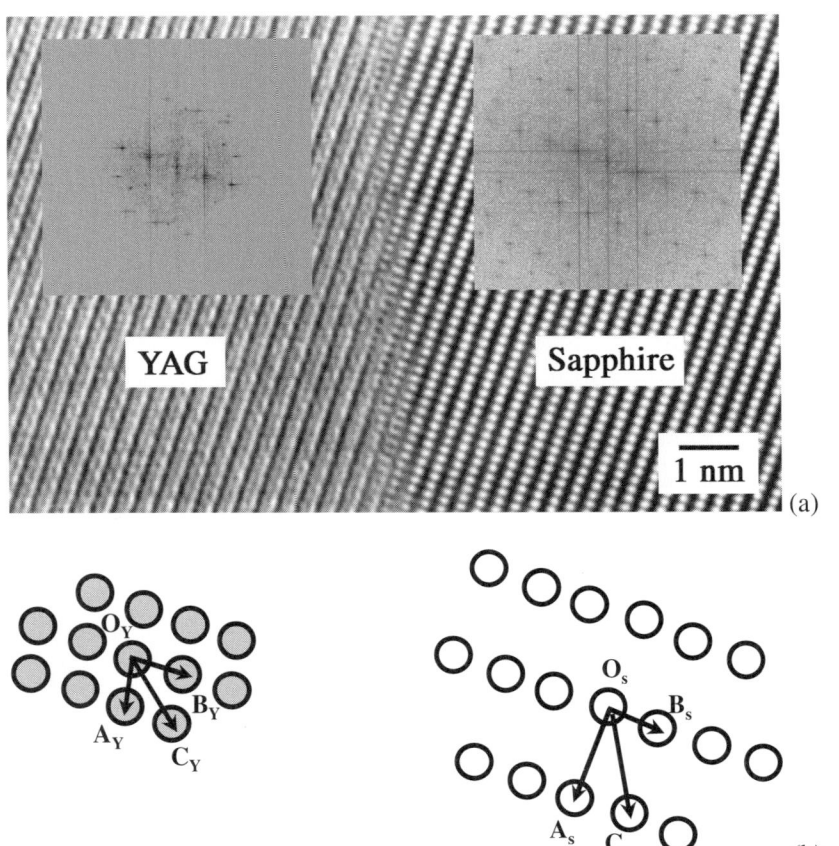

Fig. 6.9. (a) TEM image and its Fourier transformation images of YAG/sapphire eutectic. (b) Key diagram of (a): black circles are from YAG, white circles from sapphire.

Table 6.1. Various fibers grown and sapphire-YAG (SY) bulk crystals.

Material	Growth method	Structure	Tensile strength (MPa) R.T./1500 °C	Authors
fibers:				
sapphire	EFG	monocryst.	2900/200	Pollock [21]/Morscher [22]
$Fe_{75}Si_{10}B_{15}$	melt extrusion	amorphous	10000/–	Ström-Olsen et al. [18]
Al_2O_3/ZrO_2	melt extrusion	polycrystalline	2000/–	Ström-Olsen et al. [18]
$Al_2O_3/ZrO_2/TiO_2$	melt extrusion	amorphous	4000/–	Ström-Olsen et al. [18]
Al_2O_3/CaO	melt extraction	amorphous	2500/–	Allahverdi et al. [23]
Al_2O_3/CaO	melt-spinning	polycrystalline	1000/–	Wallenberger et al. [24]
Al_2O_3/YAG	μ-PD	monocryst.	927.3/576.5	Yoshikawa et al. [13]
SY bulk	Bridgman	monocryst.	200/150	Waku et al. [17]

Thermal Stability of Microstructure

Figure 6.10 shows SEM images of the microstructure for (a) as-grown fiber and after heat treatment at 1500 °C in air atmosphere ((b) 25 h, (c) 75 h). No microstructure grain growth was observed; that is, the eutectic fiber was shown to be very stable even under such extreme conditions. This stability can be explained as follows. Because the boundary between the YAG and sapphire phases is very sharp, these two phases cannot form a solid solution. Thus, there is no path for the material transport necessary for grain growth.

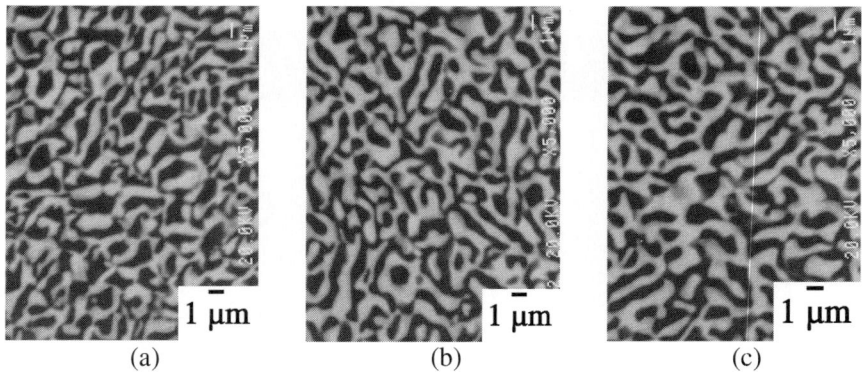

1 μm	1 μm	1 μm
(a)	(b)	(c)

Fig. 6.10. SEM images of the microstructure for (**a**) as-grown fiber and after heat treatment at 1500 °C in air atmosphere ((**b**) 25 h, (**c**) 75 h).

6.3.2 Rare Earth Para-aluminate (Garnet)/Al$_2$O$_3$ Eutectic and Rare Earth Ortho-aluminate (Perovskite)/Al$_2$O$_3$ Eutectic

Phase Identification

Oxides of rare earths from La to Pm have no eutectic point with Al$_2$O$_3$. For the eutectic forming materials, powder X-ray diffraction was measured to identify the phases. All peaks in each pattern were sharp and could be indexed. So it was found that these eutectic fibers were crystalline and contained no significant impurities. The nature of the rare-earth-containing phase changed in going from Gd to Tb. Fig. 6.11 shows powder X-ray diffraction patterns of Al$_2$O$_3$/Tb$_3$Al$_5$O$_{12}$ and Al$_2$O$_3$/GdAlO$_3$ pulverized fiber. Sm, Eu, and Gd showed Al$_2$O$_3$ and a perovskite phase, while Tb-Lu showed Al$_2$O$_3$ and a garnet phase. No three-phase eutectic was observed. This accords with and extends previous research, which had not clarified completely whether a garnet phase, perovskite phase, or a mixture the two appeared in eutectics with Al$_2$O$_3$. According to the result of this work, the materials which form eutectics with Al$_2$O$_3$ can be divided into the perovskite type and the garnet type at the boundary between Gd and Tb (see Fig. 6.12). No other type of eutectic with Al$_2$O$_3$ was observed.

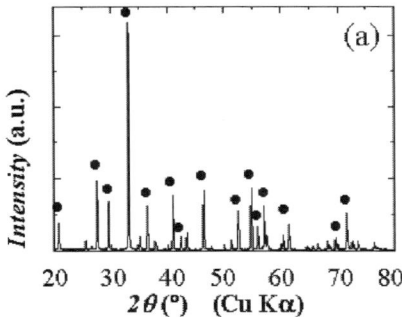

Fig. 6.11. Powder X-ray diffraction patterns of (**a**) Al$_2$O$_3$/Tb$_3$Al$_5$O$_{12}$ and (**b**) Al$_2$O$_3$/GdAlO$_3$

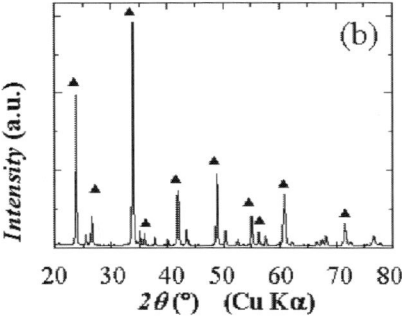

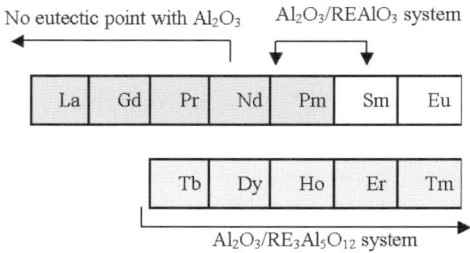

Fig. 6.12. Eutectic formation by rare-earth compounds and Al_2O_3

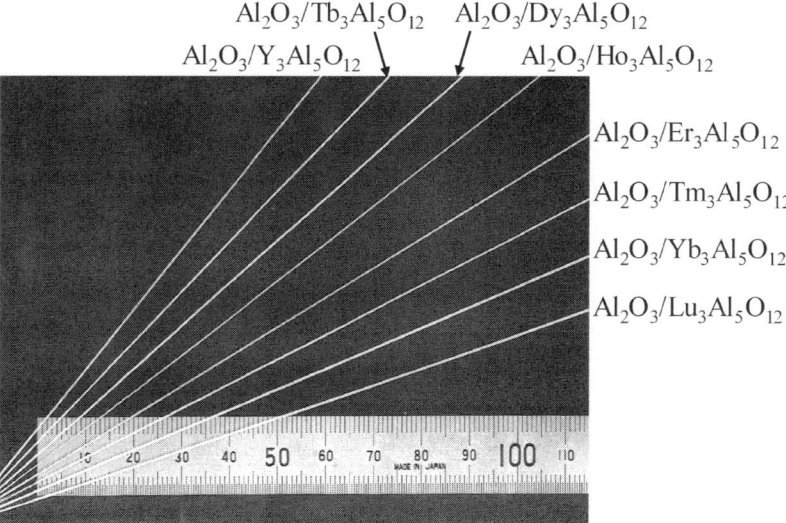

Fig. 6.13. Al_2O_3/$RE_3Al_5O_{12}$ (RE = Tb~Lu, Y) eutectic fibers.

Features of Al_2O_3/Garnet System

Fig. 6.13 shows the Al_2O_3/garnet system eutectic fibers obtained. In this system, reasonably stable growth was achieved over a wide range of pulling rates, 0.1–20 mm/min, with the highest achievable pulling rate being around 30 mm/min. The diameter of these fibers was well controlled in the range from 0.20 to 0.50 mm, and the length was 550 mm. The fibers were white or colored depending on the rare earth ion.

Fig. 6.14 shows a typical SEM image of the microstructure of Al_2O_3/$RE_3Al_5O_{12}$ eutectic fiber. In all eutectic fibers in this system, the component phases formed a three-dimensional interpenetrating network known as a Chinese script (CS) structure. It was completely homogeneous throughout the entire cross-section of the grown fibers. The size of the domains was controlled by growth speed, with a high pulling rate producing a smaller microstructure (see Fig. 6.15).

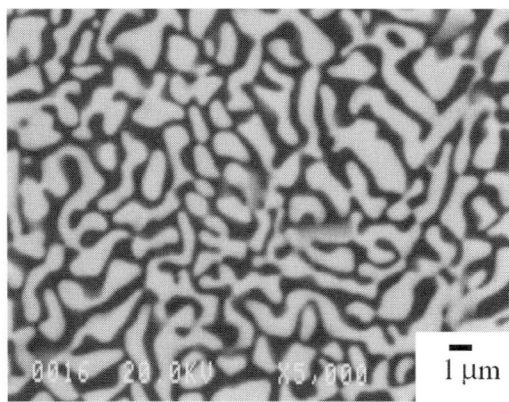

Fig. 6.14. SEM image of $Al_2O_3/Tb_3Al_5O_{12}$ eutectic fiber cross-section showing Chinese script microstructure. Other Al_2O_3/garnet materials had a similar structure (dark region: Al_2O_3 phase, bright region: garnet phase).

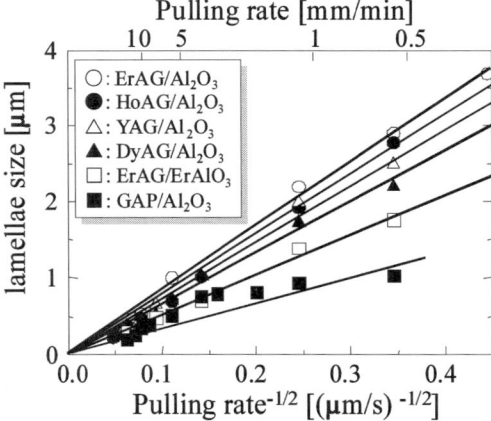

Fig. 6.15. Relation between growth velocity and lamellae size of various eutectic materials.

Features of Al_2O_3/Perovskite System

Al_2O_3/perovskite system eutectic fibers were grown using Sm, Eu, and Gd (see Fig. 6.16). These materials behaved differently from the garnet system materials, and there were some problems during growth. For example, $Al_2O_3/SmAlO_3$ easily disconnected from the melt during growth and this fiber could be grown only several centimeters in length. The maximum growth rate was 1.0 mm/min. $Al_2O_3/EuAlO_3$ exhibited more stable growth, and fibers could be grown at a pulling rate of up to 15 mm/min. Very thin fibers (around 100 µm) could be grown, but it was difficult to control the diameter, and the thin fibers easily disconnected. In the case of $Al_2O_3/GdAlO_3$, fibers could be grown at a pulling rate of up to 15 mm/min, but this material very easily bent during growth. Figure 6.17 shows typical SEM images of the microstructure of $Al_2O_3/REAlO_3$ eutectic fiber in cross-section and longitudinal section. It was found that the microstructure in this system was not uniform, but rod-shaped colonies were mixed with regions of CS. The ratio of the volumes of the rod-shape and CS structure varied and only a small number of samples from fibers grown at low growth rate exhibited uniform CS structure.

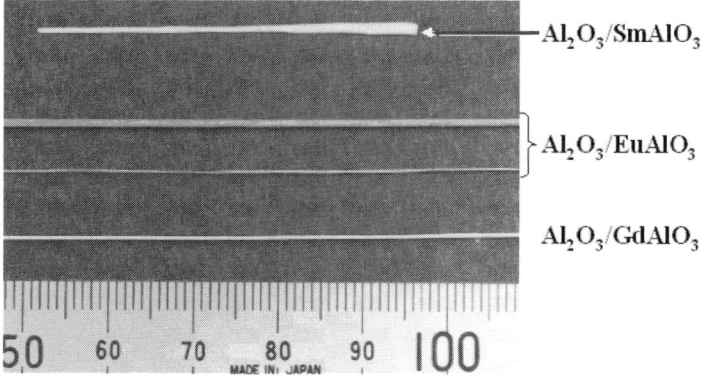

Fig. 6.16. $Al_2O_3/REAlO_3$ (RE = Sm~Gd) eutectic fibers.

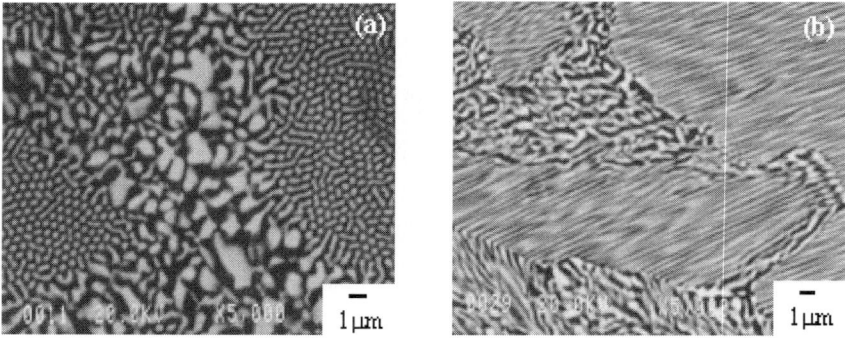

Fig. 6.17. SEM image of $Al_2O_3/GdAlO_3$ eutectic fiber (**a**) perpendicular and (**b**) parallel to the growth direction showing regions of rod-like and Chinese script type microstructures. Other Al_2O_3/perovskite fibers had similar microstructure (dark region: Al_2O_3 phase, bright region: perovskite phase).

Off-Eutectic Composition

To examine how the stability and homogeneity of the microstructure in Al_2O_3/garnet and Al_2O_3/perovskite systems are influenced by chemical composition, off-eutectic growth was attempted. Fibers were grown with chemical composition shifted from the eutectic point by several mole percent. Figures 6.18 and 6.19 show the resulting microstructure of $Al_2O_3/Y_3Al_5O_{12}$ and $Al_2O_3/GdAlO_3$ eutectic fibers. The homogeneity of the CS texture in $Al_2O_3/Y_3Al_5O_{12}$ was unchanged. In the case of $Al_2O_3/GdAlO_3$, the change in composition strongly affected the structure: a decreased fraction of Gd_2O_3 increased the amount of rod-like microstructure relative to the amount of CS microstructure. A detailed understanding of the microstructure formation is still lacking, but we can rationalize some aspects of the difference between the CS and rod-shape microstructure in the following way. As reviewed by W. J. Minford et al. [25], the basic microstructures of eutectics are correlated with the relative interfacial surface area per unit volume for the fibrous or rod-like and lamellar forms. A fibrous structure has lowest surface energy when the volume fraction of the minor phase is < 0.28. Otherwise, the microstructure tends to be lamellar. The Al_2O_3/ZrO_2 eutectic, for example, which falls near the division point, has been observed to form both rod-shape and lamellar microstructure. The change of microstructure in $Al_2O_3/GdAlO_3$ qualitatively follows this tendency. The volume fraction of $GdAlO_3$ in the $Al_2O_3/GdAlO_3$ eutectic, as measured on digitalized images, was 0.38 in rod-shape regions, but approached 0.50 in CS regions of fibers grown from Gd_2O_3-rich composition. The crossover volume fraction is very different from that for the rod-lamellar transition, presumably for the same reason that these materials form Chinese script rather than simple lamellar microstructures. We note that $Al_2O_3/Y_3Al_5O_{12}$ formed CS structure even when the volume fraction of $Y_3Al_5O_{12}$ to Al_2O_3 was over 0.62. It seems that other factors, for example facet or nonfacet behavior of the material, can also affect the type of microstructure.

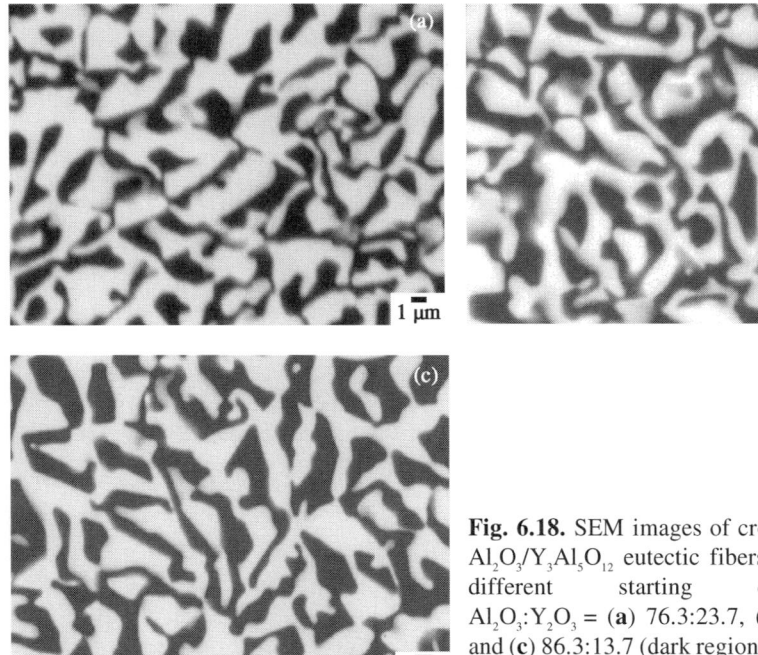

Fig. 6.18. SEM images of cross-section of $Al_2O_3/Y_3Al_5O_{12}$ eutectic fibers grown from different starting compositions. $Al_2O_3:Y_2O_3 =$ (**a**) 76.3:23.7, (**b**) 81.3:18.7, and (**c**) 86.3:13.7 (dark region: Al_2O_3, bright region: $Y_3Al_5O_{12}$).

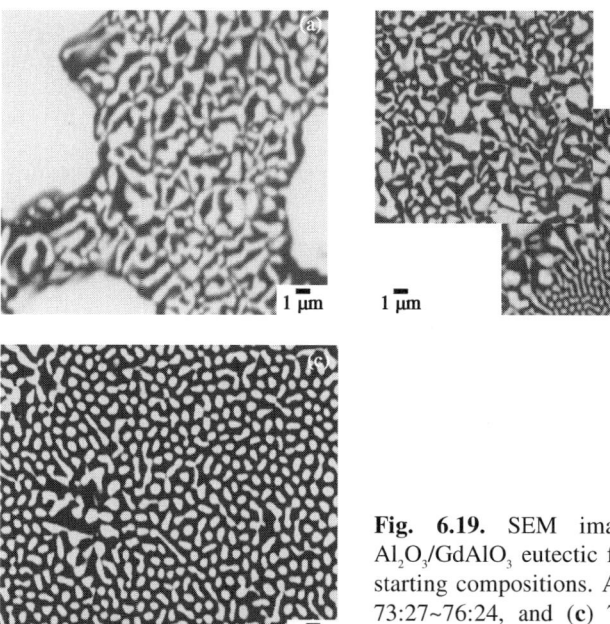

Fig. 6.19. SEM images of cross-section of Al_2O_3/GdAlO$_3$ eutectic fibers grown from different starting compositions. Al_2O_3:Gd$_2$O$_3$ = **(a)** 72:28, **(b)** 73:27~76:24, and **(c)** 77:23 (dark region: Al_2O_3, bright region: GdAlO$_3$).

Tensile Strength

To investigate the mechanical characteristics of grown eutectic fibers, tensile stress-displacement tests were carried out. Tensile strength data for eutectic fibers grown at 15 mm/min are listed in Table 6.2. Al_2O_3/garnet system eutectic fibers remained strong at 1500 °C. Al_2O_3/Tm$_3$Al$_5$O$_{12}$ (624 MPa) was the strongest among these, and was more than three times stronger than bulk Al_2O_3/Y$_3$Al$_5$O$_{12}$ material. The Al_2O_3/perovskite system eutectic fibers, which had a colony-type microstructure, had about half the strength of Al_2O_3/garnet eutectic fibers. A uniform microstructure has been shown to be necessary for the best mechanical properties; it is possible that Al_2O_3/perovskite system eutectics would be stronger if the microstructure could be made uniform.

6.3.3 ZrO$_2$/Al$_2$O$_3$ Eutectic

The eutectic composition found in this work, 62 mol% Al_2O_3/38 mol% ZrO$_2$ was consistent with Fischer et al. [26] and Cervales [27], and the eutectic temperature of 1870 ± 30 °C is consistent with Schmid [28].

Figure 6.20 shows the as-grown eutectic fibers. We were able to control the eutectic fiber diameter from approximately 0.2 mm to 2 mm. The full length was about 500 mm, limited by the apparatus. Stable growth was obtained in the range

0.1~15 mm/min. The grown fibers were completely white and had a diameter constant to within 10%.

Figure 6.21 shows a powder XRD pattern of crushed Al_2O_3/ZrO_2 eutectic fiber. In this figure, we can observe that both Al_2O_3 and ZrO_2 are crystalline and the dominant ZrO_2 phase is monoclinic with a trace of the cubic phase.

BEI of the perpendicular cross-section microstructures of Al_2O_3/ZrO_2 eutectic fibers are shown in Fig. 6.22. The eutectic microstructure was composed of white and black phases and oriented to a common direction. By EDS analysis, the white phase was shown to be zirconia and the dark one alumina. The white (zirconia) phase was formed on the black (alumina) matrix. This eutectic microstructure changed dramatically with the pulling rate. At lower pulling rates, below 0.5 mm/min, a rod-shaped structure was formed as shown at Fig. 6.22(a), but at the intermediate range of pulling rate from 1 mm/min to 5 mm/min, the microstructure shows an irregular but homogeneous broken lamellar structure (Fig. 6.22b). At higher pulling rates of over 5 mm/min, a cellular structure appeared with a broken lamellar pattern within the cells (Fig. 6.22c), which had an ellipsoidal shape and were oriented along a common direction.

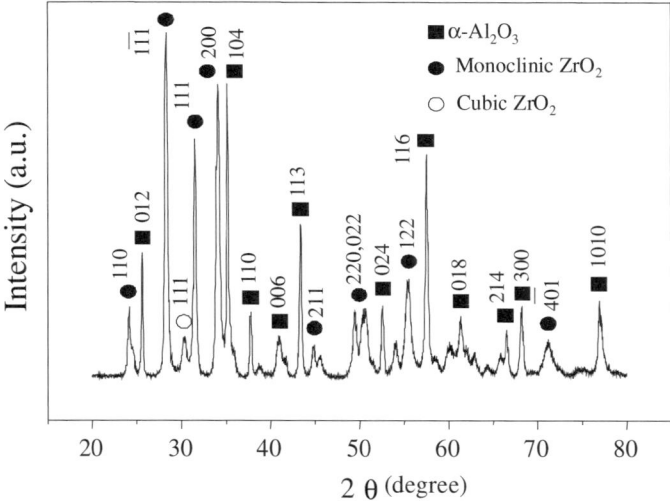

Fig. 6.20. As-grown Al_2O_3/ZrO_2 eutectic fibers.

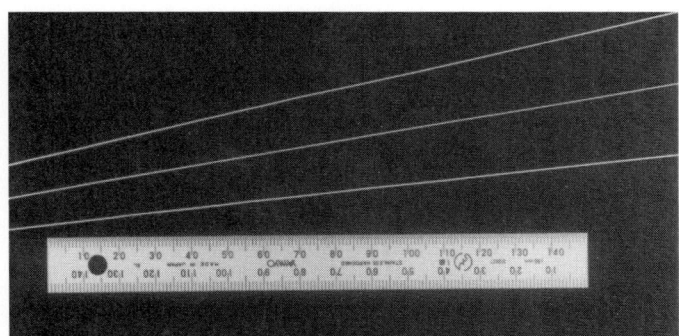

Fig. 6.21. Powder XRD pattern of crushed Al_2O_3/ZrO_2 eutectic fibers.

These "broken lamellar" and "cellular lamellar" structures are quite different from the colony-like structure of Y_2O_3-doped Al_2O_3/ZrO_2 eutectic, which was described by Schmid et al. [28], Echigoya et al. [29] and Borodin et al. [6]. While Y_2O_3-doped Al_2O_3/ZrO_2 eutectic cellular structures have a circular shape, each cell of the undoped system of this experiment was filled with thin lamellae.

The longitudinal cross-section micrographs in Fig. 6.23 also show the same trend of microstructural change as a function of the growth rate. Lamellar structures grown at the low pulling rate of 1 mm/min show nonuniform lamellae grown paralled to the growth direction in the length of about 30–50 μm. But the structure grown at a higher pulling rate of 10 mm/min has a typical columnar structure with irregular thickness and lamellae tending to tilt toward the Al_2O_3-rich regions between cells.

In order to examine the reproducibility of these microstructures, we tried to grow several fibers by growing at the same growth condition and to characterize them with SEM, and indeed, we could confirm this trend. On the other hand, in the case of the off-eutectic compositions, for example 64 mol% Al_2O_3/36mol% ZrO_2 or 60mol% Al_2O_3/40mol% ZrO_2, almost the same lamellar structures were observed, but that lamellar structures were more sensitive to thermal gradient than the eutectic composition, than those grown at pulling rates above 2 mm/min, where the lamellar structure changed to a cellular structure. The volume fraction of alumina to zirconia was 0.67±0.03 for all types of structures, which is in good agreement with the theoretical value for this eutectic composition of 0.66, assuming the ZrO_2 exists in the monoclinic form.

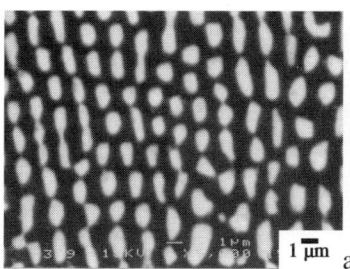
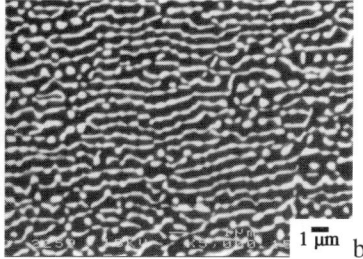

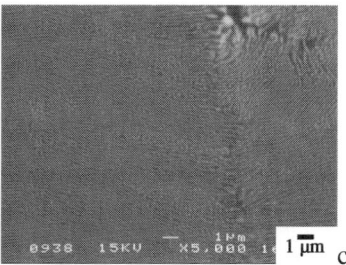

Fig. 6.22. SEM perpendicular cross-sectional images of Al_2O_3/ZrO_2 eutectic fibers at different pulling rates: (a) 0.1mm/min, (b) 1mm/min, (c) 10mm/min.

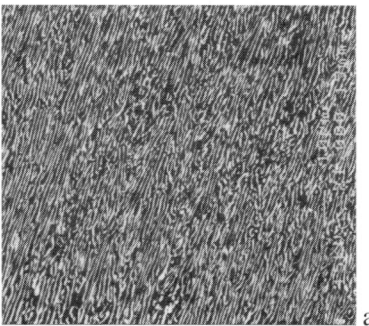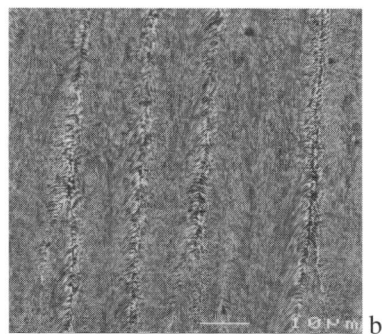

Fig. 6.23. SEM longitudinal cross-sectional images of Al_2O_3/ZrO_2 eutectic fibers at different pulling rates: (**a**) 1 mm/min, (**b**) 10 mm/min.

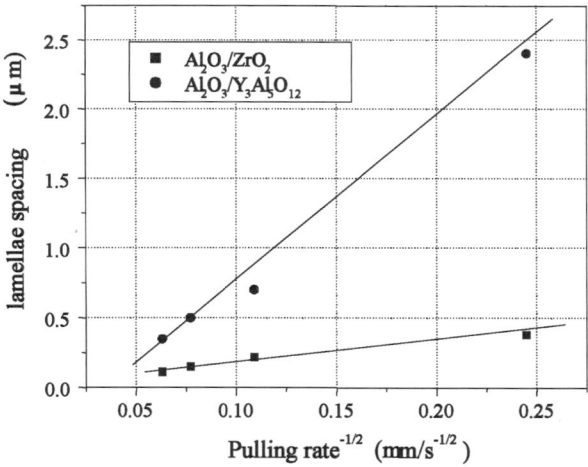

Fig. 6.24. Lamellae spacing of Al_2O_3/ZrO_2 eutectic fiber against the pulling rate.

The lamellar thickness of the zirconia phase was found to be uniform for each cross-section investigated and decreased from 380 nm to 110 nm as the pulling rate increased from 1 mm/min to 15 mm/min, as plotted comparatively with $Al_2O_3/Y_3Al_5O_{12}$ in Fig. 6.24. The general relation $\lambda \sim v^{-1/2}$ where λ is the interlamellar spacing and v is the solidification rate, can also be applied to the lamellar structure of Al_2O_3/ZrO_2 eutectic. The proportional constant is closed to 1, if λ has dimension in μm and v is in $\mu m/s$. This value is small relative to the value of 10 for $Al_2O_3/Y_3Al_5O_{12}$ and $Al_2O_3/GdAlO_3$ eutectic systems [9, 29]. The nanometer scale of lamellar thickness is very thin compared with Al_2O_3/YAG eutectic fiber having micrometer-ordered lamellar thickness [12].

The Vickers hardness value increased from 11 to 13.1 GPa as the pulling rate increased from 1 to 15 mm/min. The tensile strength at room temperature was also increased from 470 to 900 MPa as the pulling rate increased from 1 mm/min to 10 mm/min. We regard this tendency as the microstructure effect resulting from the decrease of the interlamellar spacing at the increasing pulling rate.

6.3.4 Y_2O_3 Stabilized ZrO_2/Al_2O_3 Eutectic

The basic eutectic composition used in this work was 62 mol.% Al_2O_3 and 38 mol.% ZrO_2 as decided in a previous paper [31], and the eutectic temperature of 1870 ± 30 °C was slightly decreased as the doping amount of Y_2O_3 was increased.

Fig. 6.25 shows the as-grown eutectic fibers. The grown fibers were completely white and had a diameter constant to within 10 %. It was possible to control the eutectic fiber diameter from approximately 0.3 mm to 2 mm. The full length was about 500 mm, limited by the apparatus. When the Y_2O_3 doped less than 7 mol.%, stable growth was obtained in the range 0.1–15mm/min, but compositions having over 7 mol.% of Y_2O_3 maximum stable growth rate was decreased to 12 mm/min.

Fig. 6.26 shows a powder XRD pattern of crushed Y_2O_3 doped Al_2O_3/ZrO_2 eutectic fiber. In this figure, it can be seen that these eutectic fibers were composed of only two phases of alumina and zirconia, and there were no traces of the other phases such as YAG. The dominant zirconia phases in undoped and the low content of Y_2O_3 below 1mol% had the monoclinc form, but in a higher content over 3mol% the monoclinic changed to a cubic phase via a tetragonal phase.

BEI of the perpendicular cross-section microstructures of Y_2O_3 doped Al_2O_3/ZrO_2 eutectic fibers are shown in Fig. 6.27. The eutectic microstructures composed of the white and black phase were oriented to a common direction. Under EDS analysis, the white phase was shown to be zirconia and the dark one is alumina. The white zirconia phase was formed on the black alumina matrix, and yttria was detected only in the zirconia phase. This fact confirms the report by Echigoya et al. [30] that added Y_2O_3 dissolves only in the ZrO_2 phase and almost not at all in alumina.

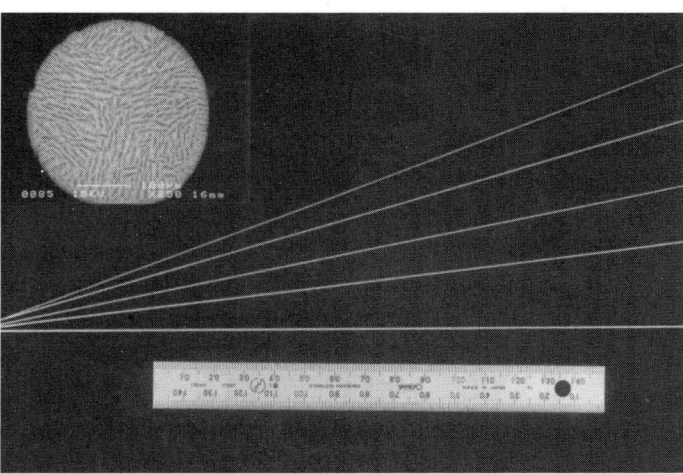

Fig. 6.25. As-grown Y_2O_3 doped Al_2O_3/ZrO_2 eutectic crystal fibers and typical cross-sectional morphology of 3 mol.% doped one grown at the rate of 10mm/min.

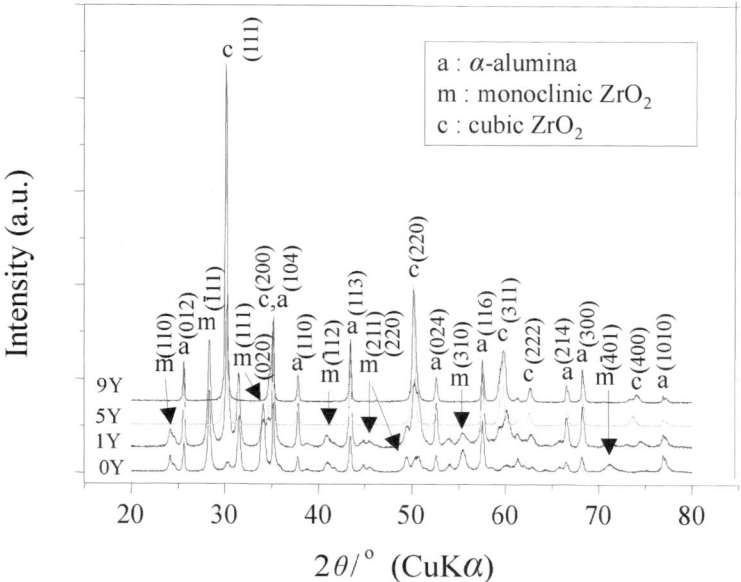

Fig. 6.26. Powder XRD patterns of crushed Y$_2$O$_3$ doped Al$_2$O$_3$/ZrO$_2$ eutectic crystal fibers.

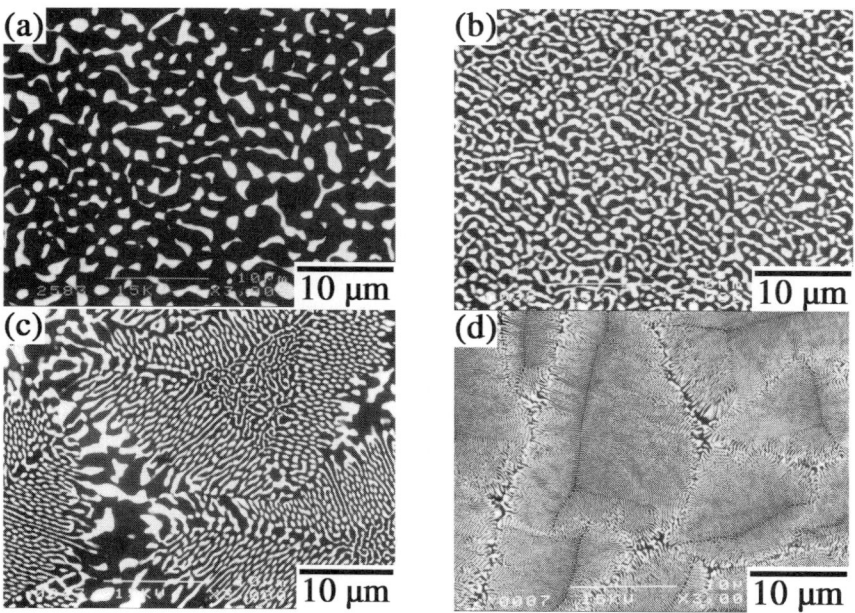

Fig. 6.27. BEI of perpendicular cross-section for 3 mol.% Y$_2$O$_3$ doped Al$_2$O$_3$/ZrO$_2$ eutectic fibers at different pulling rates: (**a**) 0.1 mm/min, (**b**) 1 mm/min; (**c**) 5 mm/min; (**d**) 10 mm/min

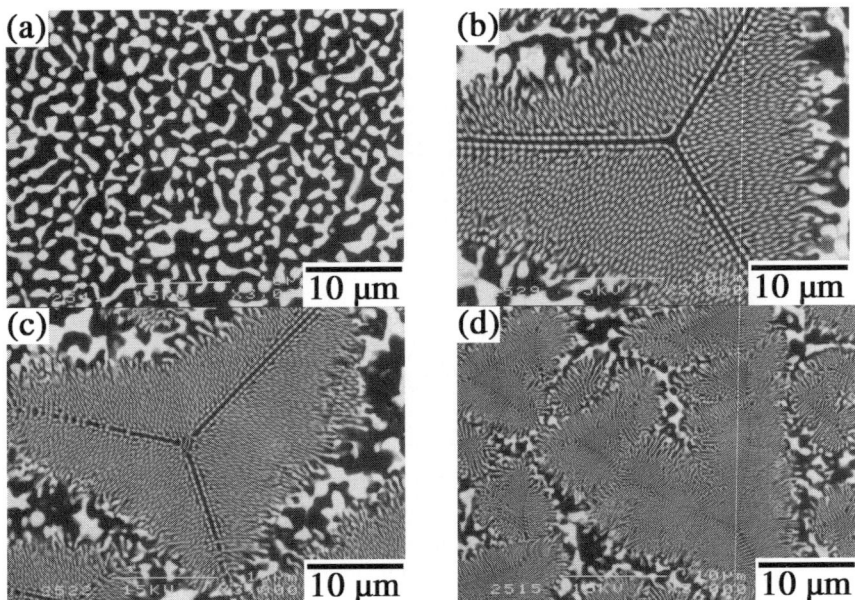

Fig. 6.28. BEI of perpendicular cross-section for 9 mol.% Y_2O_3 doped Al_2O_3/ZrO_2 eutectic fibers at different pulling rates: (**a**) 0.1 mm/min, (**b**) 1 mm/min, (**c**) 5 mm/min (**d**) 10 mm/min.

The eutectic microstructure changed dramatically with pulling rate, and the morphology and the rate of transformation were affected by the amount of Y_2O_3. In the case of below 3 mol.% Y_2O_3 doped system, there were irregular broken lamellar structures formed at lower pulling rates of below 1 mm/min as shown in Fig. 6.27a, but in the pulling rate of about 5 mm/min, the microstructure was changed to an irregular arrowhead shaped cellular structure (Fig. 6.27b), and then at higher pulling rates, over 10 mm/min, an elongated to an ellipsoidal cellular structure was formed.

In the composition having more than 3 mol.% Y_2O_3, the transformation rate became fast, then the irregular lamellar structure of the initial stage changed to a cellular one when the rate was below 1 mm/min (Fig. 6.28a, b). There was one more transformation of a circular type cellular structure, which occurred at the early intermediate pulling rate of around 1 mm/min, and then the structure changed to a highly oriented triangular cellular type separated by triply faceted joints (Fig. 6.28c). Then at a rate of over 10 mm/min an ellipsoidal shaped cellular structure having a double-sided thin lamellar pattern within the cells occurred (Fig. 6.28d). This series of microstructural changes are particular and very different from undoped Al_2O_3/ZrO_2 eutectic fibers [30] and from previous reports [16, 29, 32, 33]. There was only one report on the triangular cellular structure reported by Borodin et al. [5] but they observed only a partially and not regularly distributed and serial transformation. No trend in cell size change could be observed, but as shown in Fig. 6.29 the intercellular spacing was increased from 1.21 μm to 2.28 μm for the fibers grown at the pulling rate of 10 mm/min as the Y_2O_3 doping amount increased from 1 to 9 mol.%. It was postulated that there was an occurrence of Y_2O_3 segregation during solidification. In fact, EDS analysis showed higher Y_2O_3 content at the cell boundaries than inside, which confirmed Echigoya's report [30]. Longitudinal cross-section micrographs also showed the same trend of microstructural transformation. A lamellar structure grown at a low pulling rate of under 1 mm/min showed a regular lamellar structure developed parallel to the growth direction in the length of about 20~30 μm. But the structure grown at higher pulling rates of 10 mm/min had a typical columnar structure with irregular thickness and doubly faceted lamellar tending to tilt toward intercellular boundary regions (see Fig. 6.30).

The maximum Vickers hardness value was recorded approximately to be 20 GPa for the specimen grown at the rate of 15 mm/min and containing 9 mol.% Y_2O_3. A tensile strength test was performed at room temperature (R.T.) and at a high temperature of 1500 °C. Room temperature tensile strengths were tested for the specimens grown at the rate of over 10 mm/min, and the high-temperature tensile test was carried out for the specimen which showed the highest R.T. tensile strength. The highest strength was recorded as 2018 MPa at R.T. and 560 MPa at 1500 °C for the specimen doped with 3 mol.% Y_2O_3 in Al_2O_3/ZrO_2 eutectic fiber grown at the rate of 15 mm/min. These strength values are much higher than for undoped Al_2O_3/ZrO_2 eutectic fiber [31] which had a value of the tensile strength of 900 MPa at R.T. These values are also higher than Borodin's result [5] of 420 MPa obtained by the bending test. It is assumed that the very high strength of these eutectic fibers is due to both the stabilizing effect of Y_2O_3 and microstructural development. It also seems that the mechanical properties in these eutectic crystals are affected by the interlamellar size within the cell rather than by the cell shape and size.

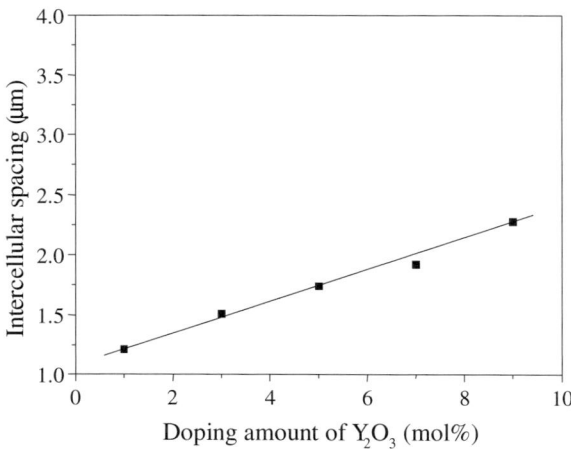

Fig. 6.29. Intercellular spacing of Y_2O_3 doped Al_2O_3/ZrO_2 eutectic fibers grown at the pulling rate of 10 mm/min as a function of the doping amount of Y_2O_3.

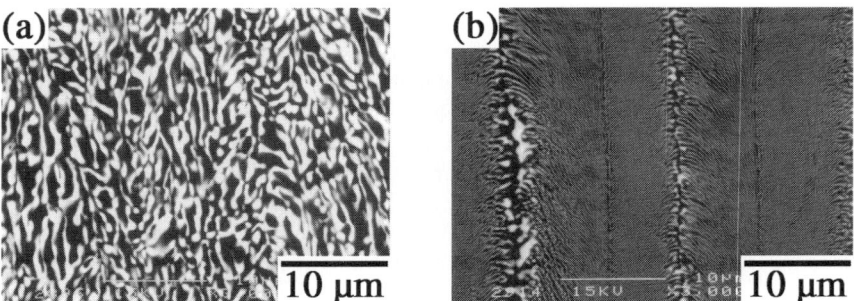

Fig. 6.30. BEI of longitudinal cross-section for 3 mol.% Y_2O_3 doped Al_2O_3/ZrO_2 eutectic fibers: (**a**) 1 mm/min, (**b**) 10 mm/min.

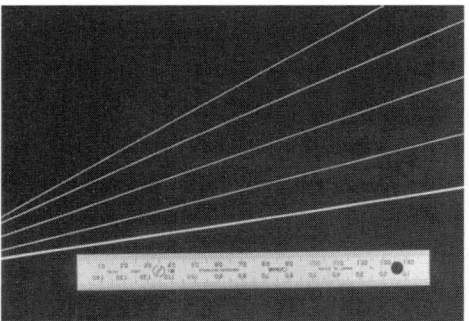

Fig. 6.31. As-grown $Al_2O_3/YAG/ZrO_2$ ternary eutectic fibers.

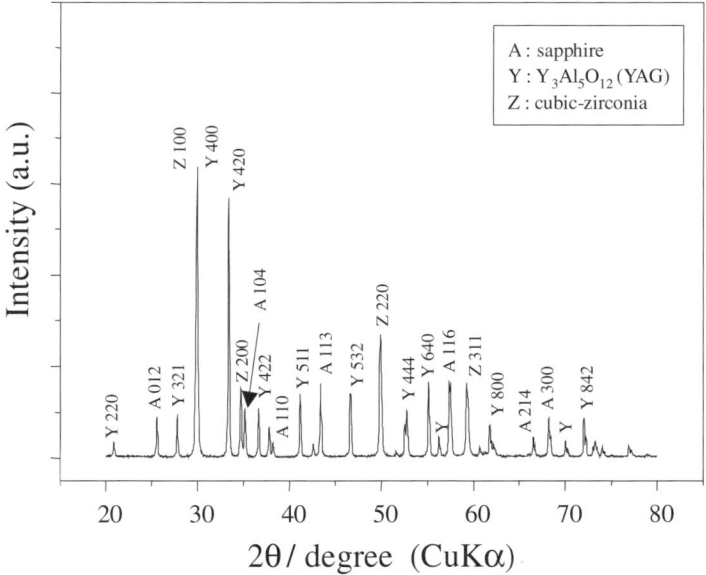

Fig. 6.32. Powder XRD pattern of crushed $Al_2O_3/YAG/ZrO_2$ ternary eutectic fibers.

6.3.5 Al₂O₃/YAG/ZrO₂ Eutectic

The real melting temperature of Al_2O_3/YAG/ZrO_2 ternary eutectics measured in this experiment was 1720 ±30 °C. As-grown Al_2O_3/YAG/ZrO_2 ternary eutectic fibers had an almost white color as shown in Fig. 6.31 and looked just like Al_2O_3/ZrO_2 binary eutectic fibers [35]. Stable growth was obtained in the range of 0.1~15 mm/min. It was possible to control the fiber diameter from approximately 0.3 mm to 2 mm within 10 % of diameter stability. The maximum length was about 500 mm, limited by the apparatus.

Fig. 6.32 shows a powder XRD pattern of crushed Al_2O_3/YAG/ZrO_2 ternary eutectic fiber. All phases were composed of crystalline α-alumina, YAG and cubic zirconia, and there was no trace of other phases such as $Y_4Al_2O_9$ (Y_2A) or $YAlO_3$ (YA). But the peaks of the cubic zirconia phase were shifted slightly as some Y_2O_3 was dissolved. On the basis of qualitative results under EDS, some Y_2O_3 was contained in the ZrO_2 phase. For the solubility limit of Y_2O_3 in ZrO_2, Echigoya [32] reported 13.3 mol.%, but Lakiza [36] wrote 19.2mol%. Although quantification of dissolved Y_2O_3 was not carried out in this work, it was clear that a small amount of Y_2O_3 dissolved in the ZrO_2 phase.

Fig. 6.33 shows the typical phase distribution of Al_2O_3/YAG/ZrO_2 ternary eutectics grown at low pulling rate. The eutectic microstructure was composed of three phases distinguished by their different shapes and colors. By EDS analysis, the black matrix was shown to be Al_2O_3, and the gray phase was YAG. ZrO_2 regions were distributed on the periphery of the YAG phase as relatively small particles that appear white in the micrograph. Both the YAG and ZrO_2 phases in the ternary eutectics showed different morphology than in their respective Al_2O_3-based binary eutectic systems [9, 12, 13, 31, 35]. In particular the particle-shaped ZrO_2 grains had a very different shape from their lamellar or cellular pattern in Al_2O_3/ZrO_2 eutectic fibers [31, 35]. It could be observed that neighboring YAG grains were connected to each other, but the ZrO_2 particles were scattered.

Zirconia   Sapphire   YAG

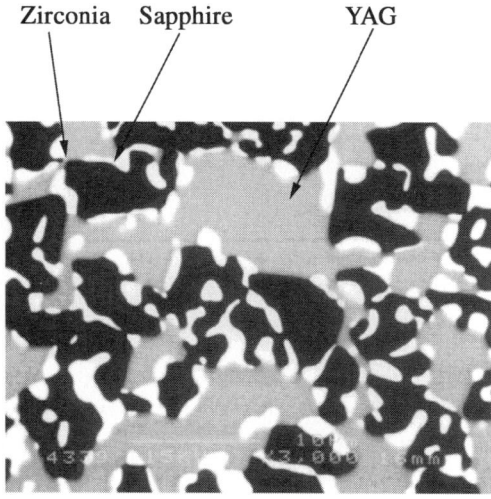

Fig. 6.33. Typical phase distribution of Al_2O_3/YAG/ZrO_2 ternary eutectic.

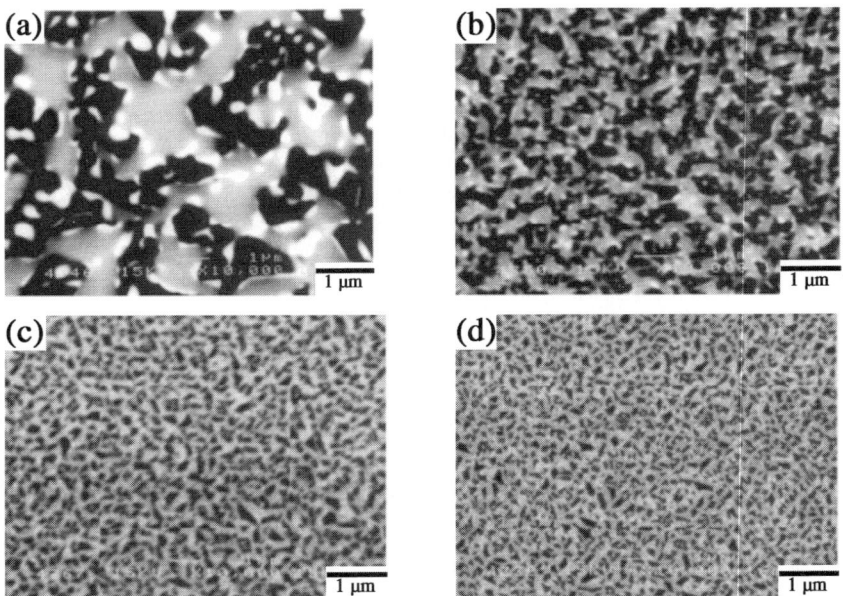

Fig. 6.34. BEI of perpendicular cross-sections of Al$_2$O$_3$/YAG/ZrO$_2$ ternary eutectic fibers grown at various pulling rates of (**a**) 1 mm/min; (**b**) 5 mm/min; (**c**) 10 mm/min; and (**d**) 15 mm/min.

BEI of the microstructures of Al$_2$O$_3$/YAG/ZrO$_2$ ternary eutectic fibers in perpendicular cross-section are shown in Fig. 6.34. The microstructure changed its size with pulling rate, but not its shape, which was different from observations in Al$_2$O$_3$/ZrO$_2$ binary eutectics. At lower pulling rates, below 0.5 mm/min, the YAG phase showed an irregular shape with varying size, as shown at Fig. 6.33, but at pulling rates over 1 mm/min, the interconnection between YAG grains became thick and the microstructure changed to Chinese-script lamellar pattern with regular shape and size as in Fig. 6.34. The script size of the YAG phase was found to be uniform for each cross-section investigated.

The interlamellar spacing decreased from 1.7 µm to 200 nm as the pulling rate increased from 1 mm/min to 15 mm/min, as plotted in comparison with Al$_2$O$_3$/YAG and Al$_2$O$_3$/ZrO$_2$ binary eutectics in Fig. 6.35. The general relation $\lambda \sim v^{-1/2}$ where λ is the interlamellar spacing and v is the solidification rate, could also be applied to the script structure of Al$_2$O$_3$/YAG/ZrO$_2$ ternary eutectics. The proportionality constant is close to 8, if λ is in µm and v is in µm/s. This value is intermediate between 10 for Al$_2$O$_3$/YAG [12, 13] and 1 for Al$_2$O$_3$/ZrO$_2$ binary eutectics [31, 35].

BEI micrographs of longitudinal cross-section in Fig. 6.36 showed that the script structure of the ternary eutectic connected three-dimensionally in both of the microstructures developed at both low and high growth rates.

In order to examine the mechanical properties of the Al$_2$O$_3$/YAG/ZrO$_2$ fibers, tensile strength tests were performed for the eutectic fibers grown at a pulling rate of 15 mm/min at various temperatures from room temperature to 1500 °C. As shown in Fig. 6.37, the highest strength was 1100 MPa at room temperature; the strength diminished only slightly with temperature until it decreased drastically to 280 MPa at 1500 °C, due to the proximity of its melting temperature.

The measured strengths of $Al_2O_3/YAG/ZrO_2$ ternary eutectic fibers are much higher than the Al_2O_3/YAG binary fiber over the whole range of tested temperature except for 1500 °C. It is assumed that higher strength of the ternary eutectic fibers over Al_2O_3/YAG results from both the strengthening effect of ZrO_2 and the smaller script size. Vickers hardness values also increased as the pulling rate increased. Maximum hardness was recorded at 17.4 GPa for the specimen grown at a pulling rate of 15 mm/min. Hardness values of the ternary eutectics the intermediate between the 16.3 GPa for Al_2O_3/YAG and the 20 GPa for Al_2O_3/ZrO_2 binary eutectics [35].

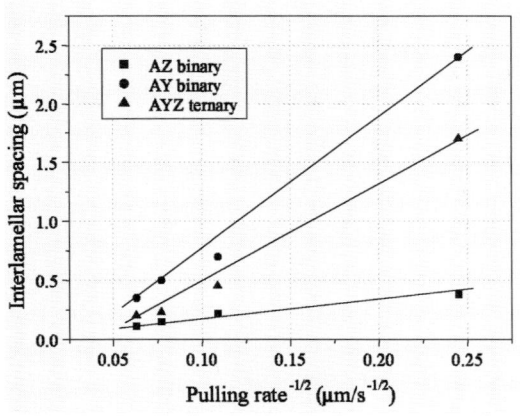

Fig. 6.35. Interlamellar spacing of $Al_2O_3/YAG/ZrO_2$ ternary eutectic fibers grown at the various pulling rates.

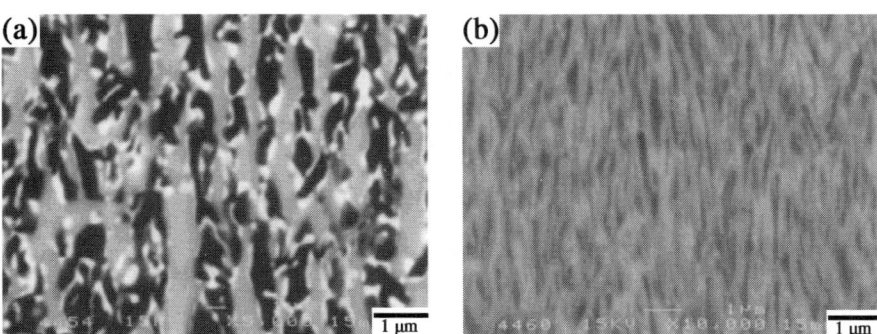

Fig. 6.36. BEI of longitudinal cross-sections of $Al_2O_3/YAG/ZrO_2$ ternary eutectic fibers grown at various pulling rates of (**a**) 1 mm/min and (**b**) 15 mm/min.

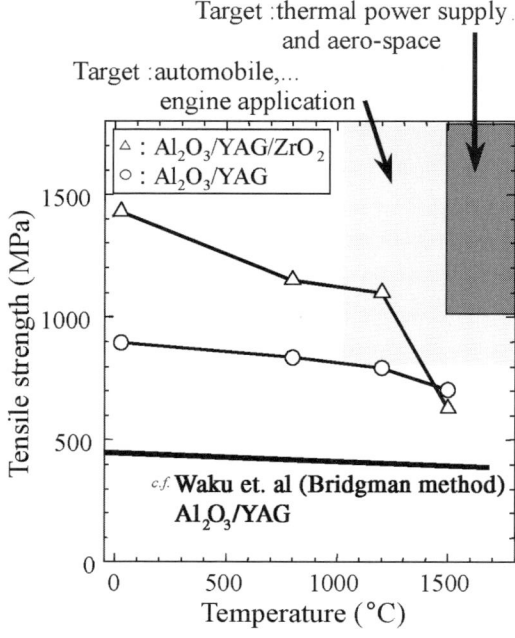

Fig. 6.37. Tensile strength of $Al_2O_3/YAG/ZrO_2$ ternary eutectic fibers at various temperatures.

References

1. V. S. Stubican and R. C. Bradt, Ann. Rev. Mater. Sci. 11 (1981) 267.
2. W. A. Tiller, Liquid Metals and Solidification (1957) 279, Cleveland Ohio: Am. Soc. Met.
3. J. D. Hunt and K. A. Jackson, Transactions of the Metallugical Society of AIME 236 (1966) 843–852.
4. D. Viechnicki and F. Schmid, J. Mater. Sci. 4 (1969) 84.
5. V. A. Borodin, A. G. Reznikov, M. Yu. Starostin, T. A. Steriopolo, V. A. Tatarchenko, L. I. Chernyshova and T. N. Yalovets, J. Cryst. Growth 82 (1987) 177–181.
6. V. A. Borodin, M. Yu. Starostin and T. N. Yalovets, J. Cryst. Growth 104 (1990) 148–153.
7. F. L. Kennard, R. C. Bradt and V. S. Stubican, J. Am. Ceram. Soc. 62 (1979) 154.
8. R. S. Feigelson, Mat. Sci. Eng. B1 (1988) 67.
9. A. Yoshikawa, B. M. Epelbaum, K. Hasegawa, S. D. Durbin and T. Fukuda, J. Cryst. Growth 205 (1999) 305–316.
10. D. H. Yoon, I. Yonenaga, N. Onishi, and T. Fukuda, J. Cryst. Growth 142 (1994) 423.
11. B. M. Epelbaum, K. Shimamura, S. Uda, J. Kon and T. Fukuda, J. Cryst. Res. Technol. 31 (1996) 1077.
12. B. M. Epelbaum, K. Inaba, S. Uda, K. Shimamura, M. Imaeda, V. V. Kochurikhin, and T. Fukuda, J. Crystal Growth 176 (1997) 559.

13. A. Yoshikawa, B. M. Epelbaum, T. Fukuda, K. Suzuki and Y. Waku, Jpn. J. Appl. Phys 38 (1999) L1623.
14. B. M. Epelbaum, A. Yoshikawa, K. Shimamura, T. Fukuda, K. Suzuki, and Y. Waku, J. Cryst. Growth 198/199 (1999) 471.
15. D. Viechniki and F. Schmid, J. Mater. Sci. 4 (1969) 84.
16. Y. Waku, N. Nakagawa, T. Wakamoto, H. Ohtsubo, K. Shimizu, and Y. Kohtoku, J. Mater. Sci. 33 (1998) 1217.
17. Y. Waku, N. Nakagawa, T. Wakamoto, H. Ohtsubo, K. Shimizu, and Y. Kohtoku, Nature 389 (1997) 49.
18. J. O. Ström-Olsen, G. Rudkowska, P. Rudkowski, M. Allahverdi and R. A. L. Drew, Mat. Sci. Eng. A179/A180 (1994) 158.
19. P. Wu and A. D. Petron, J. Alloys Compd. 179 (1992) 259.
20. B. Baretzky, B. Reinsch, U. Taeffner, G. Schneider and M Ruehle, Z. Metallkd., 87 (1996) 332.
21. J. T A. Pollock and J. S. Bailey, J. Mat. Sci. 9 (1974) 323.
22. G. N. Morscher and H. Sayir, Mat. Sci. Engin. A190 (1995) 267.
23. M. Allahverdi, R. A. L. Drew, P. Rudkowska, G. Rudkowski and J. O. Ström-Olsen, Mat. Sci. Engin. A207 (1996) 12.
24. F. T. Wallenberger, N. E. Weston, K. Motzfeld and D. G. Swartzfager, J. Am. Ceram. Soc. 75 (1992) 629.
25. W. J. Minford, R. C. Bradt and V. S. Stubican, J. Amer. Ceram. Soc., 62 (1979) 154.
26. G. R. Fischer, L. J. Manfredo, R. N. McNally, R. C. Doman, J. Mater. Sci., 16 (1981) 3447.
27. V. G. Cervales, Ber. Dtsch. Keram. Ges., 42 (1968) 216.
28. F. Schmid and D. Viechnicki, J. Mater. Sci., 5 (1970) 470.
29. A. Yoshikawa, K. Hasegawa, J. H. Lee, S. D. Durbin, B. M. Epelbaum, T. Fukuda and Y. Waku, J. Cryst. Growth 218 (2000) 67–73
30. J. Echigoya, Y. Takabayashi, H. Suto and M. Ishigame, J. Mater. Sci. Lett. 5 (1986) 150.
31. J. H. Lee, A. Yoshikawa, K. Hasegawa, S. D. Durbin, D. H. Yoon and T. Fukuda, J. Cryst. Growth 222 (2001) 791–796
32. J. Echigoya, Y. Takabayashi, K. Sasaki, S. Hayashi, H. Suto, Trans. Jap. Inst. Metals, 27 (1986) 102.
33. M. Yu. Starostin, B. A. Gnesin, T. N. Yalovets, J. Crystal Growth, 171 (1997) 119.
34. V. A. Borodin, A. G. Reznikov, M. Yu. Starostin, T. A. Steriopolo, V. A. Tatarchenko, L. I. Chernyshova and T. N. Yalovets, J. Crystal Growth, 82 (1987) 177.
35. J. H. Lee, A. Yoshikawa, K. Lebbou, H. Kaiden, T. Fukuda, D. H. Yoon and Y. Waku J. Cryst. Growth 231(2001) 179–185.
36. S. M. Lakiza and L. M. Lopato, J. Am. Ceram. Soc. 80 (1997) 893.

7 Oxide Fiber Crystals Grown by μ-PD and LHPG Techniques

Kheirreddine Lebbou and Georges Boulon

Single crystal fibers are very attractive for optical applications as active and passive elements. The research into single crystal fibers became active 30 years ago, especially when it was found that the crystal fibers had a high crystalline perfection and near theoretical strength. They can be produced by a variety of methods, the most versatile of which are the laser-heated pedestal growth and micro-pulling down techniques. We will describes these two techniques and will present some original results based on these two process.

7.1 Introduction

The many properties of single crystal fibers and their geometry makes them highly useful in a variety of optical devices. The fiber application for miniaturization and operation of laser sources are a big example, especially for telecommunications and photonics. There is a big interest in the development and improvement of solid state materials for the generation, transmission, detection and conversion of optical signals over a broad range of wavelengths and power levels. Single crystals have an important role in most of these applications either in the form of bulk crystals (three-dimensional) or epitaxial thin film (two dimensional). The single crystal fiber geometry cal largely be oriented in one dimension during growth and could have the potential for use in a variety of fiber optic devices such as laser sources, electro-optics modulators, switches, couplers, isolators, transmission lines, remote sensors, etc [1]. Quite often, for the desired application, it is important that the grown fibers have a specific length, diameter and orientation. There are different techniques to produce single fiber crystals. But not all of them allow the parameters cited to be controlled. This is a reason why it is impossible to use, for example, vapor and solution growth techniques to prepare single crystal fibers. The optimal growth or preparation technique to get fibers are the following: (1) EFG technique [2], (2) pulling through a die [3], (3) floating zone (pedestal) growth [4–7], solidification in a capillary tube [8–10], capillary drawing [11] and pressurized capillary-fed growth [12]. Generally, the choice of the method is strongly depended on the physical and the chemical properties of the materials to be grown. Optical loss can be causes due to imperfect bulk crystals, index of refraction variations, diameter fluctuation and surface defects. Diameter fluctuations depends on the azimuthal dependence and spatial frequency of the perturbation. We estimate losses on the order of 25% for 1% random diameter fluctuation on a 5-cm length with a fiber diameter of 25μm. The optical losses can be reduced by using a diffused cladding if the diffusion depth is large compared to the scale length of the diameter variation. Diffused cladding can be accomplished in different ways.

The proposed article shows the basics of fiber crystal growth and will be focused on laser heated pedestal growth LHPG and micro-pulling down μ-PD techniques.

7.2 Single Crystal Elaboration by Floating Zone Technique Using a Laser Source as Heater

7.2.1 The Principle of Zone Melting

Melting zone techniques were reported by Kapitza et al. [13] for bismuth single crystal growth in 1928. This technique became a general method for chemical purification of semiconductors metals [14]. The germanium was obtained, from 1953 onwards, with very low impurity concentrations (<10–11at.%)[15]. Figure 7.1 shows the technique in principle. A horizontal furnace is moved along the rod. The moving furnace gives a melting in one direction and crystallization in the opposite direction. The sample is contained horizontally in the crucible. Different heating modes were used, resistive heater, inductive, infrared radiation heating, electronic bombardment or discharge using an electrical arc.

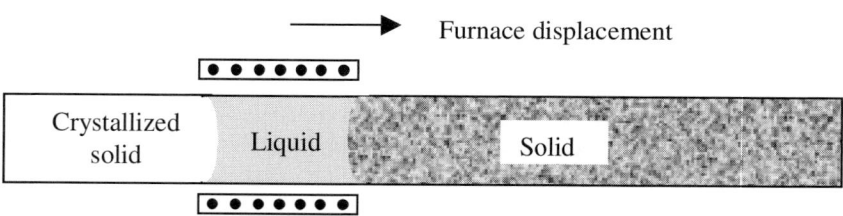

Fig. 7.1. Principal illustration of melting zone techniques

7.2.2 Species Distribution and Pfann Relation

During the motion through the melting zone, the melting and phenomena of growth are based on the equilibrium diagram, the kinetic phenomena and the moving rod speed during the heating process. The distribution of impurities in the liquid phase and the recrystallized solid in the crystallization front is controlled by the distribution coefficient k. This coefficient is defined by the ratio of the concentration of impurities in the crystallized phase (C_{sol}) and melting zone (C_{liq}):

$$k = C_{sol}/C_{liq}. \tag{7.1}$$

In the case of ideal diluted binary mixture, the liquidus and solidus curve can be considered linear, the distribution coefficient is a constant and could be calculated directly from the equilibrium diagram. Figure 7.2 gives an example in the case of purification, where k is less than one. If the equilibrium conditions are realized, for the concentration C_0 of the liquid phase, the crystallized solid composition is kC_0.

The species distribution along the rod prepared by the melting zone technique is described by the Pfann relation [14]. The calculated distribution coefficient is based on the following hypothesis:

– the melting zone is a conservative process, but in reality a small amount of the material can be vaporized in the melting zone;
– the calorific capacities and the enthalpy of the phase transition are constant along the rod. This means that the melting zone length must be constant;
– the liquid is homogeneous;
– solid diffusion is negligible;
– solid liquid interface is planar;
– the volume mass of the solid and liquid are equivalent.

If the transversal section of the rod is S, l the melting zone length and n_x the species mole number in the melt, when the melting zone extends between x and $x+l$ (Fig. 7.3), the concentration in the melt is:

$$C_{liq,x}=n_x/S.l \qquad (7.2)$$

When the melting zone moved dx along the rod of C_0 initial composition, we can write the material analysis of i species in the melt as,

$$dC_{liq}.S.l=C_0Sdx - C_{sol,x}Sdx \qquad (7.3)$$

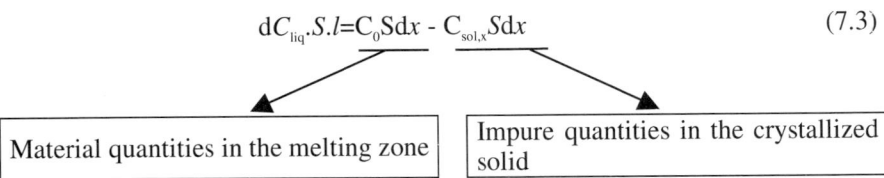

| Material quantities in the melting zone | Impure quantities in the crystallized solid |

$C_{sol,x}$ is the concentration in the crystallized solid
Following the definition of the distribution coefficient k, $C_{liq}=C_{sol,x}/k$

We deduce:

$$(1/k)\, dC_{sol,x}=C_0dx-C_{sol,x}.dx \qquad (7.4)$$

$$\Leftrightarrow \quad (dC_{sol,x}/dx) + (k/l)C_{sol,x} -(k/l)C_0=0 \qquad (7.5)$$

for $x=0$, the melting zone composition is C_0 and the crystallized solid composition will be $C_{sol,x=0}=kC_0$, we can write the Pfann composition after integration of the equation (7.5) as

$$C_{sol}(x) = C_0 \{1-(1-k)e^{-kx/l}\}. \tag{7.6}$$

Fig. 7.4 shows the impurity concentration evolution as a function of the rod length. The rod composition is C_0 and $k<1$. This figure shows:

- in the domain for which the crystallized solid composition varied from kC_0 to one maximal limit value $C_{sol,lim}$ if $k<1$ and minimal if $k>1$.
- A stationary domain for which the composition of the crystallized solid is constant and equal to $C_{sol,lim}$. This region is also labeled (orientation zone) and allows the growth of a single crystal.

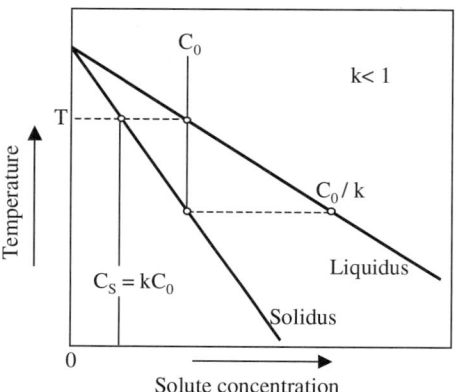

Fig. 7.2. Definition of segregation coefficient (k)

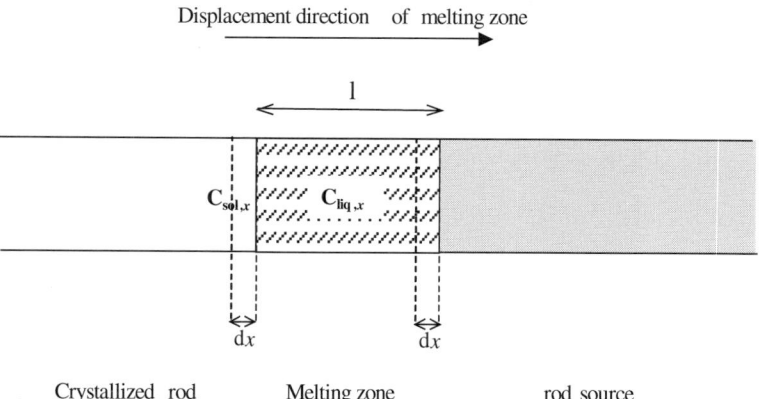

Fig. 7.3. Melting zone description during pulling

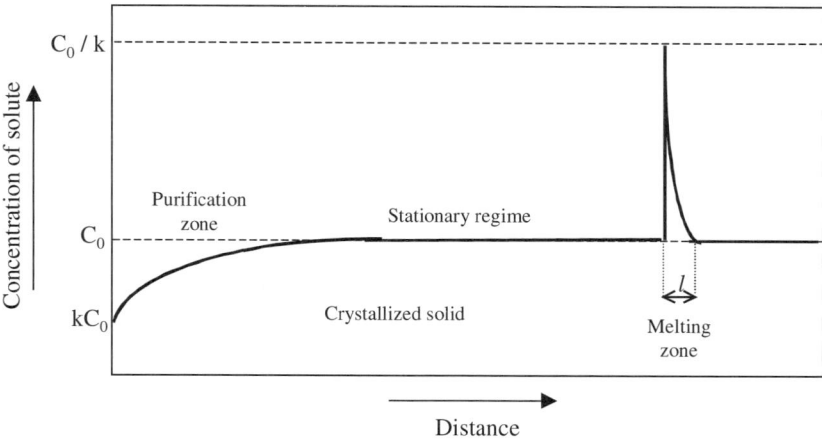

Fig. 7.4. Dopant repartition along the rod after melting ($k<1$)

For a slow displacement of the melting zone (slow pulling rate), the thermodynamic equilibrium solid⇔liquid is obtained at the crystallization interface. The distribution coefficient is assumed to be the coefficient k_0 deduced from liquidus and solidus curve.

If the melting zone displacement is slightly slow, a diffusion layer at the crystallization interface is observed. The diffusion layer corresponds to:

– bad diffusion of impurities rejected by the crystallized solid ($k<1$);
– a poor impurities melting zone region comparing with the average concentration ($k>1$).

Burton, Prim, and Slichter defined an effective distribution coefficient (k_{eff}), for which the cooling rate and diffusion in crystallization interface are involved.

$$k_{eff} = k_0 \, / \, [k_0 + (1-k_0) \, e^{-V\delta D}] \tag{7.7}$$

δ is the diffusion layer thickness,
V is the rate of melting zone displacement, and
D is the diffusion coefficient in the liquid.

For a fast pulling rate, the exponential part is 0, in this case the segregation coefficient is 1. The crystallized solid has the same composition as the melting zone.

7.2.3 LHPG (Laser Heated Pedestal Growth) Technique

Poplawski [16] was the first one who has initiated crystal growth using the pedestal growth design process based on the melting materials by energy created by an image furnace. Then Haggerty [17] et al. developed the LHPG technique in 1972 and later this was improved on by Feigelson [18] at Stanford University.

A schematic illustration of LHPG is given in Fig. 7.5. It is based on the utilization of CO_2 IR laser beam emitting at 10.06 μm. The laser power and beam are most stable if the laser is allowed to run at constant high power.

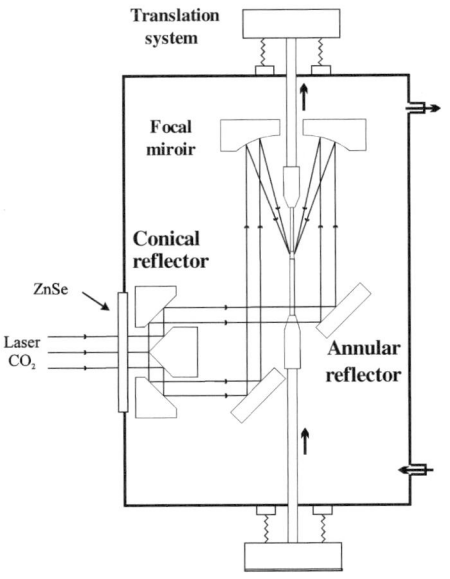

Fig. 7.5. Schematic illustration of the laser heated pedestal growth technique

Fig. 7.6. Inside the growth LHPG chamber

Fejer et al [19] used a novel optical system focusing the laser beam onto the fiber in a 360° axially symmetric distribution to prevent cold spots in the growth zone. The power is computer controlled by a function proportional controller. The laser beam enters into a hermetically sealed chamber. This chamber prevents starry air currents from causing diameter variations and allows the use of a reactive or inert atmosphere and also supports a good vacuum (Fig. 7.6).

The amount of power necessary to melt a compound depends on its melting point, the diameter of the source rod and the optical absorption coefficient of the material. 10.06 μm can easily melt the materials with a melting temperature of around 3000°C. The remarkably important application for single crystal fiber processing through the LHPG technique is the growth of the new materials in single crystal form for physical, chemical and crystallographic evaluation. It is also a powerful technique for growing crystals of solid solutions as well as incongruently melting materials and crystals with different dopant concentrations. This method is ideal because single crystals can easily be produced in a much shorter time then by conventional methods and at much lower cost. This method is crucible free, it does not require furnace insulation (Fig. 7.7) and contamination problem are completely minimized. The source material can be easily manufactured in the form of either round or square cross section rods from a polycristalline material which has been fabricated from solidified melts pressed and sintered powders or from hot pressed samples.

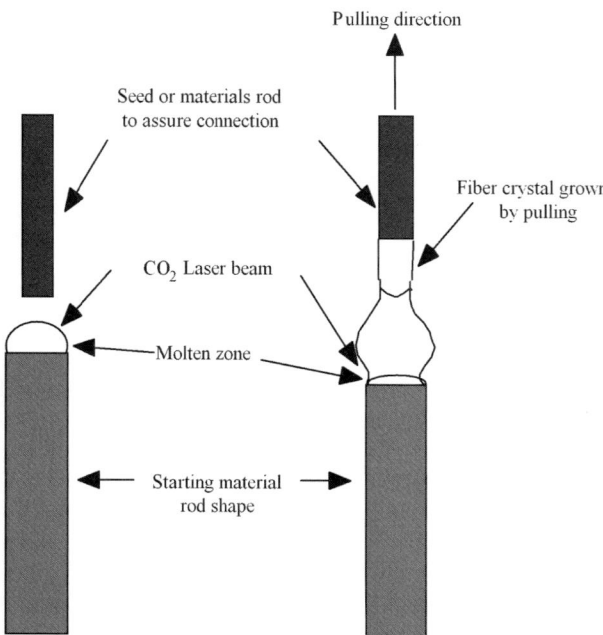

Fig. 7.7. Fiber crystal growth using LHPG technique (crucible free and the heated source is a CO_2 laser); left: before dipping; right: during growth.

For good diameter control, the source rods should have a round cross section with a very uniform diameter. Rods can be densified by partially melting with the laser beam before growth. The small diameter of the crystals grown allows easy grain selection. A single grain will dominate the cross section after only a short amount of the growth. To initiate growth the use of either an inert wire, a single crystal seed or a polycristalline rod have been found equally effective. Defects, such as twins and dislocations, unless they propagate exactly parallel to the growth direction will eventually grow out of the crystal leaving the fiber with a lower defect density than normally found in bulk crystals. Fibers can easily be grown along specific crystallographic directions. Controlled fiber growth requires a stable molten zone.

The stability of the molten zone results from a steady-state heat flow at the solid -melt interface according to the relationship:

$$Q_s = Q_f + Q_m = A\rho\Delta H_f(dx/dt) + Ak_l(dT/dx)_l = Ak_s(dT/dK)_s = \text{Constant} \qquad (7.8)$$

Q_s is the heat flow in the crystal away from the growth interface,
Q_m is the heat flow from the melt toward the interface,
Q_f is the latent heat of the crystallization,
A the area of the interface,
ρ_s density of the solid,
ΔH_f the latent heat,
K_l and K_s the thermal conductivity of the liquid and the solid,
$(dT/dx)_s$ and $(dT/dx)_l$ are the temperature gradient in the solid and liquid.

The material which melts congruently, has a low vapor pressure at its melting point and evaporates congruently, will generally be easy to grow in single crystal form by the LHPG method. Also the floating zone and pedestal growth methods are both excellent techniques for producing single crystals from incongruently melting compounds. If we start with a source rod of such a material, the melt which is formed will produce a crystalline phase which is initially different from that of both the melt and the source rod and the crystal will have the same composition. If the melt is incongruent, this means that one or more components are lossy, it is possible to grow good quality fiber single crystals. This can be realized by the incorporation onto the source material of an excess of the volatile component. LHPG technique has been used to grow a lot of materials. Table 7.1 summarize the different materials grown by the LHPG technique.

As it is shown in Table 7.1, LHPG technique has been used to grow a wide variety of materials. But there are some materials that LHPG techniques is not suitable to be used as a growth technique for, due to thermodynamic or kinetic considerations. Many IR materials melt at too low a temperature and exhibit a significant vapor or dissociation pressure at their melting temperature. In this case other single crystal fiber techniques which avoid the dissoziation such as growth in a capillary tube or die, might be more appropriate.

Table 7.1. Fiber grown by LHPG technique

Materials	Melting temperature (°C)	Diameter (μm)	Application
SBN	1480	100-700	ferroelectric devices
BNN	1450	100-800	ferroelectric devices
LiNbO3	1250	50 –900	electro-optics
Nd:LiNbO3	1250	50-900	lasers
Nd:YAG	1940	50-800	lasers
YIG	1560	100-700	isolators
GGG	1750	150-700	magnetic devices
Nd-YVO4	1830	100-800	lasers
BaTiO3	1620	200-700	ferroelectric devices
Bi2Sr2CaCu2Ox	incongruent	100-1000	superconductivity
Eu -Y2O3	2410	100-800	laser spectroscopy
Gd2O3	2450	100-800	crystallographic study
CaF2	1360	200-800	IR guiding
BaF2	1280	200-800	IR guiding

7.3 Pulling Technique from a Meniscus Using the Micropulling Down Technique (μ-PD)

Single crystal fiber growth using μ-PD method was developed further at the Fukuda Laboratory of the Tohoku University in Sendai Japan [20]. Figures. 7.8a and 7.8b show a schematic illustration and the seed preparation used in this technique. The micro-pulling system is based on a resistive or radio frequency heating system. The fiber growth method involves growing a fiber through a micro nozzle by pulling in the downward direction, as shown in Fig 7.8a. The growth equipment consists of a Pt or Ir crucible as a function of melting temperature. The crucible is directly heated resistively or by radio frequency. An after-heater made from a Pt wire or an Ir tube is used to reduce the temperature gradient. A second annealing furnace is also used to anneal crystal in order to reduce the thermal stress concentration. A micro X-Y stage allows lowering of the mechanism.

The raw materials is melted in the crucible and allowed to pass through the micro nozzle or the crucible hole in the case of the Ir crucible. The fiber is formed by attaching the seed to the tip of the micro nozzle and then slowly pulling it downward with a constant velocity. The growth procedure is observed directly during the process using a camera. The crucible shape and the geometry are a function of melting temperature and wetting angle (Fig. 7.9). Congruent melting components with high melting temperature require an Ir or Pt/Rd crucible, but the crucible's thickness and geometry including the size is a function of melting temperature.

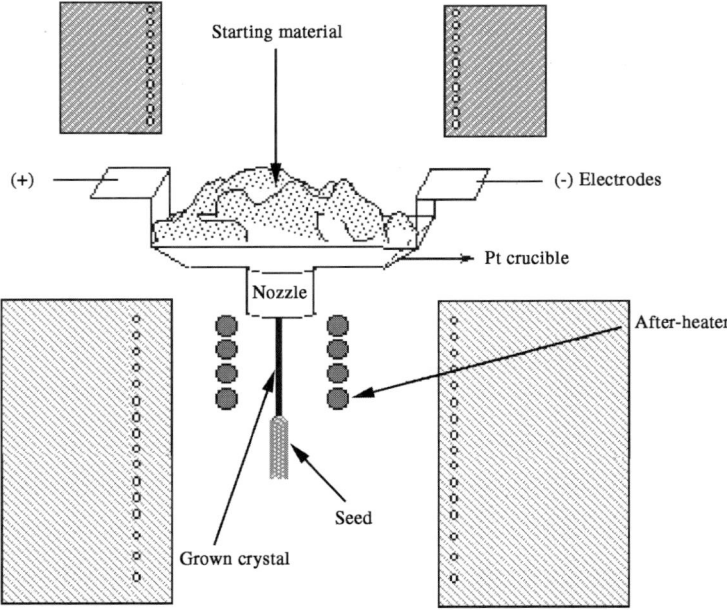

Fig. 7.8.a Schematic illustration of the resistive micropulling down technique

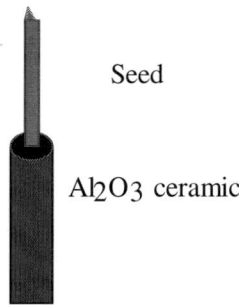

Seed

Al$_2$O$_3$ ceramic

Fig. 7.8b Seed preparation

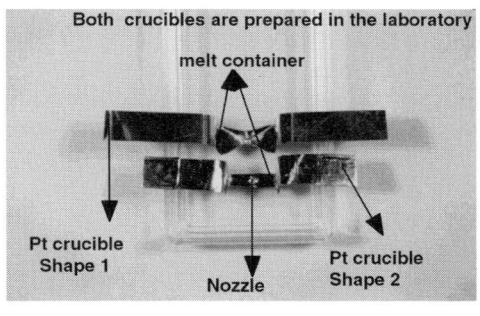

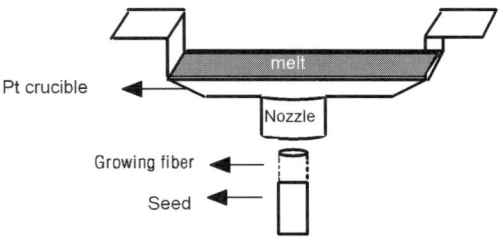

Fig. 7.9. Crucible conception design for the low temperature μ-PD technique

The utilization of a Pt/Rh crucible decreases the crystal quality and some coloration could be observed in the crystals because of the Rh presence. This phenomena was observed, especially for $LiNbO_3$ and SBN niobate crystal growth using Pt/Rh10%. Table 7.2. summarize the comparison between LHPG and μ-PD technique. It is clear that the absence of the crucible in the LHPG method is a certain advantage including the final crystal qualities which will not be contaminated during the growth process.

Table 7.2. Comparison between μ-PD and LHPG techniques

	μ-PD	LHPG
Crucibles	Pt, Pt/Rh, Ir,...	-
Starting materials	melt	rod
After-heater	Pt, Pt-10Rh	-
Annealing furnace	yes	-
Atmosphere	controlled	controlled
Contamination	yes (crucible effect)	no
Temperature	crucible function	very high
Pulling rate	fast	fast
Seed adjustment	yes	yes
Fiber length	~1meter	limited by the cell sizes
Interface	one	two
Melt qualities	congruent and incongruent	congruent and incongruent
Segregation coefficient	close to one	function of pulling Rate
Heating system	resistive or RF	laser
Power regulation	could be stable	function of laser stabilities

7.4 Illustration of the Fiber Crystal Growth Potential

In the following selected optical materials, superconductors and fluoride crystals grown by the LHPG and μ-PD the technique will be presented.

7.4.1 Nonlinear Niobate Fiber Crystals LiNbO3

Lithium niobate is a well known synthetic single-crystal oxide material. Its total production now amounts to about 50–100 tons per year, driven by its successful use in several field of application such as surface acoustic wave (SAW) devices, non linear optics, electro-optics and guided-wave devices. Single crystal lithium niobate (LN) is typically produced using the Czochralski technique. Traditionally LN single crystal boules are of congruent composition (as opposed to the stoichiometric $LiNbO_3$). This is because the composition of the melt does not change during solidification of the stochiometric composition. Even though a lot of progress was done on the technology of crystal growth, a lot of problems exist during the growth process and the composition problem remains unresolved. The properties of stochiometric $LiNbO_3$ single crystals are better than the congruent LN crystal, this is a reason why new research is needed to investigate more closely the crystal growth and more study is needed to correlate the crystal growth process, chemical properties and optical application. Using LHPG technique we have grown $LiNbO_3$ and $Yb-LiNbO_3$ single crystal fibers with good qualities. Figure 7.10, shows the morphology of a fiber. The fiber is transparent defect free and homogenous.

The distribution of Yb^{3+} along the fiber grown at 70 mmh^{-1} was obtained by monitoring the intensity of the Ytterbium luminescence under transversal and punctual excitation along the fiber. We have found, from the plotted data, that the Yb^{3+} concentration is not homogeneous and tends towards a stationary value corresponding to the Yb^{3+} concentration of the source rod, as predicted by the Pfann equation:

$$C_t(x)/C_s = (R^2v_R\rho/r^2v_t\rho)-[(R^2v_R\rho/r^2v_t\rho) - k]*\exp[(-\pi r^2/V)v_t kt] \qquad (7.9)$$

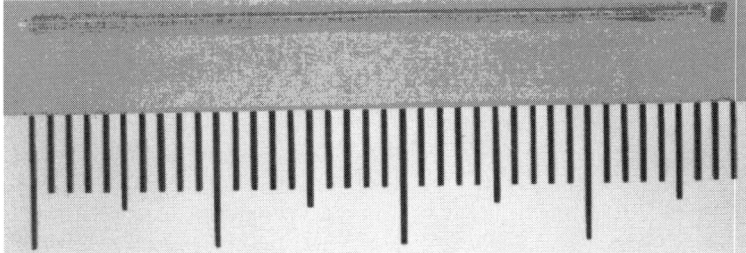

Fig. 7.10. LiNbO3 Fiber grown by LHPG technique (scale mm)

where $C_i(x)$ is the Yb concentration at position x of the fiber, C_s the concentration of the feed, k the apparent distribution coefficient, r the radius of the fiber, R the radius of the feed, V the volume of the molten zone, v_r the speed of the crystal, v_R the speed of the feed and ρ' the compacity of the feed and ρ the theoretical compacity. We found that the Yb distribution coefficient was larger than 1. No attempt was made in the fibers to localize the Yb^{3+} ions inside the lattice but in bulk $LiNbO_3$ crystals, Rutherford back-scattering spectrometry/channeling techniques have located Yb^{3+} ions in the Li^+ octahedron of the unit cell, shifted from the Li^+ regular position as previously reported [21–22]. In order to investigate the ability of the ytterbium doped fibers to self-double their σ laser emission in the green spectral range near 530 nm, we have measured the wavelength for non-critical birefringent phase matching at room temperature. We have found that the room temperature the phase matching wavelength is strongly composition dependent, allowing an exact adjustment to the laser emission wavelength, expected near 1060 nm. The fiber giving the shortest wavelength for phase matching was a Li-rich fiber (49.47% mol Li_2O) grown from a 50% mol Li_2O feed rod at a pulling rate of 120 mm h^{-1}.

The fiber is prepared by micro-pulling down have high qualities and are close to LHPG materials. Uda et al. [23] have grown a Mn: $LiNbO_3$ fiber using micro-pulling down process. They have observed a temperature gradient inversion in the molten zone and which is specific to the μ,-PD method. This behavior reduces the radial influence of the interface electric field leading to a rather homogeneous distribution of the solute in the crystal with mid to large diameter. Yoon et al. [24] have grown a dislocation-free $LiNbO_3$ with small diameter. The majority of the fiber grown by micro-pulling down and LHPG techniques are closed to the stochiometric $LiNbO_3$ compound, comparing to other techniques such as Czochralski and Bridgman techniques. This is related to the molten zone dimension which is smaller than in other techniques including the convection phenomena which are drastically minimized compared with the others techniques.

7.4.2 Ba$_2$NaNb$_5$O$_{15}$ (BNN)

One of the major problems during the elaboration of BNN single crystals by conventional Czochralski technique is the occurrence of cracks during the cooling process of the crystals growth. Cracking is due to the large thermal expansion of the c-axis at the ferroelectric transition temperature [25–28]. Moreover, BNN crystals exhibits micro-twins because of the exchange of a and b axes of the orthorhombic crystalllographic structure. As the twins disturb the optical properties, crystals should be detwinned by cooling under a compressive stress and simultaneously, poled under a direct-current voltage. The necessity of such a procedure which sometimes leads to breaking of the crystals has also constituted an impediment for commercial development. The crystal growth of BNN is accompanied by striation development [29] perpendicular to the growth axis which perturb the optical properties and changes the refractive index value (Fig. 7.11).

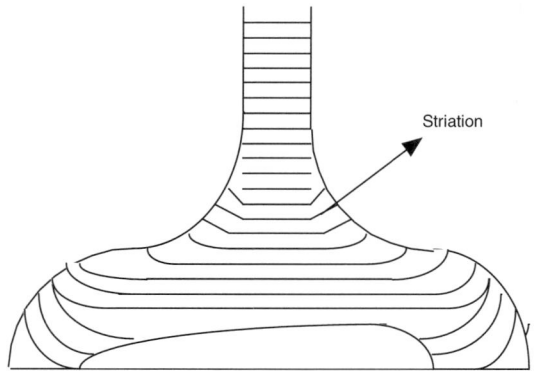

Fig. 7.11. Striation propagation during a grown BNN crystal

Using the μ-PD method and LHPG techniques, $Ba_2Na_{1-3x}Yb_xNb_5O_{15}$ (0.02 •x•
0.06) and $Ba_2Na_{1-3x}Gd_xNb_5O_{15}$ were grown from the stoichiometric melt [30–31]. In
the case of undoped BNN materials, cracks appear and are directly observable will
the eyes during the growth process [29], but for Yb-and Gd doped compounds, no
cracks were observed during growth (Fig. 7.12). The crystals were colorless and
transparent independently of the melt composition, with uniform shape and free of
macroscopic defects such as cracks, bubbles or inclusions.

The cracks and diameter variation were observed only for stoichiometric BNN
[29]. The typical size of crystals was 0.4–0.8 mm in cross section, depending on
nozzle diameter and a few centimeters in length. In the main, about 90 vol % was
crystallized into single crystals . It was also possible to grow BNN single crystals
with a high pulling rate of about 1.5mm/min. Phase homogeneity of the crystals
was studied by x-ray diffraction (XRD) powder analysis at room temperature. The
crystals were BNN-Yb and BNN-Gd single phase with tetragonal structure
(JCPDS 40-1463). Figure 7.13 shows the crystal composition as a function of distance along the growth axis. The uniformity of the composition of the crystals was
found to be high (effective distribution coefficients close to 1). The deviation from
the average data for Ba, Na, Yb, Gd and Nb concentrations within the same crystals did not exceed about 1%, which corresponds to the experimental error of the
EPMA measurements.

We have shown that slightly Nd^{3+}-doped fibers are orthorhombic and at concentration higher than 3% Nd^{3+} the structure is tetragonal with P4bm space group
[32–35]. Low temperature spectroscopy reveals that Nd^{3+} ions occupy the two
$Na^+$and Ba^{2+} crystallographic sites as can be seen in Fig. 7.14. The stimulated
cross-section near 1060 nm of the $^4F_{3/2}$-$^4I_{11/2}$ has been measured to be large enough
for laser applications (Fig. 7.14), and its branching ratio has been determined
through the McCumber formula. Polarization of the ordinary wave only needs to
be considered for self-doubling laser operations with birefringent phase-matching
[36]. The first report of laser action in a Nd^{3+}-doped BNN crystal was in the mid
1970s.

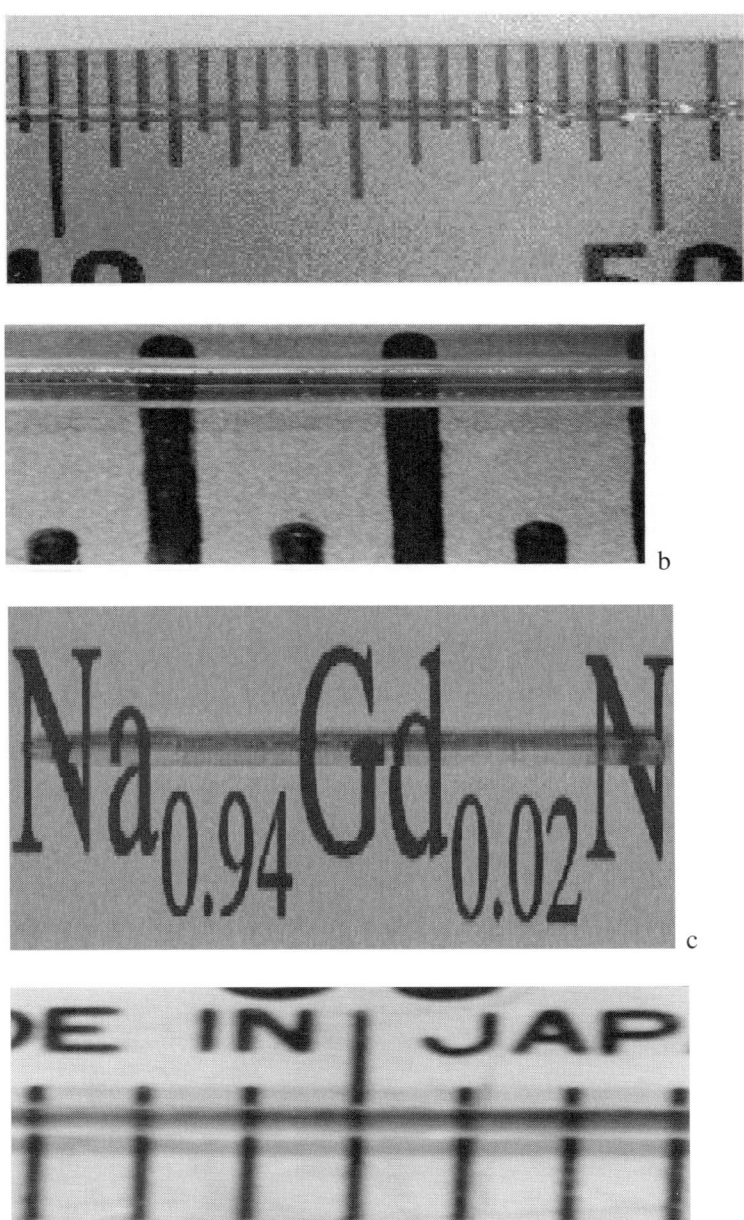

Fig. 7.12. Single crystal fibers grown from $Ba_2Na_{1-x}(Yb_x,Gd_x)Nb_5O_{15}$ melt (scale in mm) (**a**) is $Ba_2Na_{0.94}Yb_{0.02}Nb_5O_{15}$ and (**c**) is $Ba_2Na_{0.94}Gd_{0.02}Nb_5O_{15}$, (**b**) is $Ba_2Na_{0.82}Yb_{0.06}Nb_5O_{15}$ and (**d**) is $Ba_2Na_{0.82}Gd_{0.06}Nb_5O_{15}$

Recently, continuous wave laser action in the near infrared region and self-frequency doubling has been reported [37]. Multi-wavelength frequency conversion was also reported in the same work by means of diffuse quasi-phase matching due to the aperiodic ferroelectric domains present in nonpoled samples. In order to provide a full understanding of the laser properties related to this promising material, the Judd-Ofelt intensity parameters of Nd^{3+} ions in the nonlinear $Ba_2NaNb_5O_{15}$ crystal have been determined, presenting values of 2=1.91×10-20, 4=2.61×10-20 and 6=2.36×10-20 cm^2 [38]. These values were used for the first time to calibrate the excited state excitation measured around the main laser radiation at 1060 nm in terms of absorption cross-section units (cm^2). The experimental results, using a two-beam experiment, show a low value of the excited state absorption cross-section ($<1\times10^{-20}$ cm^2) in comparison with the luminescence emission cross-section (about 4×10^{-20} cm^2) [38].

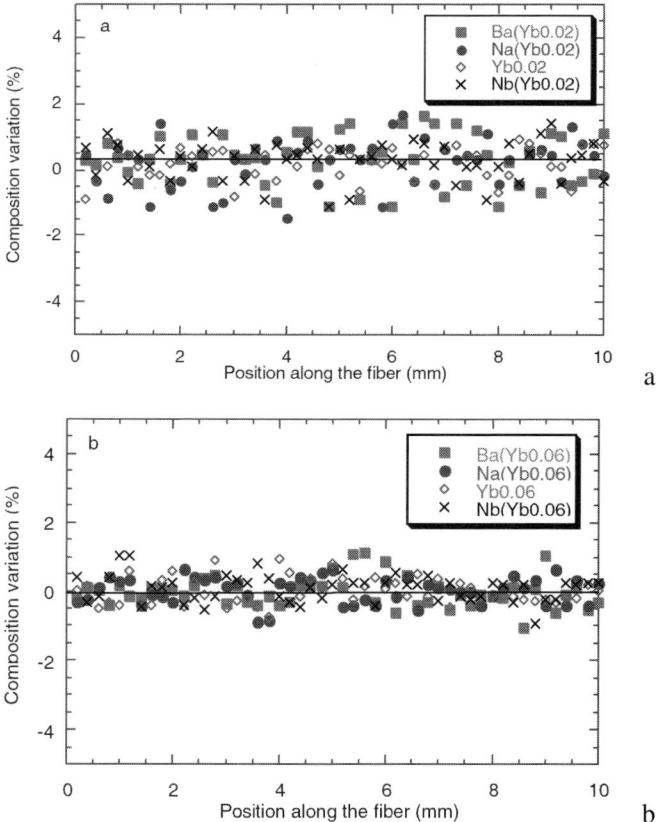

Fig. 7.13. Variation of the Ba_2Na_{1-3x} $Yb_xNb_5O_{15}$ crystal composition along the growth axis (**a**) is $Ba_2Na_{0.94}Yb_{0.02}Nb_5O_{15}$, (**b**) is $Ba_2Na_{0.88}$ $Yb_{0.06}Nb_5O_{15}$

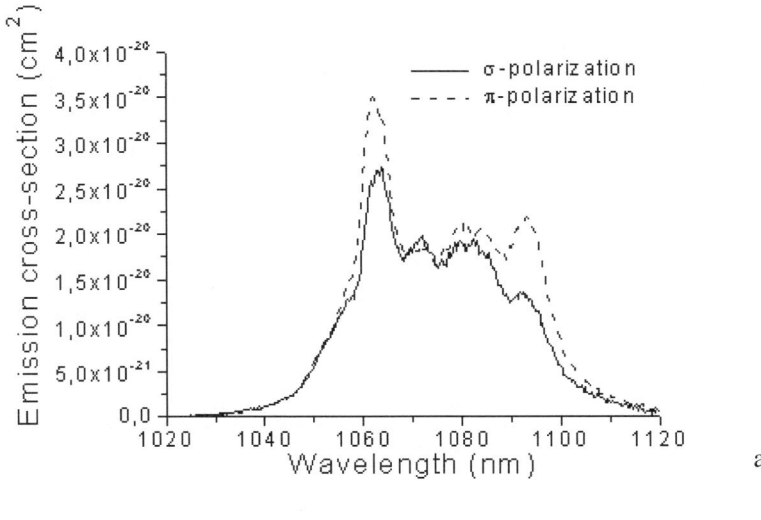

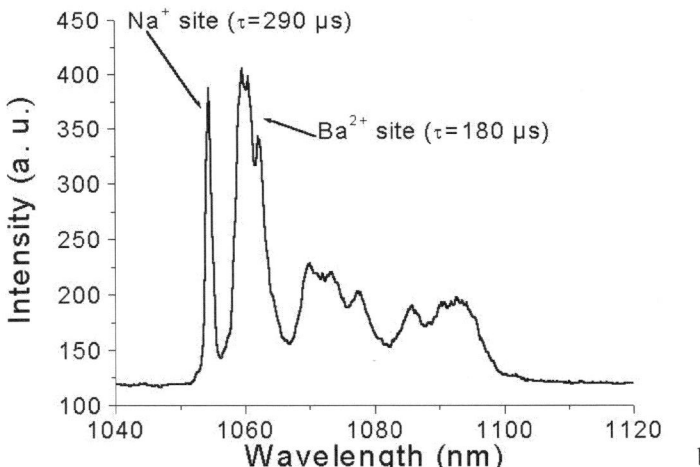

Fig. 7.14. $^4F_{3/2}$-$^4I_{11/2}$ emission transition of Nd^{3+}-doped BNN at room temperature (right) and at low temperature (left) under broad band excitation into the $^4F_{3/2}$. The two types of Nd^{3+} sites have been interpreted as mentioned

7.4.3 $Sr_xBa_{1-x}Nb_2O_6$ (SBN)

SBN materials ($Sr_xBa_{1-x}Nb_2O_6$) are ferroelectric with a tungsten bronze structure (TB) over a wide solid solution range. The belong to the same niobates as BNN. These compounds have a lot of interest for the field of opto-electronics device applications. They have large pyroelectric coefficients and excellent photorefractive properties. Even these materials were grown in bulk single crystal form by the

Czochralski technique, Stepanov method, as single crystal fibers by LHPG technique and in thin film form by a variety of techniques. A lot of problems are observed during growth process. The main problem in the practical application of this important material is the difficulty in growing large single crystals with a good optical homogeneity. The diameter instability is also an unavoidable problem. The apparition of striations perpendicular to the growth axis affects the optical properties and causes some inhomogeneities of the refractive index along the pulling axis. Some tentative steps have been taken in order to improve the crystal quality by changing the growth conditions especially the determination of the congruent composition. In order to investigate more closely the growth conditions, the SrO-BaO-NbO$_{2.5}$ systems have been studied using the crystal growth by the μ-PD method. The Sr$_x$Ba$_{1-x}$Nb$_2$O$_6$ compounds were prepared by solid state reaction from mixtures of SrCO$_3$, BaCO$_3$, Nb$_2$O$_5$, powders (99.99% purity), cold pressed under 1kgcm^{-2} into discs of 15 mm diameter which were placed into an alumina crucible. The pellets were sintered at 1000°C for 10h and at 1200°C for 20h in an air atmosphere twice. The formation of the SBN phases was confirmed by X-ray powder diffraction at room temperature. The melt behavior and the phase transitions were studied by DTA/TG in an air atmosphere. Using the μ-PD method, SBN was grown from the melt. The crystals were transparent, independent of the melt composition, with uniform shape and free of macroscopic defects such as cracks, bubbles or inclusions (Fig. 7.15).

The typical size of crystals was 0.4–0.8 mm in cross section, depending on the diameter of the nozzle and a few centimeters in length. In the main, about 90% of the volume was crystallized into single crystals.

The crystals were SBN single phases. Regular striations with periodic distribution perpendicular to the growth axis are observed. The optimization of the growth condition is based on the equilibrium diagram. Figure 7.16 shows the evolution of the growth conditions as a function of the compound position in the ternary SrO-BaO-NbO$_{2.5}$ equilibrium diagram. By moving along the BA direction, crystal growth conditions become uncontrolled and the quality of the crystal decreases dramatically, even by using a slow pulling rate. Both LHPG and μ-PD techniques have given the same results and the crystal quality was comparable.

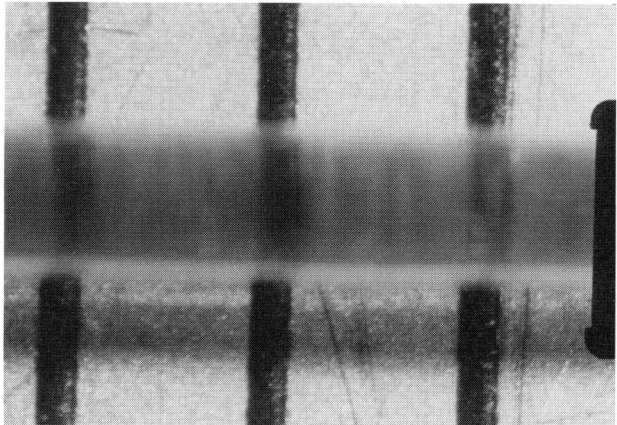

Fig. 7.15. Periodic striations perpendicular to the growth direction in a SBN crystal

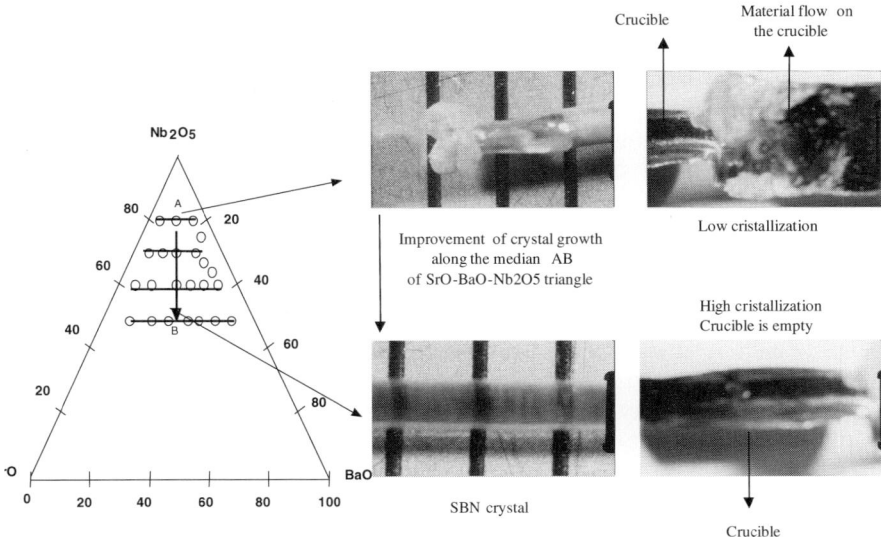

Fig. 7.16. Optimization of SBN crystal composition using μ-PD technique and equilibrium phase diagram. The growth conditions are improved along AB direction. At the end the crucible is empty

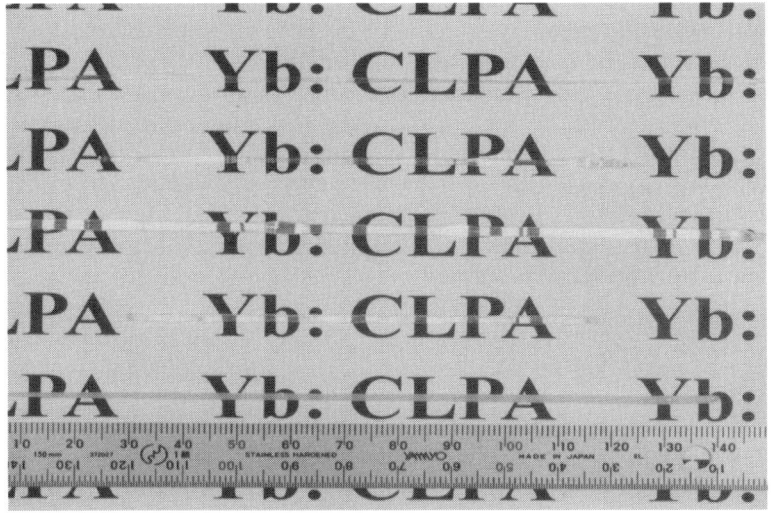

Fig. 7.17. Yb^{3+}-doped oxyapatite single crystal prepared by the μ-PD technique

7.4.4 Laser Fiber Crystals : Yb^{3+}-doped $Ca_8La_2(PO_4)_6O_2$ Oxyapatite

Single crystals with the apatite structure (space group $P6_3/m$) are effective luminescent host and laser materials. Among the many synthetic crystals, compounds of the fluoroapatite family are of practical interest as efficient laser-host materials when doped with neodymium or ytterbium, because of their high crystal field and good thermal, mechanical, and chemical properties. However, the investigation of oxyapatites as laser materials has not progressed far. This is related to the difficulties which are observed during synthesis and especially the crystal growth process. To resolve this technological problem, we have used the crystal fiber growth techniques. First of all, we present results from the μ-PD technique. The crystals were grown at pulling rates of 0.1–15.0 mm/min and were 0.5–3.0 mm in diameter (Fig. 7.17). The typical length of the molten zone was about 0.1 mm.

The melt wets Ir very well, so the control of the fiber diameter is rather difficult. Best reproducibility corresponded to a pulling rate of 3.0 mm/min. From $x = 0.00$ to 0.06 of $Ca_8La_{2-x}Yb_x(PO_4)_6O_2$, the fiber crystals were transparent and colorless [39]. Transparency was lost when $x \cdot 0.08$. As we use <0001> single crystal as a seed, it is clear that the preferred growth axis is controlled to be <0001>. The crystal grown from the $x=0.06$ melt was analyzed for the distribution of Yb^{3+} along the growth axis. The result shows homogeneous distribution of the cations along the growth direction. We could not find any detectable variation of Ca, La or P concentration along the growth axis. But, for Yb, there was a detectable change. Though the monophased field as determined from the sintering experiments extended from $x= 0.0$ to 0.2, the Yb^{3+} concentration could be incorporated into a single crystal by the μ-PD growth technique up to 0.06. A white impurity phase appeared at the fiber surface when the solidification fraction was higher than 0.3. This result is very successful for the growth of these family of materials, especially for fiber shape. These materials present a fast decomposition at the liquid's temperature because of high kinetic decomposition and losses of phosphor oxide. But by using the μ-PD with relatively high pulling rate, it has been possible to get an oriented single crystal fiber along the seed direction. Research is underway to find new compositions containing much higher Yb^{3+} concentrations than those introduced in $Ca_8La_2(PO_4)_6O_2$ oxyapatite suitable for large boule crystal growth. Growths of oxyapatites have also been succeeded with the LHPG technique on calcium borohydroxyapatite [40], oxyphosphate-silicate apatite [41] and more recently on oxyboroapatites [42].

7.4.5 Superconductors : Bi2212

After the discovery of high critical temperature (T_c) superconducting materials [43], an important effort has been devoted to improving the critical temperature and critical current. The main objective was to produce materials with high (T_c) for tape applications, and to understand the different phenomena involved in preparation and growth [44–50]. But high-T_c materials are complex. The superconducting systems are metastable and substitution is necessary to stabilize them [51–55]. Many attempts to grow crystals by different methods have been made such as the self-flux method [56], floating zone [57], top-seeded solution [58], and Bridgman methods [59]. But reproducibility has been poor, due to the complexity of the growth process and the limits of the techniques. Using the μ-PD technique, we have successfully grown superconducting Bi2212 fiber from an incongruent melt composition and we have optimized the best composition which allows a high percentage of crystallization with high critical temperature. The fiber morphology grown at 0.4 mm/min is shown in Fig. 7.18: the length is about 6 cm, the diameter about 1.5 mm. The fibers are hard, difficult to break, without cracks, and black in color. No blackish-copper color was observed in any fiber. The surface structure is fibrous. The fibers show plate-like crystallites with highly aligned morphology along the growth direction [60–61].

The fiber length depends on the quantity of raw materials in the Pt crucible. About 90% of raw material was crystallized. No diameter variation along the growth direction was observed for $Bi_{2.2}Sr_2CaCu_2O_x$ and $Bi_{2.4}Sr_2CaCu_2O_x$, which had bismuth-rich stoichiometry. During growth, small facets on the surface of the molten zone indicating the formation of the large plate-like structure were observed in Fig. 7.19. The melt was stable and did not boil or decompose during growth.

According to these first results of Bi 2212 crystal growth by micro-pulling down growth is most rapid along directions in the (a-b) plane, and so this orientation was selected by the crystal itself, even when we did not use an oriented seed. The crystal lattice parameters are calculated by a program for unit cell refinements and are a=5.384(8) Å, b = 5.397(6) Å and c= 30.809(7) Å. These results are in good agreement with the investigation of Bonfigt et al. [62].

Bi-2212 was also grown using the LHPG technique. Similar results were observed including physical properties [63–64].

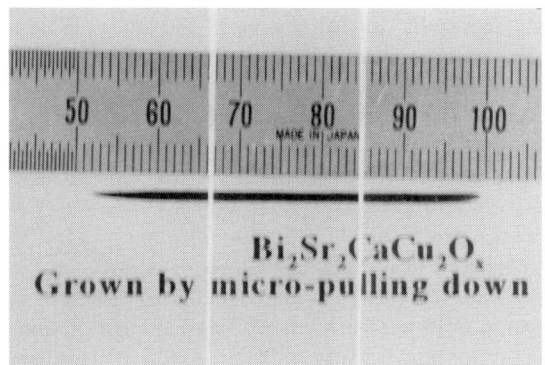

Fig. 7.18. Bi2212 grown by the μ-PD technique

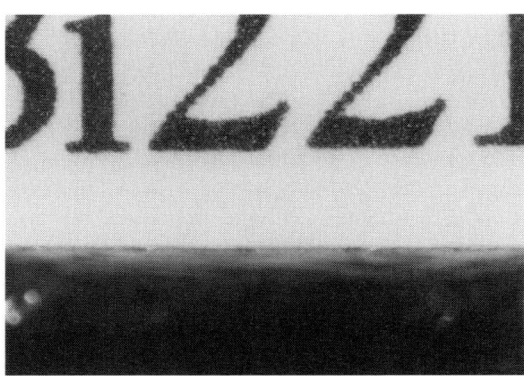

Fig. 7.19. Bi2212 fiber with presence of facets showing a good texturation of the material

7.4.6 Refractory Sesquioxide Fiber Crystals Grown by the LHPG Technique and Proposition of a New Combinatorial Chemistry Approach

We have demonstrated the possibility to grow yttria materials under the shape of small crystal fibers by using LHPG apparatus [65]. Yttria is an attractive laser crystal for several reasons: it is a refractory oxide with a melting point of 2380 °C and it has a very high thermal conductivity, (κ_{Y2O3}= 27 W/mK, twice YAG's, κ_{YAG}= 13 W/mK). In addition, this crystal has the body centered cubic structure with the space group Ia3 (T^7_h) and finally it is optically isotropic, with a refractive index of 1.91. Another interesting property allowing high radiative transition probabilities between electronic levels is that the dominant phonon energy is 380 cm^{-1}, which is one of the smallest phonon cut-offs among oxides [66]. It is obvious that the high melting point of this oxide is an obstacle for classical crystal growth techniques as the growth by the Czochralski method. However, the Bridgman process has been performed in an expensive rhenium crucible [67].

We would like to mention we have focused our research on original methods for the creation of a new and original combinatorial approaches based on both synthesis and in situ exploitation of concentration gradient samples. Recently, combinatorial chemistry has been adapted to inorganic compounds as well as being increasingly applied in the study for new advanced materials. This systematic, efficient and fast method was applied to investigate a large field of ternary, quaternary and high order solid state compounds and was tried successfully for superconducting [68], magneto-resistive [69], photoluminescent [70] and catalyst [71] materials. The combination of automated thin film deposition and physical masking techniques was used for parallel synthesis and has generated a spatially addressable library. As an example, an unusual blue white phosphor (Sr_2CeO_4) with an unexpected structure was discovered from a library of more than 25000 members [72]. Despite the great number of samples constituting the combinatorial library, this method remains discontinuous. A simultaneous chemical and structural characterization is not associated to the physical properties measured, and the given composition corresponds to the bulk composition without taking into account whether or not the sample is monophased. A new combinatorial approach where a rod or a fiber, single crystal or not, with continuous composition gradient is prepared by using a melting zone technique or a floating zone method [73]. The method is based on the synthesis and investigation of "concentration gradient fibers or rods" where composition changes continuously from one end to the other between two well-defined compositions C_A and C_B. If a solid solution exists, a single crystal can be grown with composition varying continuously between C_A and C_B. The advantage of such samples is that every point of the fiber can be considered as a single crystal where composition and physical properties can be correlated by means of in situ measurements.

This combination has previously been applied to the study of Yb^{3+} monodoped sesquioxides and to the $Yb^{3+}-Er^{3+}:Y_2O_3$ systems where composition and spectroscopic propertiessuch as decay times and initial risetimes of fluorescence have been correlated on monocrystalline fibers [74]. $Yb^{3+}- Er^{3+}$. Codoped systems are studied for the achievement of all solid-state lasers generating eye safe IR radiation of the Er^{3+} at 1541 nm. These systems have to be optimized in order to prevent back energytransfer $Er^{3+} \rightarrow Yb^{3+}$ and up conversion in Er^{3+} excited state processes that limit the laser efficiency. In such systems, the $^4I_{13/2}$ lifetime is controlled by the contents of the activator (Er^{3+}) and sensitizer (Yb^{3+})[75]. Gradient concentration fibers have been used to study, as an example, the influence of the sensitizer (Yb^{3+}) and the activator (Er^{3+}) concentrations in Y_2O_3 crystal to optimize the Er^{3+} infrared emission at 1541 nm.

Gradient concentration fibers are grown from a special ceramic rod which already has a variation of concentration. The rod is composed of two rods cut along a slant and placed side by side as shown in Fig. 7.20. During the melting zone, there is an homogeneous mix of the species in the liquid: the amounts of the solubilized species vary continuously when the molten zone moves and a composition gradient is induced in the crystallized rod. By using the gradient concentration ceramic rod, the concentration of the liquid varies depending on the position of the molten zone.

The concentration of the crystallized solid is directly linked with the concentration of the melted zone through the segregation coefficient k:

$$k = C_S / C_L \qquad (7.10)$$

with C_S and C_L representing the concentration of the solid and the concentration of the liquid respectively. There is an equilibrium segregation coefficient k_0 which is directly obtained from the phase equilibria diagram, however, due to the melting zone being rarely applied under equilibrium conditions, an effective segregation coefficient k_{eff} is currently used. The relationship between k_0 and k_{eff} has been expressed by Burton, Prim and Slichter in the following equation (7.11) [76]:

$$k_{eff} = \frac{k_0}{k_0 + (1 - k_0)\exp(-v\delta / D)} \qquad (7.11)$$

v is the freezing rate, δ is the thickness of the boundary layer and D is the diffusion coefficient in the molten zone. Equation (7.11) shows that for high a freezing rate the exponential term tends to zero and consequently, the effective segregation coefficient tends to unity. Thus for a high pulling rate, the composition of the crystallized solid is the same as those of the liquid. When the molten zone is fed with a rod possessing a progressive variation of its concentration, gradient concentration fibers can be grown.

A previous study has shown that for a pulling rate of the order of 50 mmh^{-1} no segregation appears in the crystallized fiber and we shall admit that for such pulling rate $C_S = C_L$.

The gradient concentration fiber is an unique tool for making the correlation between composition and physical properties. Composition and physical properties have been measured in situ on the same sample in relation to the distance from the top of the crystallized rod. The two curves were correlated by using a reference made of a homogeneous monocrystalline fiber with well-defined composition and physical properties. All crystals prepared by this technique were transparent and crack free as it is shown in Fig. 7.21.

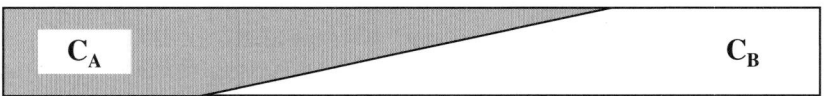

Fig. 7.20. Scheme of a ceramic feeding rod used for the growth of concentration gradient fiber

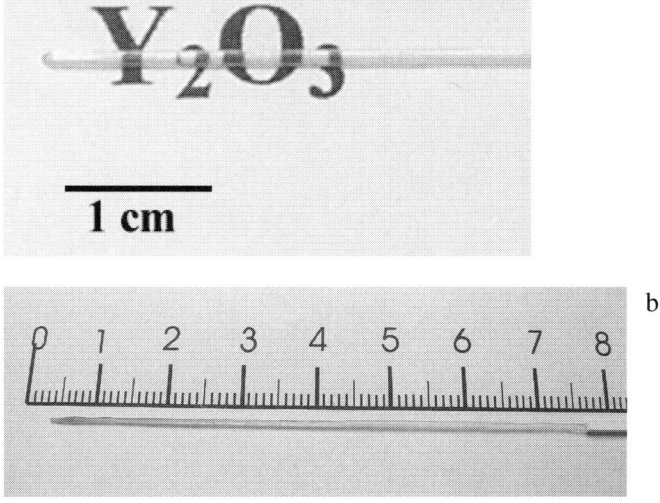

Fig. 7.21. Undoped Y_2O_3 (**a**) and Ho-doped Y_2O_3 (**b**) fiber single crystals prepared by LHPG technique. Ho-doped Y_2O_3 has been selected in this case to show Ho gradient composition by the absorption change in the visible along the growth direction

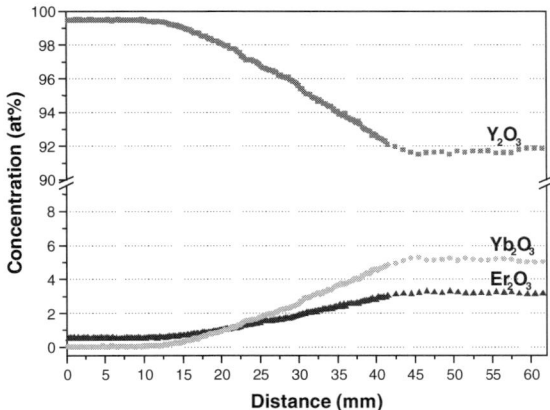

Fig. 7.22. Chemical analysis of a sample in which concentration varies between C_A (0% Yb, 0.5% Er) and C_B (5% Yb, 3% Er)

We have applied this new combinatorial approach for the optimization of some properties in relation to the composition. The application of this approach to laser materials is based on the growth of single crystals with composition gradients shown in Fig. 7.22 with in-situ measurements of composition and spectroscopic properties [74].

The originality of the new procedure proposed here relies on the investigation of a monocrystalline fiber as a "combinatorial library" with continuous data. If a total solubility domain exists in the range of investigated concentration, the library stretches over the entire field of the solid solution and the method constitutes a good, fast and cheap way to optimize the system. If, in the studied composition range, a limit of solubility exists for one or more elements then this approach is useful for determining the value of these limits. In the present case we investigated a three-constituent system, but there is no limit on the order of the system studied. With the development of fully automated analysis apparatus, gradient concentration samples are the key for fast and thorough knowledge in materials science.

7.4.7 Example of Application of the Growth of Rare Earth-doped Concentration Gradient Crystal Fibers: Analysis of Dynamical Processes of Laser Resonant Transitions in Yb^{3+}-doped Y$_2$O$_3$

The development of reliable InGaAs laser diode pump sources emitting in the 900-980-nm spectral range is strongly influencing the field of lasers based on Yb^{3+}-doped solid state crystals. Ca$_5$(PO$_4$)$_3$F (C-FAP) and S-FAP (Sr$_5$(PO$_4$)$_3$F) have been recognized as favorable hosts for Yb^{3+} lasing in the nanosecond pulse regime. This fact has been supported by an evaluation of the spectroscopic properties of several Yb^{3+}-doped crystals useful for laser action [77]. This evaluation was based on two parameters known from spectroscopy, the emission cross-section at the laser wavelength and the minimum pump intensity required to achieve transparency at the laser wavelength. There is a need for a new evaluation of Yb^{3+}-doped crystals in order to predict the laser efficiency in a more realistic manner in different kinds of regimes. A new evaluation [78-79-80], based on a quasi-three level laser model, leads to a comparison of all known Yb^{3+}-doped crystals in a two-dimensional diagram considering the laser. Such research programs in our laboratory involve the LHPG technique mainly on oxoapatite, niobate and sesquioxide crystals.

As an example, new and original crystalline samples, grown by the LHPG technique, having a continuous longitudinal concentration gradient have been used to study dynamical processes of resonant transitions in rare earth-doped sesquioxides. This fast and simple method allows the measurement of the influence of radiation trapping and impurity quenching on the excited-state lifetime as a function of the dopant concentration, which plays an important role in solid-state laser gain media.

Resonant radiative energy transfer is well known for acting as a radiation trap. Actually resonant transitions permit a long-range energy diffusion between identical ions by successive reabsorption/reemission processes as shown in Fig. 7.23. This serial mechanism affects the experimental excited state decay since, after the initial emission occurs, each subsequent reabsorption event acts as a time reset on the relaxation of this excitation. Consequently, a lengthening of the fluorescence lifetime measured over the volume of the sample relative to the lifetime of a single isolated ion should be observed. Then the accurate determination of the radiative lifetime which is occurring in the stimulated emission cross-section value is required.

Strong spectral overlap between the fluorescence and absorption spectra enhances fluorescence reabsorption. The most relevant examples are those of the resonant laser transitions such as $^2E \rightarrow ^4A_2$ of Cr^{3+} in ruby at 694 nm, and several rare earth ions in lasing hosts such as $5I7 \rightarrow 5I8$ in Ho^{3+} at 2100 nm, $^3F_4 \rightarrow ^3H_6$ in Tm^{3+} at 2000 nm, $4I_{13/2} \rightarrow ^4I_{15/2}$ in Er^{3+} at 1540 nm and $^2F_{5/2} \rightarrow ^2F_{7/2}$ in Yb^{3+} at 980 nm. Examples in Yb^{3+}, Er^{3+} and Ho^{3+} rare earth ions have already been presented [75]. The case of Yb^{3+}-doped Y_2O_3 is shown in Fig. 7.24 where the total overlap of the resonant transition is clearly seen.

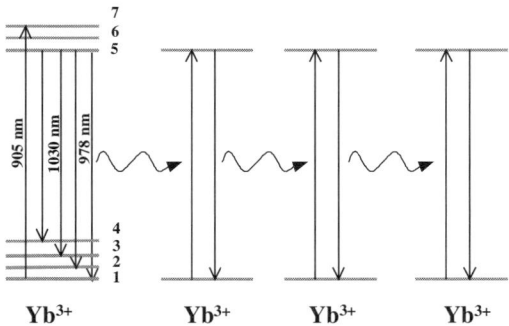

Fig. 7.23. Schematic representation of the radiative energy transfer (self-trapping) between Yb^{3+} ions corresponding to the increase of τ_{exp} as can be seen from Fig. 7.25

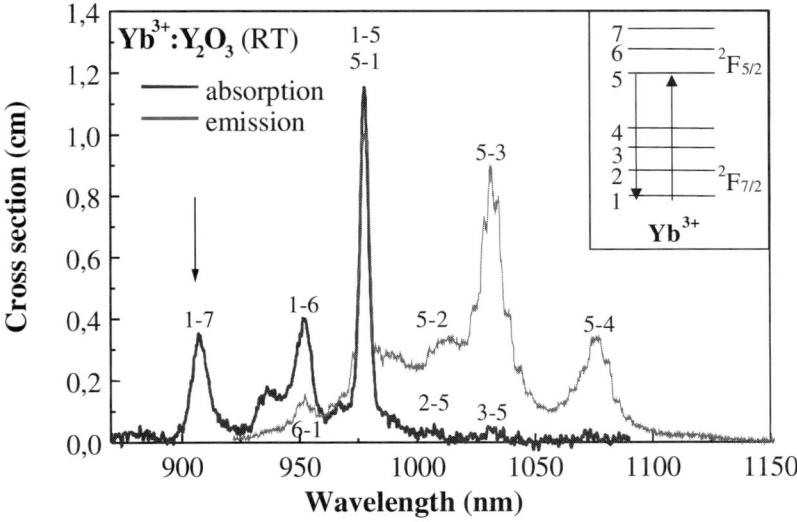

Fig. 7.24. Absorption and emission spectra of Yb^{3+}-doped Y$_2$O$_3$. The total overlap of the 1 ↔ 5 lines can be seen

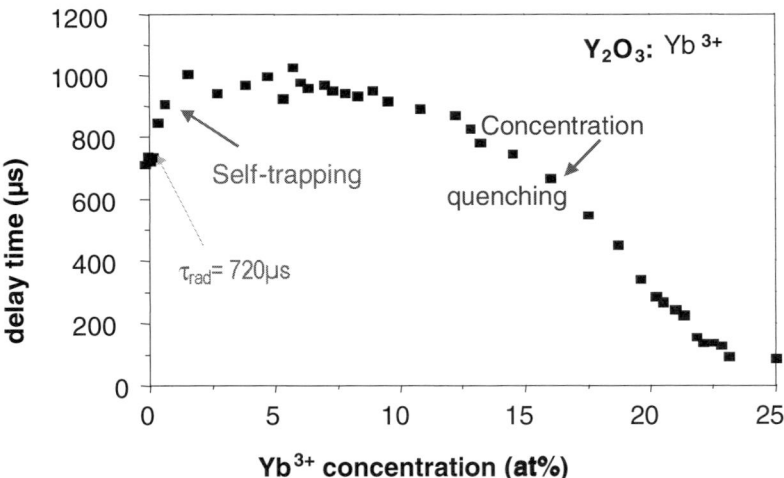

Fig. 7.25. Experimental decay times of the $^2F_{5/2}$ excited state as a function of the Yb^{3+} concentration in Y$_2$O$_3$

The size and the concentration of the sample are critical factors since the probability of a reabsorption/reemission event depends on the distance that the emitted light has to travel in the sample [1,2]. Moreover it has been shown that the geometric configuration of the sample and its refractive index play an important role since a fraction of the light emitted can be trapped by successive total internal reflections (TIR) at the sample/air interface. The higher the refractive index of the material is, the smaller the critical angle for TIR. A high refractive index, associated with a high symmetry of the sample, leads to an increase in the path length of the light and therefore enhances reabsorption effects.

In order to prevent overestimating the radiative lifetime, measurements can be performed after reducing the samples into splinters. Other nondestructive methods have been used in order to limit the TIR. Each consists of the elimination of the media/air interface by including the sample in a transparent material of identical refractive index. We intend to demonstrate both radiative trapping, quenching processes and radiative lifetime measurements by using concentration gradient fibers. The originality of concentration gradient samples is that one fiber contains a collection of single crystals of different concentration spread along its axis. Such crystalline fibers present a dopant concentration which varies continuously between two chosen compositions.

Figure 7.25 shows the experimental results of the observed lifetime as a function of the dopant concentration .

Actually, from a microscopic point of view, the radiative transfer probability for a given system is strongly dependent on the distance as R^{-2} between two involved neighbor ions. Measured lifetime variations clearly demonstrate the existence of energy transfer mechanisms. The energy transfer probability can be easily deduced from the effectively observed lifetime (exp) and the purely radiative one (rad) through the following simple equation:

$$W(\text{transfer}) = 1/\tau(\text{exp}) - 1/\tau(\text{rad}) \qquad (7.12)$$

By performing a statistical calculation over 100 lattices, we have calculated the average distance between two dopant ions as a function of the doping rate in the Y_2O_3 matrix. In this calculation we have considered that the two different sites included in the Y_2O_3 structure are involved in the energy transfer processes.

Purely radiative lifetime has been evaluated by extrapolation of the curve giving a lifetime for the Yb^{3+} lowest concentration: $\tau(\text{rad})=0.72$ ms.

Identifying the origin of the observed lifetime variations provides an understanding of excited state dynamics. In the case of Yb^{3+}-doped Y_2O_3, the curve in Fig. 7.25 can be divided into two regimes:

– in the lowest concentration range, the experimental lifetime increases as the dopant concentration increases;
   for the higher concentration range, measured lifetime decreases when the doping rate increases.

In the intermediate region, competitions occur between the two trends and consequently they compensate each other leading to a constant value of the measured lifetime.

We assign these two behaviors respectively to:

- fluorescence reabsorption so-called self-trapping process between resonant transitions giving radiative energy transfer (see Fig. 7.23 in the case of 1↔5 transitions of Yb^{3+} ions);
- because of the lack of other excited levels in Yb^{3+}, the decreased lifetime in the high dopant concentration-region can be assigned to energy transfer by up-conversion (ETU) leading to the de-excitation of the emitting level of unexpected Er^{3+} (and Tm^{3+}) impurities according to the model drawn in Fig. 7.26. Yb(5%)-doped Y$_2$O$_3$ excited under 930 nm, exhibits a bright green anti-Stokes luminescence visible to the naked eye. Indeed, the presence of rare earth impurities is certain. All starting materials used in this study have a purity of 99.99%. Because rare earth elements are chemically related, it is difficult to separate them from each other. Thus impurities are inevitable. Moreover, looking at the Dieke diagram one can see that many resonant energy transfers are possible between trivalent lanthanide ions. In particular in the 10000 cm^{-1} energy range matching with excited state of Yb^{3+} ions, resonant energy transfers are allowed with Er^{3+}, Dy^{3+} or Pr^{3+},without taking account of phonon-assisted energy transfer ,as it is known, with Yb^{3+} - Tm^{3+} ions.

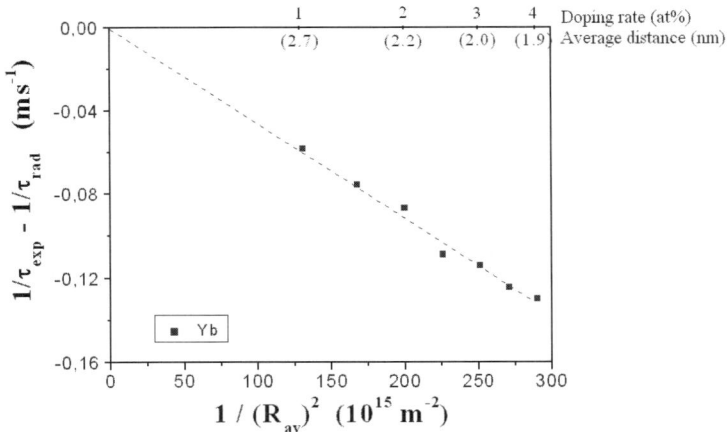

Fig. 7.26. Plot of the difference $1/\tau_{exp}$-$1/\tau_{rad}$ as a function of $1/R^2$ for the first excited level of Yb^{3+} ions

The average distance between two neighboring crystallographic sites is relatively short: in YAG, the minimum distance between dopant rare-earth ions substituting Y^{3+} atoms is 3.67 Å, just slightly higher than in Y$_2$O$_3$ (C$_2$ site– C$_{3i}$ site distance=3.51 Å and C$_2$ site – C$_2$ site distance= 3.53 Å). The two hosts should give the same type of anti-Stokes emission connected to pairing phenomenon with a lower probability in YAG due to higher distances than in yttria. The probable cooperative emission was not immediately apparent, hidden by the impurities emitting with higher fluorescence intensities in the same spectral range. However, detailed and repetitive measurements have recently shown the real presence of cooperative luminescence in a 5% Yb^{3+} sample. IR convoluted emission spectrum can be exactly overlapped with the visible part of anti-Stokes emission represented by the green part in Fig. 7.27. Consequently we detect without doubt the presence of Yb^{3+} ion pairs and probably more generally Yb^{3+} ion clustering within the Y$_2$O$_3$ host.

We wished to show these results in optics in order to illustrate the potential of the new combinatorial chemistry in growing concentration gradient fibers. The method can be used to get an overall view of the dependence on composition of a physical property since only one sample contains all possible compositions of a predefined composition field.

7.4.8 Fluoride Crystal Growth (CaF$_2$)

The choice of materials for the new generation of stepper optics is fairly limited. Obviously, a major criterion is high transmission at the application wavelength. Another criterion is a high damage threshold, or low induced absorption. Induced absorption arises mostly from point defects (structural defects often caused by impurities). Low birefringence and high optical homogeneity are also required for transferring high-resolution patterns. Any absorption experienced by the material during lithography will result in an increase in local temperature and therefore negatively effect birefringence and hence resolution. Materials meeting the transmission criteria include fused silica and several monocrystalline fluorides. Whereas much progress has been made in improving the transmission of fused silica, there is an inherent limitation due to the absorption edge at 160 nm. Fused silica is not a viable material for 157 nm lithography. Transmission tends to decrease at 200 nm and problems exist with laser resistance and compaction. Of the fluoride candidates, those with band gaps in the deep-UV region include LiF (120 nm), NaF(140 nm), MgF$_2$(150 nm), BaF$_2$(150 nm) and CaF$_2$(130 nm). LiF and NaF are hygroscopic and difficult to polish, their use is limited by problems associated with treatment and handling. MgF$_2$ is tetragonal and thus is not optically isotropic, birefringence is an issue. CaF$_2$ crystals are the material of choice as they are transparent in the UV region, optically isotropic, non-hygroscopic, and although brittle, can be handled with relative ease. CaF$_2$ does not suffer from compaction. Based on the data, CaF$_2$ looked promising for a preliminary investigation on the growth of fluoride single crystal fiber material, via the LHPG method, for some specialized IR transmission applications. CaF$_2$ formed a stable molten zone in a vacuum atmosphere (10^{-4}). The grown fibers are transparent, homogeneous and free of macroscopic defects (Fig. 7.28). The fiber crystals are bubble and inclusion free (Fig. 7.29). The transmission percentage is higher than 95%. Our grown sample qualities are very close to the commercial materials and are oriented along the initial seed direction (111).

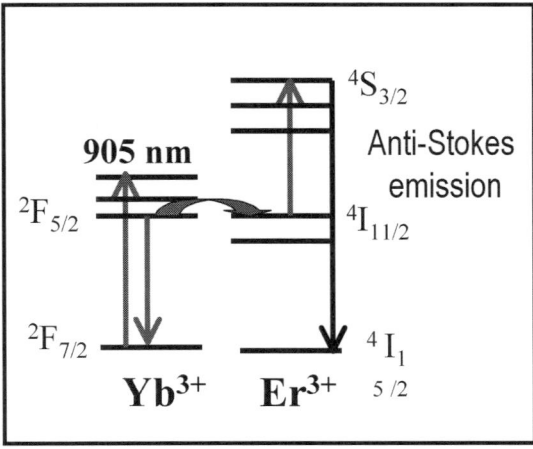

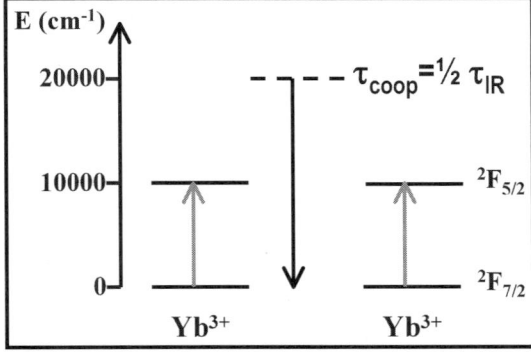

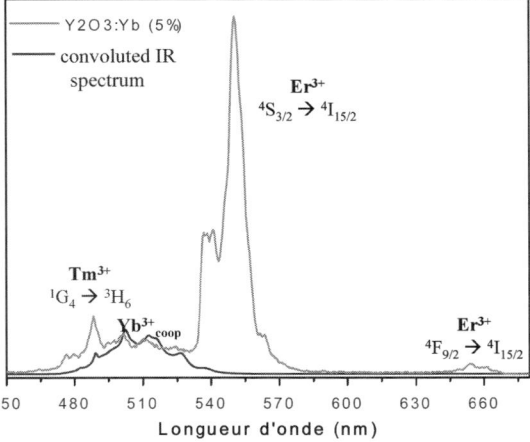

Fig. 7.27. Anti-Stokes emission in Yb [3+]-doped Y$_2$O$_3$ due to both energy transfer upconversion between Yb [3+] ions and unexpected Er [3+] (and probably Tm [3+]) ions as impurities and co-operative emission related with dimers and clustering of Yb [3+] ions

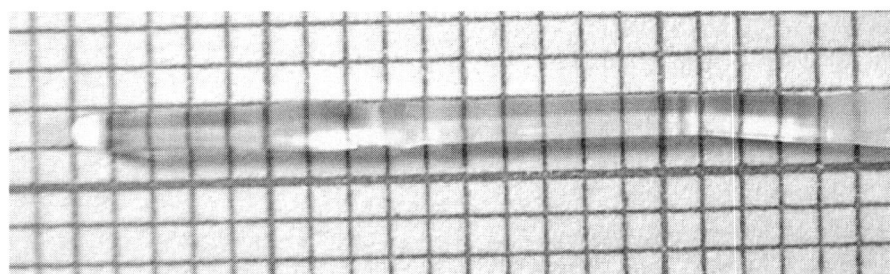

Fig. 7.28. CaF$_2$ single crystal fiber grown by LHPG technique.

7.5 Conclusion

We have shown that the interest in fibers is growing because of their properties, important for applications such as superconductors, laser sources, as well as the telecommunication and photonics field. Oriented single crystal fibers could be grown from the melt with long length, thin diameter and various rare earth dopant concentrations. Up to now, only two complementary techniques are used for fiber crystal growth, which are μ-PD and the LHPG methods complementary to the conventional Czochralski and melt zone techniques, respectively. The melt quality is important for obtaining the best crystal properties. A variety of materials have already been grown using LHPG and μ-PD techniques which are also very helpful tools for performing and improving fundamental research into materials science. These two techniques have been used to understand the different phenomena observed during crystal growth process such as convection phenomena, meniscus, interface shape and kinetics of crystal growth. Their usefulness has also been demonstrated by drawing the equilibrium diagram in complex systems, understanding phases relation in multi-compounds serials, optimizing new crystal composition with low cost and even creating a new and original combinatorial chemistry approaches by using the floating zone of the LHPG technique. The strong impact in research of new superconductors or optical materials and new physical phenomena has been shown.

Fig. 7.29. Photomicrographe of CaF$_2$ using transmitted light

Acknowledgement

This study is performed in the frame of co-operation between the Fukuda laboratory of the Institute for Materials Research at the Tohoku University of Sendai and the laboratory of Physical Chemistry of Luminescent Materials at the Claude Bernard /Lyon1 University including CNRS in France from the « Groupement De Recherche » (GDR 1148) on « Laser Materials » and the Japanese Society for the Promotion of Science (JSPS 161) in Japan. The authors thank all French and Japanese collaborators for their great effort for success in this mutual research.

References

1   P.Rudolph, T.Fukuda, Cryst.Res Technol 34 (1999) 13.
2   H.E.Labelle, Jr. and A.I..Mlavsky,Mater.Res.Bull. (1992) 184.
3   E.von Comperz,Z.Physik 8 (1992) 184.
4   J.S.Haggerty, Production of fiber by Floating Zone fiber Drawing technique, NASA Repport, 120948 (May 1972).
5   J.S.Haggerty, W.P.Menashi and J.F.Wenckus, US Patent 3,944, 460, March 16, 1976.
6   C.A.Burrus and J.Stone, Appl.Phys.Letters 26 (1975) 318.
7   R.S.Feigelson, Growth of fiber crystals in : Crystal Growth of Electronics Materials,Ed.Kaldis (North-Holland, Amsterdam, 1985) p.127.
8   J.L.Stevenson and R..B.Dyott, Electron.Letters 10 (1974) 449.
9   H.P.Weber, P.F.Liao,B.C.Tofield and P.M.Bridenbaugh, Appl.Phys.Letters 26 (1975) L269.

10   R.Cohen-Adad, Private communication.

11   Y.Mimura, Y.Okamura,Y.Komazawa and C.Ota, Japan.J.Appl.Phys.19 (1980) L269.

12   T.J.Bridges, J.S.Hasiak and A.R.Strnad,Opt.Letters 5 (1980) 1985.

13   P.Kapitza, Proc.Roy.Soc (London) A119 (1928) 358.

14   W.G.Pfann, K.M.Olsen, Phys.rev.89 (1953).

15   W.G.Pfann, Zone Melting, Wiley, New York (1959).

16   R.P.Popawsky, J.Appl.Phys.339(1962) 1616.

17   J.S.Haggerty, NASA Repport, CR-120948 (1972).

18   R.S.Feigelson, J.Crystal Growth 79 (1986) 669-680.

19   M.Fejer, J.L.Nightingale, G.Amagel, R.Byer, Rev.Sci.Instrum.55 (1984) 1791.

20   D.H.Yoon, T.Fukuda, J.Korean Ass.of Crystal Growth 4 (1994) 405.

21   A.Lorenzo, H.Jaffrezic, B.Roux, G.Boulon, L.Bausa, and J.Garcia-Sole, Physical.Rev 1996.

22   J.Garcia-Sole, T.Petit, , H.Jaffrezic, G.BoulonEuro-Phys .Letters 24 (1993) 719.

23   S.Uda et al, J.Crystal Growth 182 (1997) 403-415.

24   D.Ho Yoon, I. Yonenaga, T.Fukuda, Norio Ohnishi, J.Crystal Growth 142 (1994) 339-343.

25   P.B Jamieson, S.C.Abrahams and Bernstein, J.Chem.Phys 50 (1969) 4352.

26   R.G.Smith, J.E.Geusic, H.J.Levinstein, J.J.Rubin, S.Singh, L.G. Van Uitert, Appl. Phys. Lett 12 (1968) 308.

27   S.A.Baryshev, V.I.Pryalkin, A.I.Kholodnykgh, Sov. Phys. Lett 6 (1980) 415.

28   S.Singh, D.A.Draegert, J.E.Geusic Phys.Rev B2 (1970) 2709.

29   K. Lebbou, A. Yoshikawa, T. Fukuda, M. Th. Cohen Adad, G. Boulon, Material Rescarch Bulletin V8 (2000) 35.

30   K.Lebbou, H.Itagaki, A.Yoshikawa,T.Fukuda,F.Carillo-Romo, G.Boulon,A.Brenier, M.Th.Cohen-Adad. J.Cryst.Growth 210 (2000) 655-662.

31   K.Lebbou H.Itagaki,.A.Yoshikawa, T.Fukuda, G.Boulon, M.Th.Cohen-Adad. J.Crystal Growth 224 (2001) 59-66.

32   G.Foulon, M.Ferriol, A.Brenier, G.Boulon, S.Lecocq, Eur.J.Solid State Inorg.Chem 33 (1996) 673.

33   G.Foulon, M.Ferriol, A.Brenier, M.T.Cohen-Adad, G.Boulon, Acta Physica Polinica 90 (1996) 63.

34   G.Foulon, A.Brenier, M.Ferriol, M.T.Cohen-Adad, G.Boulon, Chem.Physics Letters 245 (1995) 555.

35   M.Ferriol, G.Foulon, A.Brenier, G.Boulon, J .Mat.Sci. 33 (1998) 1227.

36   G.Foulon, A.Brenier, M.Ferriol, M.T.Cohen-Adad, G.Boulon, Chem.Phys.letters 249 (1996) 381-386.

37   A.A. Kaminskii, D. Jaque, S.N. Bagaev, K. Ueda, J. García Solé and J. Capmany. Quant. Electron. 29 (1999) 95-97.

38   J.J.Romero,A.Brenier,L.Bausa,G.Boulon,J.Garcia Sole,A.Kaminskii, Optics Communication 191 (2001) 371.

39   G.Boulon, A.Collombet, A. Brenier, Mth.Cohen-Adad, A.Yoshikawa, K.Lebbou, J.H.Lee, T.Fukuda, Adv .Funct.Mater 11 (2001)263-270.

40   R.Ternane,G.Panczer, M.T.Cohen-Adad, C.Goutaudier, G.Boulon, N.Kbir-Ariguib, M.Trabelsi-Ayedi Optical Materials 16 (2001) 291-300.

41   R.El Ouenzerfi, G.Panczer,C.Goutaudier, M.T.Cohen-Adad, G.Boulon, M.Trabelsi-Ayedi, N.Kbir-Ariguib optical Materials 16 (2001) 301-310.

42 R.Ternane, G.Boulon,Y.Guyot, M.T.Cohen-Adad, M.Trabelsi-Ayedi, N.Kbir-Ariguib, Optical Materials (submitted on July 2002).

43 J.G.Bednorz, K.A.Muller, Z.Phys.B-Condensed Matter 64 (1986) 189.

44 S.Elschner, J.Bock, H.Bestgen, Supercon.Sci.Techn 6 (1992) 413.

45 G.Triscone, J.Y.Genou, T.Graf, A.Junod, J.Muller, Physica C 185-189 (1991) 783.

46 P.Hermann, E.Beghin, G.Bottini, C.Cotteveille, A.Leriche, T.Verhaeghe J.Bock,Cryogenics 77 (1993) 87.

47 T.Inoue, S.Hayashi, S.Miyashita, M.Shimizu, Y.Nishimura and H.Komatsu, J.Crystal .Growth 123 (1992) 615.

48 J.Zhao, M.Wu, W. Abdul-Razzaq, M.S.Seehra, Physica C 165 (1990) 135.

49 Y.Huang, M.H.Huang, K.W.Yeh, M.Y.Hong, Mat.Chem.Phy 41 (1995) 290.

50 Y.S.Sung, E.E.Hellstrom, Physica C 252 (1995) 155.

51 W.J.Yu, Z.Q.Mao, l.Yang, C.Y.Xu, M.L.Tian, L.Shi, G.E.Zhou, Y.Zhang, Physica C 244 (1995) 135.

52 G.D.Gu, K.Takamuku, N.Koshizuka, S.Tanaka, J.Cryst. Growth 137 (1994) 472.

53 H.jin, N.L.Wang, Y.Chong, M.Deng,L.Z.Cao, Z.J.Chen, J. Cryst. Growth 149 (1995) 269.

54 K. Lebbou, M.Th. Cohen Adad, R. Abraham, S. Trosset, R.E. Gladyshevskii, R. Flükiger, P. Gallez, G.W. Schuz, H.W. Weber, M. Couach, Physica C 297 (1998) 201.

55 C.N.R.RaoSupercond Sci Techn 3 (1990) 242.

56 B.Wanklyn, E.Dieguez, C.Changkamg, A.K.Pradhan, J.W.Hodby, H.Yongle, D.Smith, F.Wondre, J.Cryst.Growth 128 (1993) 738.

57 G.D.Gu, K.Takamuku, N.Koshizuka, S.Tanaka, J.Crystal Growth 130 (1993) 325.

58 P.Schätzle, G.Krabbes, G.Stöver, G.Fuchs, D.Schläfer, Supercond.Sci.Tech. 12 (1999) 69.

59 S.Kishida, E.Hosokawa, J.Cryst.Growth 192 (1998) 136.

60 F.M.Costa, R.F.Silva, J.M.Vieira, Physica C 323 (1999) 23.

61 K.Lebbou, A.Yoshikawa , M.Kikuchi, T.Fukuda, MTh.Cohen-Adad, G.Boulon Physica C 254-260 (2000) 336.

62 O.Bonfigt, H.Somnitz, K.Westerholt and H.Bach, J.Cryst.Growth 128 (1993) 725-728.

63 D.Gazit, R.S.Feiglson, J.Cryst.Growth 91 (1988) 318.

64 H.D.Brody, J.S.Haggerty, M.J.Cima, M.C.Flemings, R,L.Barns, E.M.Gyorgy, D.W.Johnson, W.W.Rhodes, W.A.Sunder and R.A.Laudise, J.Cryst.Growth 96 (1989) 225.

65 C. Goutaudier, F.S. Ermeneux, M.T. Cohen-Adad, R. Moncorge, J. Crystal. Growth 210 (2000) 694.

66 L. Laversenne, Y. Guyot, C. Goutaudier, M.T. Cohen Adad, G. Boulon, Opt. Materials 16(4) (2001) 475.

67 L. Fornasiero, E. Mix, V. Peters, K. Petermann, G. Huber, Cryst. Res. Technol 34 (1999) 255.

68 X. D. Xiang et al., Science 268 (1995) 1738-1742.

69 G. Briceno et al., Science 270 (1995) 273-275.

70 E. Danielson et al., Nature 389 (1997) 944-948.

71 S. Borman, Chem. Eng. News (1996) 37-38.

72 E. Danielson et al., Science 279 (1998) 837-839.

73 M.T.Cohen-Adad, L.Laversenne, M.Gharbi, C.Goutaudier, G.Boulon, R.Cohen-Adad J.of Phase Equilibria 22 n 4 (2001) 379-385.

74 L. Laversenne, S. Kairouani, Y. Guyot, C. Goutaudier, G. Boulon, M. Th. Cohen-Adad, Optical Materials 19 (2002) 59-66.
75 L. Laversenne, C. Goutaudier, Y. Guyot , M. Th. Cohen-Adad , G. Boulon J.of Alloys and Compounds 341 (2002) 214-219.
76 J.A. Burton, R.C. Prim, W.P. Slichter, J. Chem. Phys. 21 (1953) 1987.
77 D. De Loach, S.A. Payne, L.L. Chase, L.K. Smith, W.L. Kway and W.F. Krupke IEEE, J. Quantum Electron. 29 4 (1993), p. 1179.
78 A. Brenier J. Lumin. 92 (2001)199-204.
79 A. Brenier and G. Boulon J. Alloys Comp. 323-324 (2001) 210-213.
80 A. Brenier and G. Boulon Europhys. Lett. 55 (2001), 647-652.

8 Growth of Micro and Bulk Crystals by Modified Micro-PD and their Properties

Tsuguo Fukuda

The pulling down method has been modified to adapt to the growth of thin fibers, in-situ core-clad growth and bulk crystal growth. The concept of the modified pulling down method will be introduced in this section on the basis of the adjustable geometry of the crucible.

8.1 Thin Fiber Growth

In the regular μ–PD growth [1] the diameter of grown fiber is given by the size of the capillary nozzle. Depending on the crucible material it is relatively easy to prepare nozzles 500–1000 μm in diameter, but crucibles with thinner capillaries are much more difficult to produce. Another problem of very thin capillary channels is that they can be easily plugged during growth because of foreign phases or bubble formation. This often makes further use of a damaged crucible impossible. In the following section a geometrical modification of the μ–PD crucible by insertion on of an additional shaping element into the crucible nozzle is described for both metal and graphite crucibles used in high-temperature versions of the method [6].

8.1.1 Modification of Metal Crucibles for Oxide Growth.

The modified fiber growth assembly is shown schematically in Fig. 8.1. The change to the μ-PD crucible was made by inserting the additional shaping element into the crucible orifice, which enables one to grow thinner fibers using the same main crucible (compare Figs. 8.1.b and 8.1.c). Thin iridium wire, insert was 300 μm in diameter, was suspended with the wire holder so that the slightly sharpened insert end was below the bottom opening. This simple construction is self-aligning and flexible enough for precise adjustment of the insert extension length. Final adjustment was made after bringing the crucible with the insert to working temperature. Afterward the insert was slowly impressed into the crucible by the sapphire seed until the excess length was disposed.

The compound crucible combines two known approaches for the arrangement and feeding of a small liquid meniscus pool from which the fiber is drawn. If the meniscus is fixed on the crucible bottom (Fig. 8.1.b) "thick" fiber can be grown. The meniscus is continuously fed through the capillary split between the opening wall and insert. This is the same as in the regular μ-PD (and also in the EFG method, La Belle [2]), and actually in this case the insert plays no role, being shorter than the meniscus. If the meniscus is fixed on the insert tip (Fig. 8.1.c), much thinner fiber can be grown. Meniscus should be fed by melt spread along the insert surface. That approach to fiber shaping was first applied by Onishi [3] in his micro-Czochralski process for LiNbO$_3$ fiber growth.

Although our compound construction was intended mainly for growth of thin fibers, it features a few advantages prior to the direct miniaturization of the main crucible itself:

- the initial charge cannot be very small, but of reasonable size (1–5g) allowing a more accurate melt composition and a longer growth process;
- the residual melt can be quickly removed if necessary by the pulling of a "thick" fiber; there is no danger of split blockage by a precipitate of excessive or foreign phases, since the split is essentially large;
- a crucible of regular size is cheaper than a crucible having an ultrathin orifice; handling and cleaning of that crucible is much easier and safer.

The growth procedure for "thick" fibers involved routine steps such as: (i) bringing the crucible assembly to growth temperature until a bright melt surface was detectable inside the orifice, (ii) sapphire seed contact with the melt inside the orifice with formation of the melt meniscus, (iii) pulling the seed down with mutually interacting manual adjustments of both the pulling rate and the power input to the RF coil until stable meniscus conditions were established. Step (iii) was a trial and error procedure, essentially assisted by the self-stabilization of the meniscus (Surek [4], Tatarchenko [5]).

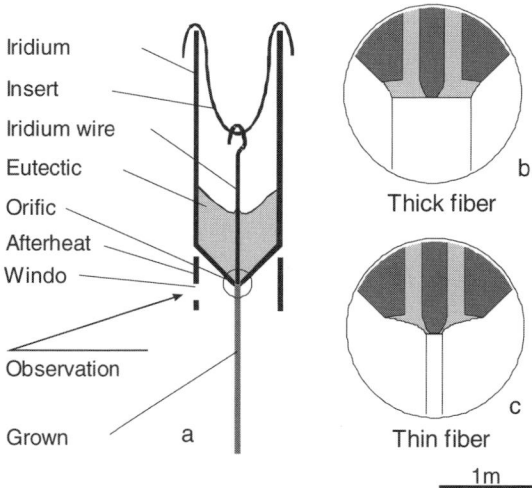

Fig. 8.1. A schematic diagram of the micro pulling down assembly modified with a wire insert

The growth procedure for thin fibers involved the same steps, but the accomplishment of step (iii) required special care. The drawback at that point originated from the limited wetting of iridium by many oxide melts. For instance, contact angles measured by us in the images of melt droplets on the iridium surface were $30 \pm 5°$ for all alumina based eutectics under investigation (see Chap. 6). In contradistinction to the LiNbO$_3$ melt in the micro-CZ experiments of Onishi [3], the eutectic melt did not spread spontaneously along the iridium surface. The seeding operation cannot be completed on the insert tip without forced melt film extension along the insert until it is all covered with the melt. What this means is that the seeding should be started from the melt near the insert and only then can the proper position be achieved. During the initial stage it is very important to prevent the fiber from undergoing enlargement and transition to "thick" growth. The necessary elements of step (iii) were as follows.

(iii a)   a few minutes were allowed for thermal equilibrium in the meniscus area to be reached and then a single temperature decrease was made immediately up to specified supercooling;

(iii b)   strictly within the period of reaching the new equilibrium state (usually 2–3 min) the pulling rate was increased so that the fiber steadily became slightly thinner but does not pull free from the insert;

(iii c)   after the specified temperature drop was compensated with the increase in heat of solidification the pulling rate was kept constant.

In this way the manual adjustment was performed using only 'one-way' variation of the pulling rate. Attempts to control the process of 'thin' growth in the usual way by tailoring both the pulling rate and temperature were mostly unsuccessful, since if the enlargement of a thin fiber was started once, it was impossible to turn it back.

Using the modified μ-PD method reasonably stable growth of eutectic fibers 150–300 μm and 800–900 μm in diameter and up to 500 mm in length was achieved over a wide range of pulling rate 0.05–20.00 mm/min. The highest achievable pulling rate was 24 mm/min (Fig. 8.2). The maximal length of a separate fiber was limited by the pulling mechanism. However, as the growth of 500 mm fiber was completed, it was separated from the orifice, removed from the furnace without lowering the temperature and the seeding procedure was repeated. The total length of fibers produced in the course of a single growth experiment was up to 3 meters. The diameter of each individual thin fiber in the range 200–300 μm and "thick" fibers were constant within 5–10% accuracy but most of thinnest fibers (150-200 μm) were not uniform in diameter demonstrating peculiar surface waviness (see Fig. 8.2).

Control of diameter uniformity can be achieved only by better control of the meniscus shape, which is dictated by the balance of stresses caused by surface tension, hydrostatic pressure, dynamic pressure and viscosity. The conventional assumption is that capillary pressure dominates over hydrodynamic stresses in small-scale capillary growth systems (Tatarchenko [5]). In our case this assumption is definitely valid for menisci remote from the insert ("thick" growth in Fig. 8.1), but it is not so in the case of thin fiber growth close to the insert tip, as will be discussed later. The equations governing the static meniscus shape were presented in detail in the Sect. 3.2, and here only the peculiar solution relevant to the present study will be considered.

The shape and dimensions of the actual crucible orifice with inserted iridium wire are shown in Fig. 8.3. The coordinate system is based on the lowest point of the crucible opening (the anchoring point for the menisci), see Fig.8.3. Theoretical meniscus shapes $Z(r)$ for a number of different fiber radii R in Fig. 3 were calculated using the equation

$$Z(r) = R\sin(\pi/2 - \phi)\left[\operatorname{arch}\frac{r_0}{R\sin(\pi/2 - \phi)} - \operatorname{arch}\frac{r}{R\sin(\pi/2 - \phi)}\right] \qquad (8.1)$$

where r_0 is the crucible bottom radius and ϕ is the melt/crystal growth angle.

The broken line $Z(r)$ represents all possible pairs of static values for variation of the crystal radius R with meniscus height Z. This was calculated according to

$$Z(r) = R\sin(\pi/2 - \phi)\left[\operatorname{arch}\frac{r_0}{R\sin(\pi/2 - \phi)} - \operatorname{arch}\frac{1}{\sin(\pi/2 - \phi)}\right]. \qquad (8.2)$$

The melt/crystal growth angle ϕ was set equal to 20° in our calculations, since measured values for sapphire are 17° for <0001> and 35° for <1010> directions [33]. In (8.1) and (8.2) the linear parameters (r_0, R) are dimensionless by using the capillary constant $a = \sqrt{\dfrac{2\sigma}{\rho g}}$ as the reference length, but in Fig. 8.3 they are adjusted to the actual crucible/insert geometry (a was set equal to 4 mm) and plotted in micrometers for ease of visual examination.

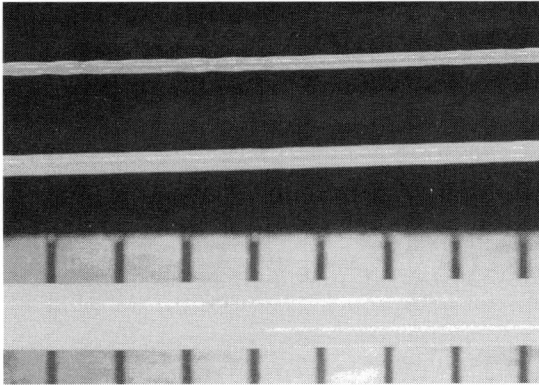

Fig. 8.2. As-grown eutectic Al$_2$O$_3$/YAG fibers 150, 220 and 900 μm in diameter. Note the diameter instabilities in the 150 μm fiber

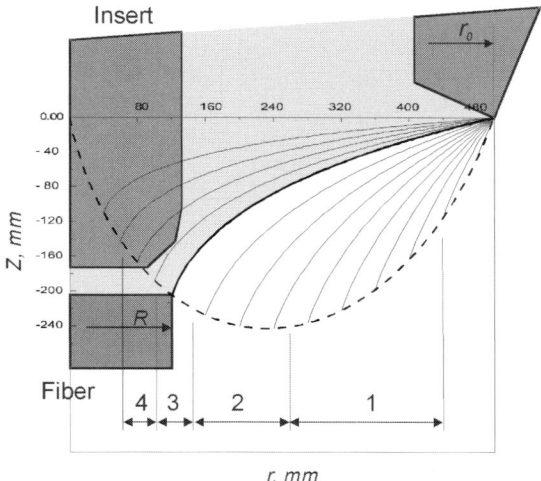

Fig. 8.3. Calculated meniscus shapes for actual crucible/insert geometry and stability areas

The analysis made by Surek [4] with the use of a static approach shows that the fiber shape will be stable if the anchoring-defining radius r_0 does not exceed approximately twice the radius of the fiber R. This self-stabilization area is marked as section 1 in Fig. 8.3. It corresponds to "thick" growth and is in full agreement with experimental practice. If the fiber is getting smaller and comes into the portion of the $Z(r)$ curve where $dZ/dr < 0$ the growth becomes unstable. From the unstable section 2, fiber tends either to get thicker and return back to the stable section 1 or gets steadily thinner. In the regular μ-PD process (without an insert) the second case leads the fiber to pull free from the melt. In the presence of the insert the diminishing meniscus is subjected to hydrodynamic pressure which increases strongly when the melt film thickness between the meniscus and insert becomes sufficiently small (see the flow streamlines in Fig. 8.5). This brought the additional stable section 3 into existence. Certainly the static approach for the meniscus shape is invalid near the insert but it is still useful for qualitative interpretation. Figure 8.4a shows a screen capture image of the stable growth process for Al_2O_3/YAG fiber about 200 μm in diameter. The crucible bottom has an outer diameter of 1 mm which meant that growth occurs in section 3 of Fig. 8.3. Comparison with calculations is made in Fig. 8.4b by overlaying the plot from Fig. 8.3 onto Fig.8.4a. It clearly shows meniscus extension over the theoretical static value (the small asymmetry and some viewing angle aberration also should be taken into account, see Fig.8.1).

An attempt to grow still thinner fiber (below 200 μm, i.e. within section 4 in Fig.8.3) always resulted in specific diameter instability (compare Fig. 8.5b and Fig. 8.2).

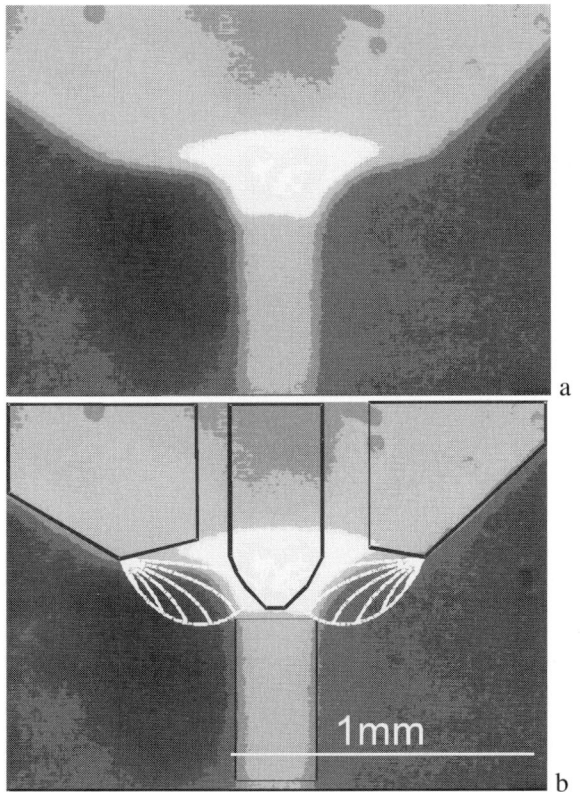

Fig. 8.4. (a)A screen capture image of the stable growth process for Al$_2$O$_3$/YAG fiber about 200 μm in diameter. The crucible bottom has an outer diameter of 1 mm. **(b)** Comparison with calculations

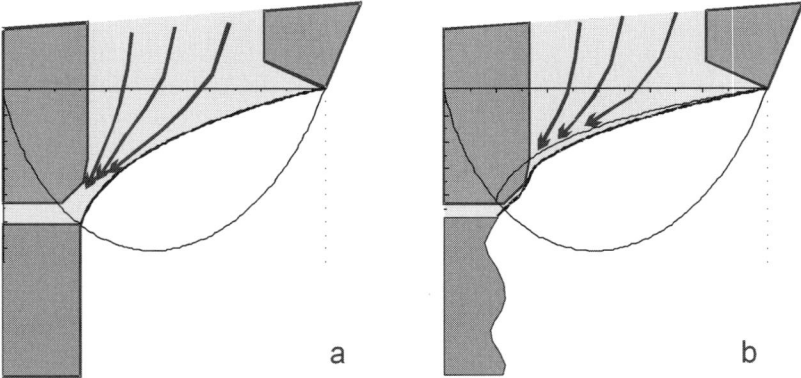

Fig. 8.5. Schematic for stable **(a)** and unstable **(b)** growth regimes for the fiber fixed on an insert tip

We suggest instabilities of that kind to be of hydrodynamic nature. As was determined experimentally the demarcation point between section 3 and 4 lies near the intersection of $Z(r)$ with the insert contour (about 100 µm in our assembly). This can be interpreted in the following way: stable fiber growth can occur within section 3 until the insert does not separate the meniscus melt pool. In section 4 separation of the meniscus pool into two parts, those connected by the melt film at the insert surface, brings up a detectable meniscus pool turbulence followed by diameter instability (see the illustration in Fig. 8.4b). The immediate implication of this result on thin fiber growth is that the insert diameter should be about 30–40% greater than the preferred fiber diameter to achieve external shape stability.

8.1.2 Adjustment of Graphite Crucibles for Thin Silicon Fiber Growth

A great amount of work has been done on crystal-size enhancement technology for silicon single crystals. Large 12 diameter crystals are now produced commercially. On the other hand, Si fiber crystals having a large surface to volume ratio are also attractive for special device applications. The present research was carried out to clear up the possibility of growing semiconductor crystals of small diameter and to investigate the operating limits of the process.

The silicon melt has low viscosity, high surface tension and high chemical activity. The chemical activity of the Si melt makes graphite the only suitable material for preparation of wetted crucibles for silicon processing [7]. However, the manufacture of the narrow capillary and sharp crucible end of the µ-PD assembly is rather complicated because of the poor mechanical properties of graphite.

To study the growth process of a thin silicon filamentary crystal we have designed three types of crucible-die assemblies. In all cases the assembly was set up in the same hot zone of the µ-PD puller for semiconductor crystal growth. The detailed description of this apparatus has been published by Schäfer [8]. In order to reduce the amount of silicon carbide precipitation and increase the useful life of the crucible, additional attention was paid to reduce carbon monoxide circulation inside the growth system by optimization of the argon flow. The crucible was provided with lateral holes and a small gas-permeable upper shield (Fig. 8.6).

As a result, pure argon was forced to bathe the free silicon surface in the crucible and then the meniscus, and to carry the CO away from the surface of the silicon melt. The high effectiveness of this design has been proved by the sufficiently increased operation time of the crucible.

The crucible of type I had a conventional µ-PD configuration but the geometry of the capillary channel has been changed. Instead of a cylindrical hole a flat 50 µm narrow cut was made through the crucible bottom.

The crucible of type II had a movable graphite insert inside the conical opening in the crucible bottom. Three segments were removed to provide a capillary channel between the crucible and the insert. Nevertheless feeding the meniscus pool of was only achieved with the help of the coarse surface of graphite covered by silicon carbide layer. It was possible to move the insert upwards which protected it in the case of operational error. For example, an attempt at "cold seeding" (before the temperature of the seed is high enough) was not dangerous for this assembly.

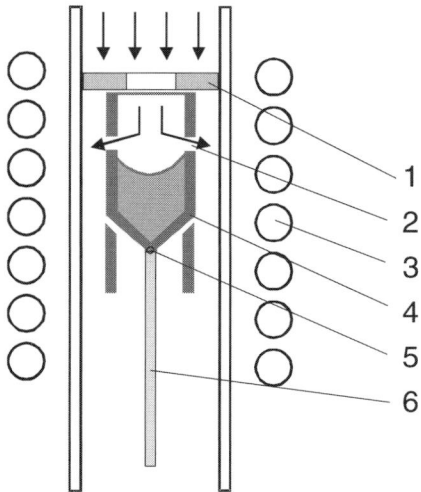

Fig. 8.6. Schematic view of improved μ-PD set-up: 1 – gas permeable shield, 2 – lateral holes, 3 – RF coil, 4 – crucible, 5 – meniscus, 6 – micro crystal

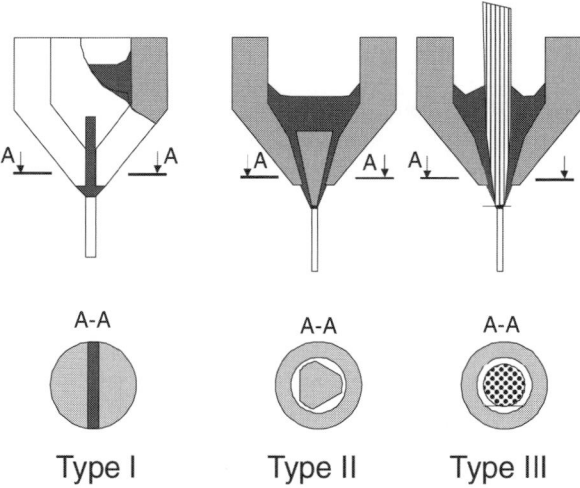

Fig. 8.7. Experimental die-crucible assemblies

In type III the insert was initially made of continuous graphite fibers 7–8 μm in diameter. After silicon melt impregnation, the graphite fibers were turned into silicon carbide almost completely and became porous "felt-pen-type" inserts. The solidified SiC-Si composite material was hard enough and the insert tip can be easily sharpened by a diamond tool up to the necessary size.

A <111> oriented rectangular Si stick prepared by cutting a 0.5 mm thick Cz silicon wafer was used as a seed. The seeds were sharpened mechanically and etched in a HF + HNO₃ acid mixture. During growth the tip of the crucible with the insert was observed by CCD camera. The liquid and crystalline silicon could be distinguished easily on the monitor screen by their difference in emissivity. This gives us the possibility of adjusting the temperature by simple visual monitoring of the solid-liquid interface motion along the melt impregnated insert which was very easy in the case of type III. The control of the growth process including the motor drive speed and the RF power adjustment was performed manually.

Applications of any of the above described types of μ-PD crucibles make it possible to grow Si fiber with a diameter from 100 to 300 μm. The diameter of each individual crystal was constant within 15–20% accuracy. As-grown crystals grown with the use of types II and III of about 100 μm diameter are shown in Fig. 8.8.

Type I gave us the possibility of growing 200 μm Si fibers at a pulling velocity of up to 1.5 mm/min. In this case successful feeding of the meniscus pool was achieved. However the fixation line of the meniscus on the sharp edges of the crucible bottom was not clearly defined as the two segmental parts were separated from each other. From time to time occasional disturbance broke the meniscus away from one segment. As a result the diameter of the crystal became smaller. One of the segmental parts was destroyed several times because of partial freezing of the crystal to the crucible bottom. The separated particle was incorporated into the fiber. The crystal shape after this had changed to oval.

Type II can be fabricated more easily than type I. The main constraint of the type II is related to the melt feeding deficiency. When the pulling velocity exceeds 0.5 mm/min the crystal was always disconnected from the crucible. Any adjustments of the RF power were ineffective.

Use of type III allowed us to increase the pulling speed up to 0.9–1.0 mm/min. A flexible and slightly inclined insert seems to be the most effective because of its ability to quench spontaneous oscillations. It was shown that type III also has the longer operation life. Assemblies of types II and III are suitable to operate with an insert top diameter less than 100 μm. In principle, the insert tip can be sharpened up to 15–20 μm. Nevertheless, in our experiments a silicon melt meniscus less than 100 μm in diameter was very sensitive to spontaneous perturbations caused by faulty operation of the pulling mechanism. The high quality growth of crystals less than 100 μm was not achieved.

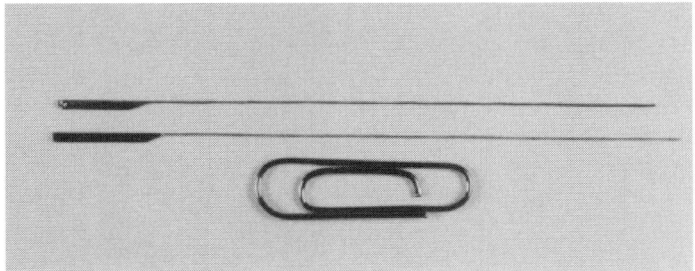

Fig. 8.8. As-grown crystals 100 μm in diameter

In all cases two important conditions of the stable growth process should be pointed out: (i) low pulling velocity in comparison with EFG silicon growth (always less than 1 mm/min); and (ii) very small meniscus height, e.g. for a 150 μm fiber it was about 20–25 μm. Calculation of the meniscus shape by solving the Laplace equation for the μ-PD growth system gives a value for meniscus height that is approximately twice as high as this one, i.e. 40–45 μm.

To clear up this difference we measured the maximum height of the meniscus, i.e. the height of the small diameter silicon liquid column at the moment of crystal separation from the crucible. The scheme of the experimental procedure is shown in Fig. 8.9.

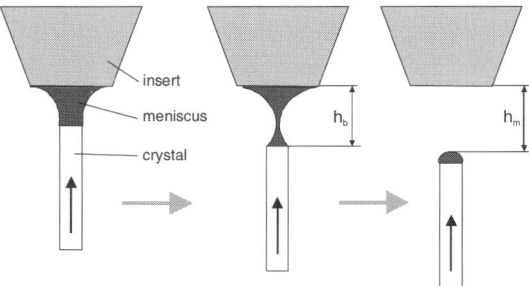

Fig. 8.9. Scheme of the measurement of μ-PD maximum meniscus height. h_b – height at the moment of separation, h_m – measured value

The fiber was raised up until it touched the overheated crucible bottom and melted. The meniscus was formed which joins the crystal and the die bottom. Because of overheating the crystal was melted further and the height of the meniscus increased. At a certain height the meniscus was broken. By applying a constant raising speed we observed a series of periodic breaks. The initial moment of contact could be easily distinguished on the LCC monitor as a bright flash. The distance H_m was determined by multiplying the average time between the flashes by the raising speed.

The simplest way to find stable values of the crystal radius R and the meniscus height h is to solve capillary and heat flow equations independently and then the functions $h(R)_{capillary}$ and $h(R)_{heat}$ obtained should be coupled, [4, 5]. This approach can be applied successfully to the μ-PD system. In our case the Bond number is about 10^{-2} and capillarity prevails completely. This can be illustrated emphatically by the comparison of the capillary rise of the silicon melt in a 100 μm wide channel equal to 75 cm with the meniscus height. For small pulling rates the heat transfer caused by crystal and melt movement can also be neglected. The micro crystal can be considered as an infinitely long object because of large length-to-diameter ratio.

Based on these assumptions the equation of heat balance on the crystallization-front can be written as follows [5]:

$$\sqrt{\frac{\beta}{R} - \frac{1}{h}} = \eta V \qquad (8.3)$$

where

$$\beta = \frac{2\mu_s \lambda_s^2 (T_0 - T_E)^2}{\lambda_L^2 (Tp - T_0)^2}, \qquad \eta = \frac{L}{\lambda_L (Tp - T_0)}$$

and μ_s denotes the coefficient of heat exchange with the environment, λ_s^2, λ_L is the thermal conductivity of solid and liquid silicon respectively, T_0 the melting temperature, T_p is the melt temperature on the die/insert level, T_E is the environmental temperature, and L is the latent heat of melting per unit volume.

Curves 1–3 in Fig. 8.10. show $h(R)$ the heat calculated for different pulling rates V from equation (8.3). To plot $h(R)_{\text{heat}}$ the refined parameters for silicon given by Kobayashi [11] were taken. The value of $(T_p - T_0)$ was assumed to be 5 K, and the environmental temperature was 300 K. The function $h(R)_{\text{capillary}}$ was calculated from the capillary equation (8.1), see Sect. 8.1.1 and plotted in Fig. 8.10. as curves 4–5.

Only simultaneous solutions corresponding to a larger radius are stable [5]. The relationship between the pulling velocity and the meniscus height for stable solutions shows that thin Si fibers are very sensitive to variations of growth rate. The conditions for stable growth could only be met if the pulling rate changed in a relatively narrow range of 0.2–2 mm/min. This peculiarity comes from the thermophysical properties of the silicon melt. For example, sapphire fiber growth is much more stable because of the effective radiative heat transfer [9].

Some additional information can be derived from comparison of the results of the meniscus height at the moment of separation with the above calculations. The criteria for the stability of a meniscus joining a vertical rod to a pool of liquid were obtained by Pitts [10]. By applying his calculation scheme to our experiment we experienced that the expanding meniscus became unstable after the angle θ between the tangent to the meniscus and the cylindrical face of a melted crystal became smaller than 80°. In the course of growth this angle was equal to 11° (growth angle for Si is 11°). Even a rough estimation shows that h_b should be at least 50 % greater than h_{max} calculated from the Laplace equation. However, the measured value of H_b was on average 30–40 % lower than the h_{max} value calculated (Fig. 8.11).

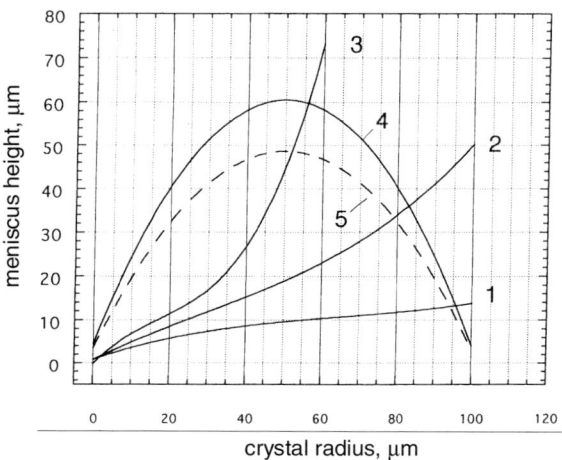

Fig. 8.10. Meniscus height dependence on fiber radius calculated from (1–3) the heat balance and (4–5) the Laplace equation: $1 - V = 0.1$ mm/min; $2 - V = 1.0$ mm/min; $3 - V = 1.5$ mm/min; 4 – capillary constant $d_{cap} = 7.6$ mm; $5 - d_{cap} = 6.2$ mm

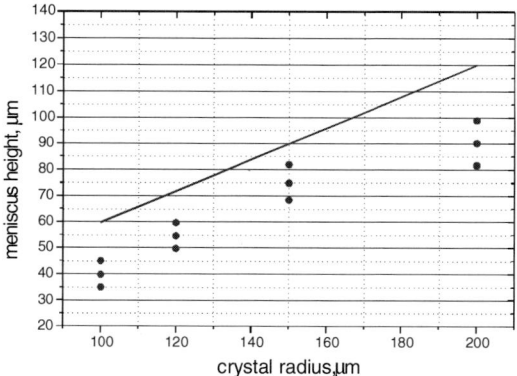

Fig. 8.11. Maximum meniscus height at the moment of separation: solid line – calculations, points – experimental

We believe that in our case the actual meniscus shape can differ noticeably from the theoretical one because of the high chemical activity of the silicon melt with respect to carbon. It is difficult to predict the exact shape of the catching line because of SiC formation. The actual magnitude of the capillary constant depends on oxygen contamination of the ambient atmosphere and seems to be actually smaller than 7.6 mm. The function $H(r)_{capillary}$ for a real system appears to be below curve 4 in Fig. 8.10. For illustration the curve 5 in Fig. 8.10 is plotted for $d_{cap} = 6.2$ mm.

In conclusion, the μ-PD method has been modified to meet the conditions for growth of Si fibers with diameter less than 150 μm. Three types of crucible-die assembly were designed and tested. Fiber crystals 100–150 μm in diameter and 70–80 mm in length were grown successfully. In spite of the high temperature gradient near the interface in the μ-PD method, the operating conditions for Si fiber growth are strongly restricted. The growth process is stable as long as the meniscus height is less than 20–25 μm and the pulling velocity is less than 1.5 mm/min. The high chemical activity of silicon melt influences the actual meniscus shape.

8.1.3 In-situ Growth of Core-Clad Fiber Structures

A reasonable way for the further development of highly efficient single crystal fibers is cladding as is commonly done with optic fiber glasses. The fiber crystal should be doped with active laser ions only in a sharply separated inner core region. The axial arrangement of a selectively doped crystal in the laser beam is very advantageous since the pumping energy is absorbed only by the central part of the crystal. The translation efficiency is increased and energy losses by heat are drastically reduced. Moreover the low level of stresses typical for fiber crystals yields a constant homogeneous refractive index distribution which is important for the stabilization of a horizontal laser mode and increasing efficiency of side pumping laser oscillations. As estimated by Rudolph [15], in the case of a laser diode with a pumping radius of about 100 μm a core diameter of 200 μm needs to be obtained. A lamp pumping laser requires a doped clad diameter of about 1.2 mm with a core of 0.6 mm.

Feigelson and his coworkers reported the investigation of the glass surface coating for Nd:YAG laser fibers and also the creation of a reduced refractive index layer in the surface of lithium niobate fibers by diffusion, ion exchange and ion implantation processes. Doping of the outside portion of the fiber was found to be preferable over outside deposition [12].

A subject of special interest is the potentiality to grow a fiber crystal doped in a separated inner core region or vice versa in a clad region immediately from the melt, that is **in-situ**. The subject of composing a well-defined concentration profile in the growing crystal has attracted considerable attention over a long period for the last 30 years. Among the experimental studies reported up to now three main approaches can be distinguished.

The first one is the application of electric, magnetic and ultrasonic fields during growth. In [16] it was shown that the electric potential changes the values of dopant distribution coefficients and the system equilibrium temperature. Doping modulation under the pulsed current was achieved. Periodic structures in crystals grown in a field of ultrasonic waves were investigated by Arakelyan [17]. The regularity in dopant modulation in crystals grown both from the melt and from solution was detected. Extremely rapid modulation of dopant segregation was observed by Lichtensteiger et.al. [18] in tellurium-doped indium antimonide single crystals. This was brought about by electric currents applied across the growth interface during Cz growth. The time constant of the modulation was found to be in the millisecond range and thus orders of magnitude smaller than that encountered in diffusion-controlled processes.

The second approach is the technique of remelting. Burrus [19–20] had grown single crystal ruby fibers by a floating-zone procedure using a CO_2 laser. Thereafter the Cr concentration in the surface layer was reduced by a factor of 100 by using the same laser to melt and regrow only the surface of the fiber. Nonetheless the possibility of coupling the fiber growth and the surface remelting into one process seems improbable.

The third idea is to supply the growing crystal from two sources of differently doped melt or even different melts. The idea for this modification of the EFG process was first proposed by Fukuda and Hirano in 1979. $LiNbO_3$ films were grown on LiTaO$_3$ substrates by the capillary liquid epitaxial technique [21]. Later Antonov [22] modified the Stepanov method of shaped crystal growth by simultaneous use of two localized dies. Each of the dies provided independent feeding of the growing crystal. A layered single crystal of LiF-LiF:Mg was grown and used as a model for the investigation of mechanical parameters and dislocation structures in composites [23]. Recently a crystal growth method for in-situ core doping was proposed by Shimamura [24]. Two crucibles combined with an outer and inner die were used for transport of independently doped melts to a single growth interface. Core doped Nd,Cr:LiNbO$_3$ crystals 5 mm in diameter with a sharply separated central doped region were grown successfully. However this technique is suitable for rod-like crystals larger than 5 mm in diameter only.

In summary, despite the promising results listed above the realization of functionally doped crystal growth remains questionable.

8.1.4 Double-die Modification of the Micro Pulling Down Method

In this part we will discuss a double-die modification of the μ-PD method which makes possible core/clad doping of fiber crystals as small as 500 μm in diameter. The method of double-die-μ-PD (dd-μ-PD) growth makes use of an assembly such as that illustrated in cross-section in Fig. 8.12. It is composed of two μ-PD crucibles that are heated resistively by passing a current through them. The upper crucible is provided with a capillary nozzle which is relatively long in comparison with the ordinary μ-PD design. The nozzle is inserted into the circular opening in the bottom of lower crucible. After melting of the charge in both crucibles the capillary tube is filled with the undoped melt spontaneously, the drop of doped melt is fixed inside the opening between the nozzle and the lower crucible bottom due to capillary tension, while the outer surface of the wetted nozzle is covered with the thin film of doped melt. As a result the nozzle is functioning as a "double-die" (dd). It arranges capillary feeding of the growing crystal simultaneously by the inner pipe channel (undoped melt) and by the outer surface (doped melt). Evidently clad or core doped growth can be achieved by changing over the position of the doped/undoped melt.

The actual design of the dd-μPD crucible is specified mainly by the melt location outside the nozzle. The geometry of the melt surface in the area of the opening is sketched in Fig. 8.13. For later analysis we will use the approach first proposed by Landau [25]. Following [25] let the liquid surface be divided into two areas: a static meniscus and a moving liquid film (see Fig. 8.13). The governing differential equation for the meniscus area is the fundamental equation of capillarity [26] known as the Laplace equation. As the governing differential equation for the film area it is possible to use the boundary layer approximation of the Navier-Stokes equation. The corresponding solutions $t_{film} = f(x)$ and $t_{men} = f(x)$ (meniscus) are valid close to and far from the nozzle surface. To match them correctly at some point $x = h$ means imposing the conditions that $t_{men} = t_{film}$, $\dfrac{dt_{men}}{dx} = \dfrac{dt_{film}}{dx}$,

$\dfrac{d^2 t_{men}}{dx^2} = \dfrac{d^2 t_{film}}{dx^2}$ near this point. The complete analytical solution based on this approach was derived in [25] for the case of a flat plate which is constantly pulled from the melt and was also proved experimentally in [27]. Here we will concentrate mainly on its application for the practical design of a miniature dd-µ-PD crucible.

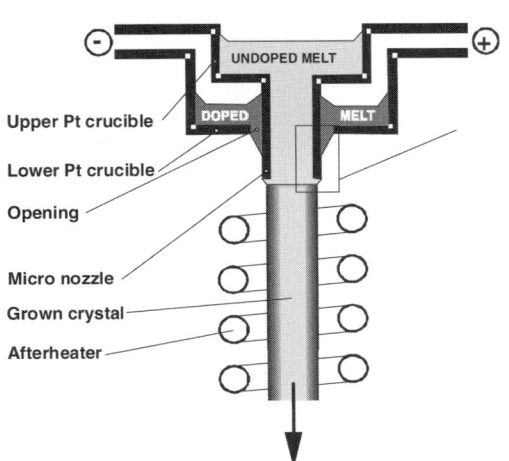

Fig. 8.12. Schematic diagram of the double-die-µ-PD process

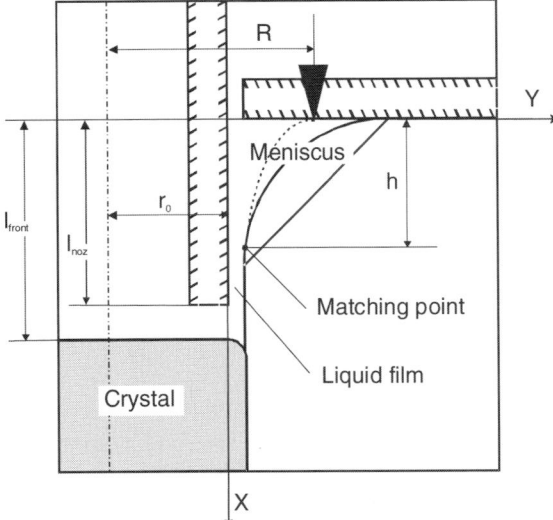

Fig. 8.13. Definition sketch for the analysis of capillarity

For the X-Y axis arrangement as shown at Fig. 8.13, when y is the dependent variable, the form of the governing equation is

$$\frac{d^2y}{dx^2} = \left[1+\left(\frac{dy}{dx}\right)^2\right]\left\{\frac{\rho g}{\sigma}\,y\left[1+\left(\frac{dy}{dx}\right)^2\right]^{\frac{1}{2}} - \frac{1}{x}\frac{dy}{dx}\right\}. \tag{8.4}$$

If the melt meets both the bottom and the nozzle surfaces at a wetting angle $\phi \approx 0$, and the nozzle diameter r_0 is small ($r_0 << a = \sqrt{\dfrac{2\sigma}{\rho g}}$, where a is the capillary constant) the analytical expression for $h = f(r_0)$ can be derived from (8.4) [28]:

$$h(r_0) = r_0\left\{\ln\left(\frac{2a\sqrt{2}}{r_0}\right) - \gamma\right\} \tag{8.5}$$

where γ is Euler's constant.

Another possible situation is the meniscus anchoring on the peculiar line at a distance of R from the nozzle axis (marked by a black triangle in Fig. 8.13). Here the analytical solution for (8.4) is also viable [21]. The meniscus shape for the pre-set fiber radius r_{cr} is given by

$$t(x) = ar_{cr}\left\{\ln\left(\frac{R}{r_{cr}} + \sqrt{\left(\frac{R}{r_{cr}}\right)^2 - 1}\right) - \ln\left(\frac{x}{r_{cr}} + \sqrt{\left(\frac{x}{r_{cr}}\right)^2 - 1}\right)\right\}. \tag{8.6}$$

The height of the meniscus fixed between the peculiar line and the nozzle is given by

$$h(r_0) = ar_0 \ln\left(\frac{R}{r_0} + \sqrt{\left(\frac{R}{r_0}\right)^2 - 1}\right). \tag{8.7}$$

Graphs for (8.5)–(8.7) are plotted in Fig. 8.14 for reasonable value of r_0 using the value of capillary constant $a = 4$ mm, as determined in [29].

Taking into consideration that the film movement is stationary or slow we can use a simplified form of the Navier-Stokes equation:

$$\frac{\sigma}{\rho}\frac{d^3t}{dx^3} + \frac{\eta}{\rho}\frac{d^2v_x}{dy^2} - g = 0. \tag{8.8}$$

The boundary condition for (8.8) is

$v_x = 0$ at $y = 0$. (8.9)

This indicates that the film is inhibited from moving down by the nozzle surface.

At the free film surface $y = t(x)$ the boundary condition is that

$$\frac{dv_x}{dy} = 0 .$$ (8.10)

The value of v_x at $y = t(x)$ is set to be equal to the pulling rate v_p. This is a matter of convention, but is undoubtedly within the range of practical purposes of our analysis.

The third condition $d^2t_{men}/dx^2 = d^2t_{film}/dx^2$ originates from the necessity to match (8.4) and (8.8) correctly.

The expression for d^2t_{film}/dx^2 was derived in [17] from the equation corresponding to (8.8)

$$d^2t_{film}/dx^2 = Kt_0^{-1}\left(\frac{\sigma}{3\eta v_p}\right)^{-2/3} .$$ (8.11)

where K is a constant equal to 0.63.

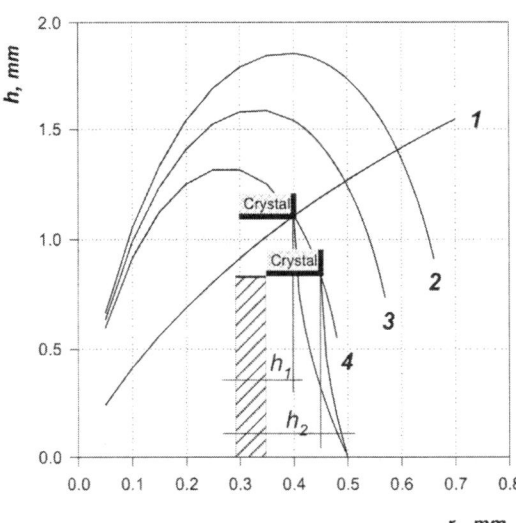

Fig. 8.14. Meniscus height via nozzle/crystal radius. 1 – meniscus wets crucible bottom, (8.5); 2–4 – meniscus is fixed at various R, (8.7); $2 - R = 0.7$, $3 - R = 0.6$, $4 - R = 0.5$

From the above Eqs. (8.4) and (8.5) it appears that near the matching point

$$d^2t_{men}/dx^2 = \frac{2r_0}{a^2}\left\{\ln\left(\frac{2a\sqrt{2}}{r_0}\right) - \gamma\right\}.$$
(8.12)

Finally equating (8.11) with (8.12) we obtain the expression for the film thickness t_0.

$$t_0 = 1.31\frac{\sigma^{1/3}\eta^{2/3}v_p^{2/3}}{\rho g}\frac{1}{r_0}\left\{\ln\left(\frac{2a\sqrt{2}}{r_0}\right) - \gamma\right\}^{-1}.$$
(8.13)

The dependence of t_0 on the pulling rate is plotted in Fig. 8.15 using viscosity data of [30]. From these curves it should be noticed that the dependence is weak and the film thickness is in any case smaller than 10 μm.

The analysis gives us the prospect of determining the main features of the proposed method and to separate two quite different approaches for its practical accomplishment.

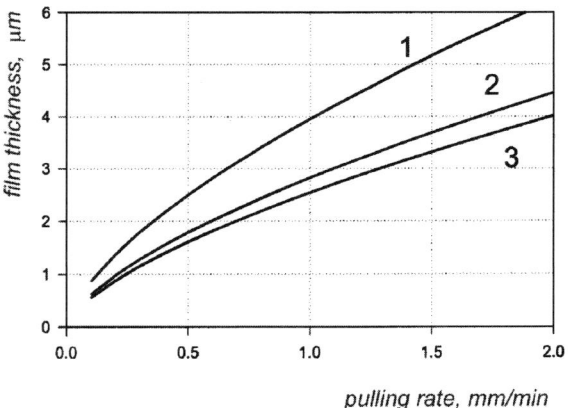

Fig. 8.15. Film thickness versus pulling rate: $1 - r_0 = 0.2$ mm, $2 - r_0 = 0.3$ mm, $3 - r_0 = 0.4$ mm

Case I. The Nozzle Length I_{noz} is Smaller than the Meniscus Height h

In this case the growth interface intersects the meniscus area. Consequently the relative amount of doped melt may be varied by arrangement of the growth interface in the vertical direction. Nevertheless an additional note is necessary: the meniscus wetting the crucible bottom is unstable against spontaneous changes of crystal dimensions akin to the Cz meniscus [5]. In order to stabilize it one should consider a fixing element of the crucible bottom for meniscus anchoring. The particular example is illustrated at Fig. 8.14 for two sizes of fiber $r_0 = 0.4$ mm and $r_0 = 0.45$ mm; note the lines of the meniscus profile. The meniscus is fixed at $R = 0.5$ mm, the nozzle radius is 0.35 mm. It can be seen that the extension of the fiber radius increases the thickness of the doped melt layer by approximately twice from h_1 to h_2. However these changes are accompanied by modification in the melt mixing conditions because the distance between the growth interface and the nozzle becomes smaller.

Case II. The Nozzle Length l_{noz} is Larger than the Meniscus Height h

In this case fiber expansion out of the nozzle is impossible and the growth process is invariantly stable. Nonetheless the thickness of the doped layer of about 5 μm is too small for practical use. The simplest method to increase the amount of doped melt is to modify the outer nozzle surface with irregularities. As a result the effective path for doped melt will become larger. In doing so the change of doped area during the growth process is impossible.

Mn-doped LiNbO$_3$ fiber has been chosen for the initial experimental verification of the proposed technique. The crucible assemblies of two types I and II (see Fig. 8.16) were made of platinum. For comparison purposes some growth experiments were performed using the ordinary μ-PD crucible design, serving as the upper crucible of assembly type I. The upper crucible size was $8 \times 2 \times 1.5$ mm^3 (rated for a charge of 70–80 mg of raw material). Nozzles of different size were applied for each of the constructions: with outer to inner diameter equal to 0.7/0.5 mm and to 0.5/0.4 mm. For assembly type II some growth experiments were performed without any mechanical treatment of the platinum surface and some others were accomplished after scratching of the outer surface of the nozzle with a sapphire pin (see the cross-section in Fig. 8.16). The temperature of the nozzle was controlled to a high degree of accuracy by the power loaded to the platinum wire afterheater. The alignment of the seed and the nozzle was adjusted by the micro X-Y stage. Visual monitoring of the process was performed using an optical microscopic tube.

Small pieces of Cz stoichiometric LN crystals were used as a charge for the upper crucible. The charge for the lower crucible was prepared by prepulling fibers containing 5 and 12 mol.% Mn by the ordinary μ-PD method.

Mn:LiNbO$_3$ crystals nearly 0.5 and 0.7 mm in diameter and 55–70 mm in length were grown in an air atmosphere along the c-axis. The pulling rate was varied in the range 0.22–1.34 mm/min. A photograph of the as-grown crystal showing the typical fiber habit and coloration is presented at Fig. 8.17. Note the dark colored heavily doped ends of the fibers which were grown under the condition of luck of undoped melt from the capillary channel. The radial Mn-distribution was analyzed by optical microscopy and by electron probe microanalysis (EPMA) using crystal sections cut perpendicular to the growth axis from the crystal mounted into a plastic holder.

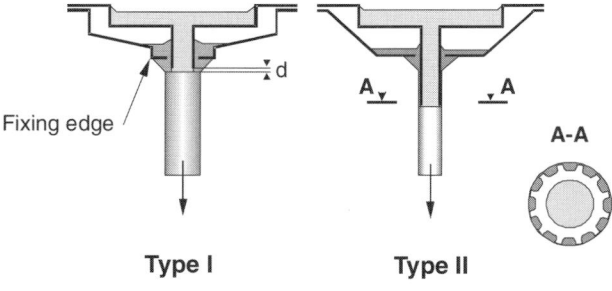

Fig. 8.16. Experimental arrangements of type I and II

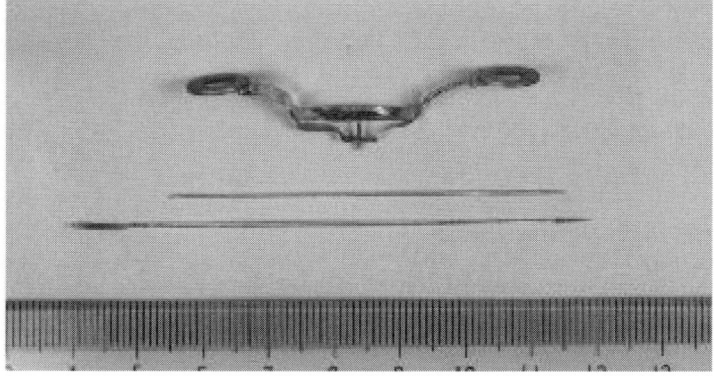

Fig. 8.17. As-grown clad-doped fibers (final part)

The radial Mn distribution curves measured by EPMA and representative for each of the applied constructions are plotted at Fig. 8.18.

It is peculiar that 5 mol.% doped LN fibers grown by the ordinary μ-PD method have a gradient in concentration across the fiber diameter (Fig. 8.18). A significant gradient appears when the fiber diameter is smaller than 700 μm, the value being stronger near the center of the fiber. This phenomena is probably due to the specific Mn distribution in the liquid which is caused by the radial interface electric field with the direction from the rim to the core, driving Mn ions to the center of the fiber (see Chap. 2). This distribution is obviously unfavorable for optical device application.

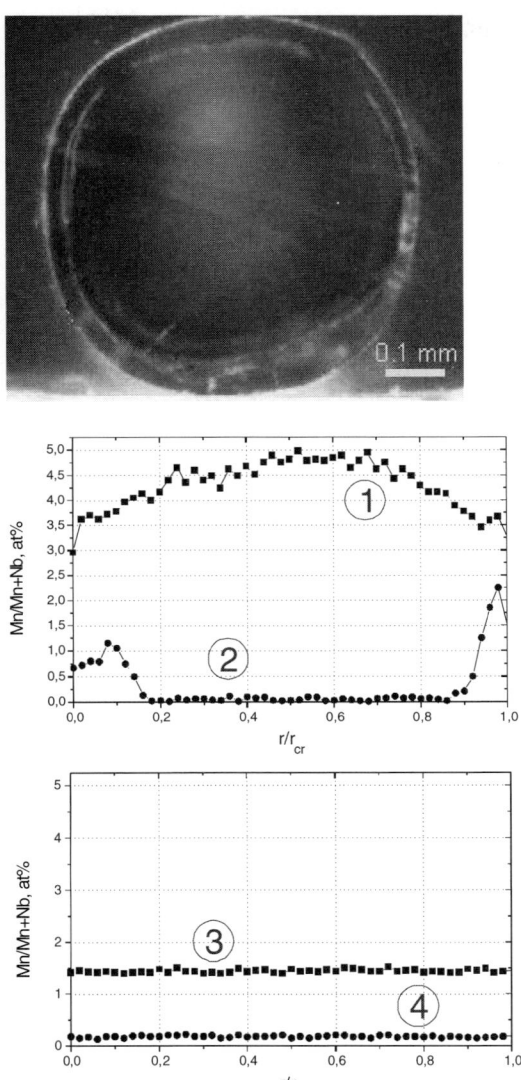

Fig. 8.18. Radial dopant distribution in grown fibers. 1 – 500 µm fiber grown at 0.8 mm/min using ordinary µ-PD; 3–4 – 500 µm fiber grown at 0.8 mm/min using dd-µ-PD type I: 3 – high meniscus, 4 – the same fiber after reducing the afterheater power; 2 – 700 µm fiber grown at 1.2 mm/min using dd-µ-PD type II

The use of assembly type I allowed us to obtain a nearly flat distribution across the fiber diameter for all tested sizes and pulling rates (Fig. 8.18). This indicates consequential mixing of doped and undoped melts in front of the growth interface. However, the possibility of pre-growth adjustment of the distance h_{front}-h_{noz}, see Fig. 8.13 , is limited. In fact it was no less than 0.3–0.4 mm for all growth experiments to prevent the fiber freezing to the nozzle. Significant variation of the dopant concentration under a change of meniscus height has been achieved (compare curve 1 and 2 in Fig. 8.18).

The fibers grown with the use of assembly type II provided with the smooth nozzle demonstrated nearly zero concentration according to the EPMA analysis. The utilization of heavily 12 mol.% doped melt gave a slight coloration of the peripheral part of the fiber detectable with an optical microscope and undetectable with EPMA.

The crystals grown with the use of assembly type II provided with the modified (scratched) nozzle at a pulling rate higher than 1 mm/min have two distinctive portions with different Mn concentration. In the peripheral portion of the fiber Mn is rich and there is a concentration peak while there is a wide area with almost zero Mn concentration (Fig. 8.18). These two portions did not look to be mixed during growth and they form clad and core, respectively. The peak in the periphery of the fiber is probably derived from the specific Mn distribution in the liquid of the same nature caused by the interface electric field as in the case of ordinary μ-PD as discussed in Chap. 2.

In conclusion, the feasibility of this novel technique for in-situ single crystal fiber cladding based on the micro pulling down (μ-PD) method has been demonstrated. The best reproducibility was achieved with the crucible provided with the long nozzle having the modified outer surface (type II). At the same time, the controlled variation of the dopant concentration under change of the meniscus height can be realized only with the use of type I. Mn-doped LN fibers 0.5 and 0.7 mm in diameter having a 40–60 μm doped outer area were grown successfully.

8.1.5 Self-Cladded Eutectic Fibers

Already in 1988 Feigelson [31] pointed out the fascinating possibility of obtaining complete phase segregation in eutectic fibers, having a diameter comparable with the size of the eutectic phases. Self-cladding structures so produced might prove very interesting for new device applications but up to now no experiments are known since plausible verification needs eutectic fibers with a diameter below 10 μm to be grown.

Deprived of the possibility of growing fibers sufficiently smaller than 120–150 μm in diameter we have made an attempt to investigate the case by applying low pulling rates in the range 0.04–0.15 mm/min (which also was near to the pulling mechanism limitation) [32]. The appearance of low-speed grown fiber and cross-sectional eutectic patterns corresponding to pulling rates of 0.05 and 0.12 mm/min are presented in Fig. 8.19. One can easily see there is the tendency of the YAG phase to concentrate in the core area and the sapphire phase occupies the surface area of the fiber. Work on small-sized fibers with a view to producing finished self-cladding structures is still in progress.

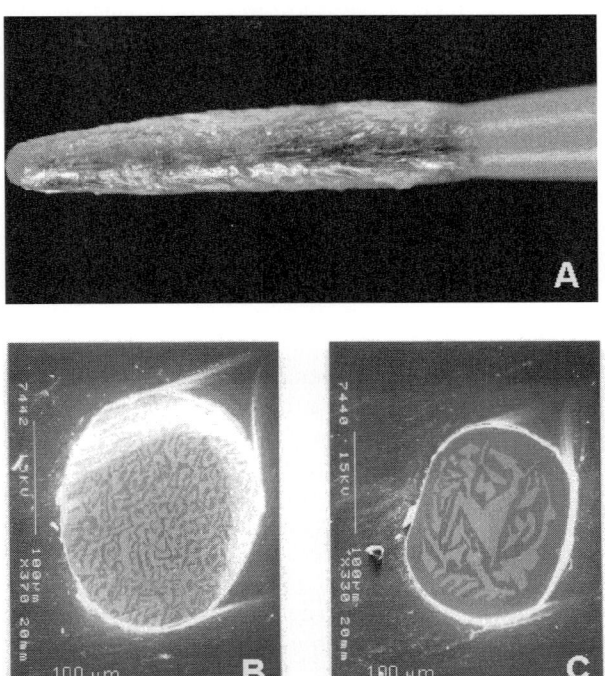

Fig. 8.19. $Al_2O_3/Y_3Al_5O_{12}$ fibers grown at 0.05 mm/min clearly show tendency for self-cladding: the YAG phase is concentrated in the center area and sapphire at the periphery

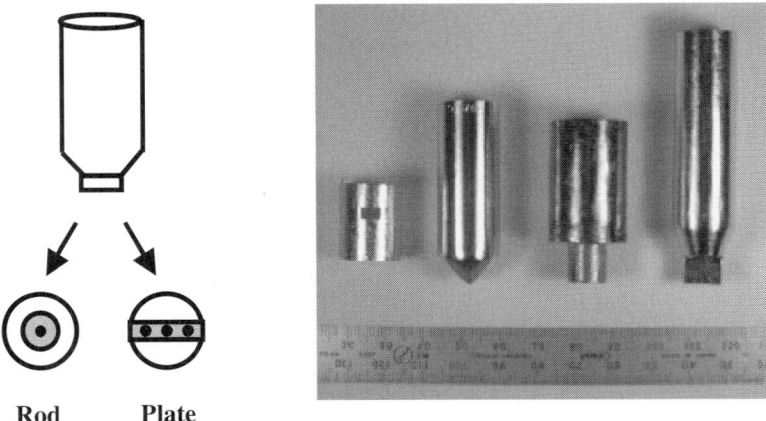

Rod   Plate

Fig. 20. The shapers for rod and plate shaped bulk crystal

8.2 Bulk Crystal Growth

The shapers for rod and plate shaped bulk crystals are shown in Fig. 8.20. The crucible bottom plays an important role for the shaper for the growing crystals. The shape and size of the bottom plane of the crucible decides those of the grown crystals. Thus several kinds of crucibles were designed for the growing rod and plate-shaped crystal.

This modified pulling down method showed credible potentiality for bulk crystal growing. As shown in Fig 8.21, Al_2O_3/YAG/ZrO_2 ternary eutectic crystals could be grown from these shapers.

5N-purity Al_2O_3 (High-Purity Chemical Co.), 4-N purity ZrO_2 (Rare Metallic Co.) and 4-N purity Y_2O_3 (Nippon Yttrium Co.) were used as the starting materials. They were mixed to the ternary eutectic composition of 65 mol.% Al_2O_3, 19 mol.% ZrO_2 and 16mol% Y_2O_3 as described in reference [13]. Al_2O_3 single crystal fibers grown by μ-PD were used as a seed. The meniscus and growing crystals were observed by CCD camera. The growth was performed under a flowing Ar gas atmosphere to prevent oxidation of the Ir crucible and after-heater. The growth process was controlled by manual adjustment of the RF power and pulling rate.

Stable growth of the bulk crystal was obtained in the range of 0.1~3 mm/min of the pulling rate somewhat lower than for the fiber crystal but much (several hundred times) faster than that of other bulk growth techniques like the Bridgman method. Rod crystals of 3 mm diameter and up to 140 mm in length could be grown successfully at up to 1 mm/min pulling rate to date. The diameter of the rod crystal was very stable within 5 % but depended strongly on the diameter of the bottom plane of the crucible (not the orifice diameter as for a fiber crystal). The only limitation to the length of the grown crystals was the volume of the crucible.

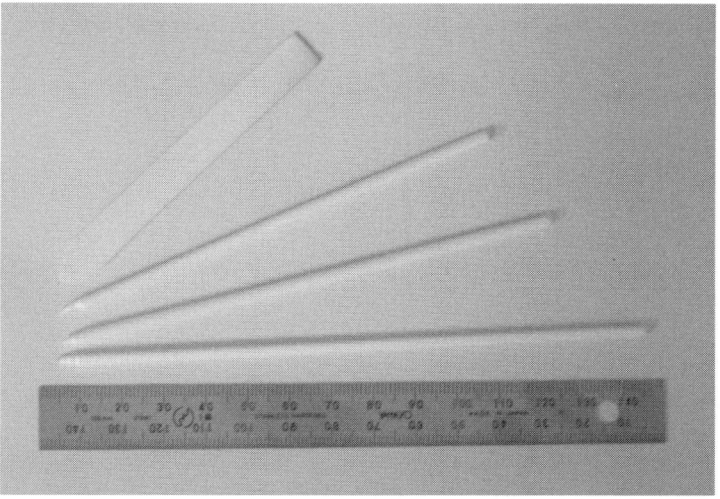

Fig. 8.21. Shaped eutectic crystals grown by modified pulling-down method. These are as-grown Al_2O_3/YAG/ZrO_2 ternary eutectic crystals

Plate-shaped crystals having a dimension of 10 mm width, 1 mm thickness and up to 140 mm length were also grown successfully as shown in Fig. 8.21. The dimension of the plate tip of the crucible strongly affected the size of the grown plate-shaped crystals. The plate crystals showed within 10% of width and thickness stability and excellent flatness. The length of grown plate crystals depended also on the volume of the crucible.

At the higher pulling rate than over 1mm/min, the diameter or width of bulk crystals decreased gradually. The meniscus shape of stable/unstable growth condition is shown in Fig. 8.22.

The size distribution of the eutectic lamellae strongly depends on the temperature distribution in the horizontal plane. Ternary eutectic rod and plate-shaped crystals grown at the pulling rate of 1 mm/min were investigated by a scanning electron microscope (SEM) using the back-scattered emission (BE) mode. The script sizes were evaluated on the chosen line on the cross-sectional micrographs perpendicular to the growth direction.

Figure 8.23 shows a typical cross-sectional view and microstructure of a Al_2O_3/YAG/ZrO_2 ternary eutectic plate grown at 1 mm/min pulling rate [34]. The BEI shows the homogeneous size distribution of the eutectic lamellae which means the homogeneous temperature distribution in the horizontal plane. The eutectic microstructure was composed of three phases distinguished by their different shapes and colors and distributed homogeneously on the whole cross-section as seen in Fig. 8.23. By EDS analysis the black matrix was shown to be Al_2O_3 and the gray phase was YAG. ZrO_2 phases were distributed on the periphery of the YAG phases as relatively small particles that appear white in the micrograph [13].

Almost the same pattern and size of microstructures were observed over the whole cross-section of the rod crystal. The microstructure of the rod crystals coincided with fiber crystals grown at the same pulling rate in spite of their larger diameter.

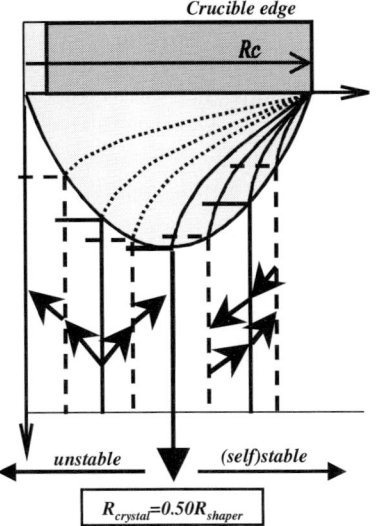

Crucible edge

Rc

unstable   *(self)stable*

$R_{crystal}{=}0.50R_{shaper}$

Fig. 8.22. The meniscus shape of stable/unstable growth condition

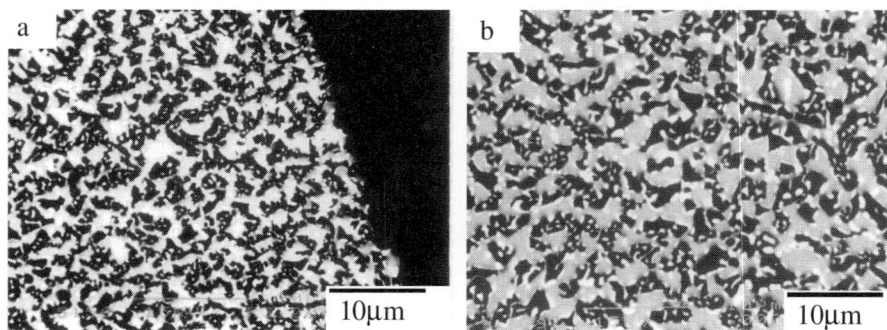

Fig. 8.23. Typical cross-sectional microstructure of marginal area (**a**) and core area (**b**) of Al_2O_3/YAG/ZrO_2 ternary eutectic plate crystal

References

1   D.H. Yoon, I. Yonenaga, N. Onishi and T. Fukuda, J. Crystal Growth 142 (1994), 423.
2   H.E. La Belle, Jr., Mat. Res. Bull. 6 (1971) 581.
3   N. Onishi and T. YAO, Jap. J. Appl. Phys. 28 (1989) L278.
4   J.Surek, Appl. Phys. 47 (1976) 4384.
5   V.A. Tatarchenko, Shaped Crystal Growth, Kluwer, Dordrecht, 1993.
6   B. M. Epelbaum, A. Yoshikawa, K. Shimamura, T. Fukuda, K. Suzuki and Y.Waku, J. Crystal Growth 198–199 (1999) 471.
7   M.N. Leipold, T.P. O'Donnell and M.A. Hagan, J. Crystal Growth 50 (1980) 366.
8   N. Schäfer, T. Yamada, K. Shimamura, H.J. Koh, and T. Fukuda J. Crystal Growth 166 (1996) 675.
9   J.P. Kalejs, H.M. Ettouney, and R.A. Brown, J. Crystal Growth 65 (1983) 316.
10 E. Pitts, J. Fluid Mech. 76 (1976) 641.
11 N. Kobayashi, J. Crystal Growth 43 (1978) 417.
12 R.S. Feigelson, J. Crystal Growth 79 (1986), 669.
13 D.H. Jundt, M.M. Fejer, and R.L. Byer, Appl. Phys. Lett. 55 (1989) 2170.
14 D. H. Yoon, M. Hashimoto and T. Fukuda Jap. J. Appl. Phys. 33 (1994) 3510.
15 P. Rudolph, K. Shimamura and T. Fukuda, J. Crystal Res. Technol. 29 (1994) 801.
16 H.G. Nalbandyan J. Crystal Growth 98 (1989) 739.
17 V.S.Arakelyan, A.G. Avetisyan and H.G. Nalbandyan, J. Crystal Growth 85 (1987) 357.
18 M. Lichtensteiger, A.F. Witt and H.C. Gatos, J. Electrochem. Soc. 118 (1971) 1013.
19 C.A. Burrus and J. Stone, Appl. Phys. Lett. 26 (1975) 318.
20 C.A. Burrus and L.A. Coldren, Appl. Phys. Lett. 31 (1977) 383.
21 T. Fukuda and H. Hirano, J. Crystal Growth 50 (1979) 291.
22 P.I. Antonov, Yu.G. Nosov, S.P. Nikanorov, Izv. Akad. Nauk SSSR Ser. Fiz 49 (1985) 2295.
23 A.V. Nikanorov, Yu.G. Nosov et. al., Izv. Akad. Nauk SSSR Ser. Fiz. 52 (1988) 2025.
24 K. Shimamura, N. Kodama and T. Fukuda, J. Crystal Growth 142 (1994) 400.
25 L.D. Landau and V.G. Levich, Acta Physicochimica URSS 17 (1942) 42.

26 L.D. Landau and E.M. Lifshits, Fluid Mechanics, Mir, Moscow, 1971, p. 289 (in Russian).
27 B.V. Deryagin and A.S. Titievskaya, DAN URSS 50 (1945) 307 (in Russian).
28 D.F.Games, J. Fluid Mech. 63 (1974) 657.
29 G.A. Satunkin, B.S. Redkin, V.N. Kurlov et. al., Izv AN USSR, Ser. Fiz. 49 (1985) 2319.
30 J. Trauth and B.C. Grabmeier, J. Crystal Growth 112 (1991) 451.
31 R.S. Feigelson, Material Science and Engineering B1 (1988) 67.
32 A. Yoshikawa, B. M. Epelbaum, K. Hasegawa, S. D. Durbin and T. Fukuda, J. Crystal Growth 205 (1999) 305.
33 A.B.Dreeben, K.M.Kim, A.Scujiko: J.Crystal Growth 50, 126 (1980).
34 J.H. Lee, A. Yoshikawa, T. Fukuda, Y. Waku, J.Crystal Growth 231 (2001) 115.

Druck: Strauss Offsetdruck, Mörlenbach
Verarbeitung: Schäffer, Grünstadt